Issues in Urban Earthquake Risk

NATO ASI Series

Advanced Science Institutes Series

A Series presenting the results of activities sponsored by the NATO Science Committee, which aims at the dissemination of advanced scientific and technological knowledge, with a view to strengthening links between scientific communities.

The Series is published by an international board of publishers in conjunction with the NATO Scientific Affairs Division

A	Life Sciences	Plenum Publishing Corporation
B	Physics	London and New York
C	Mathematical and Physical Sciences	Kluwer Academic Publishers Dordrecht, Boston and London
D	Behavioural and Social Sciences	
E	Applied Sciences	
F	Computer and Systems Sciences	Springer-Verlag
G	Ecological Sciences	Berlin, Heidelberg, New York, London,
H	Cell Biology	Paris and Tokyo
I	Global Environmental Change	

NATO-PCO-DATA BASE

The electronic index to the NATO ASI Series provides full bibliographical references (with keywords and/or abstracts) to more than 30000 contributions from international scientists published in all sections of the NATO ASI Series.
Access to the NATO-PCO-DATA BASE is possible in two ways:

– via online FILE 128 (NATO-PCO-DATA BASE) hosted by ESRIN,
Via Galileo Galilei, I-00044 Frascati, Italy.

– via CD-ROM "NATO-PCO-DATA BASE" with user-friendly retrieval software in English, French and German (© WTV GmbH and DATAWARE Technologies Inc. 1989).

The CD-ROM can be ordered through any member of the Board of Publishers or through NATO-PCO, Overijse, Belgium.

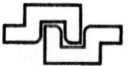

Series E: Applied Sciences - Vol. 271

Issues in
Urban Earthquake Risk

edited by

Brian E. Tucker
GeoHazards International,
San Francisco, California, U.S.A.

Mustafa Erdik
Boğaziçi University,
Kandilli Observatory and Earthquake Research Institute,
Istanbul, Turkey

and

Christina N. Hwang
GeoHazards International,
San Francisco, California, U.S.A.

Kluwer Academic Publishers

Dordrecht / Boston / London

Published in cooperation with NATO Scientific Affairs Division

Proceedings of the NATO Advanced Research Workshop on
An Evaluation of Guidelines
for Developing Earthquake Damage Scenarios for Urban Areas
Istanbul, Turkey
October 8–11, 1993

A C.I.P. Catalogue record for this book is available from the Library of Congress.

ISBN 0-7923-2914-7

Published by Kluwer Academic Publishers,
P.O. Box 17, 3300 AA Dordrecht, The Netherlands.

Kluwer Academic Publishers incorporates the publishing programmes of
D. Reidel, Martinus Nijhoff, Dr W. Junk and MTP Press.

Sold and distributed in the U.S.A. and Canada
by Kluwer Academic Publishers,
101 Philip Drive, Norwell, MA 02061, U.S.A.

In all other countries, sold and distributed
by Kluwer Academic Publishers Group,
P.O. Box 322, 3300 AH Dordrecht, The Netherlands.

Printed on acid-free paper

All Rights Reserved
© 1994 Kluwer Academic Publishers
No part of the material protected by this copyright notice may be reproduced or utilized in any form or by any means, electronic or mechanical, including photocopying, recording or by any information storage and retrieval system, without written permission from the copyright owner.

Printed in the Netherlands

Table of Contents

Preface vii

Overview ix

B. Tucker, J. Trumbull and S. Wyss *Some Remarks Concerning Worldwide Urban Earthquake Hazard and Earthquake Hazard Mitigation* 1

S. Mattingly *Socioeconomic and Political Considerations in Developing Earthquake Damage Scenarios in Urban Areas* 11

B. Bolt *Seismological Information Necessary for Beneficial Earthquake Risk Reduction* 21

W. Finn *Geotechnical Aspects of the Estimation and Mitigation of Earthquake Risk* 35

C. Rojahn *Estimation of Earthquake Damage to Buildings and Other Structures in Large Urban Areas* 79

G. Mader *Creating the Scenario and Drafting Earthquake Hazard Reduction Initiatives* 103

J. Fernandez, J. Valverde, H. Yepes, G. Bustamente, and J.-L. Chatelain *The Quito Project: Technical Aspects and Lessons* 115

T. Abdo *Governmental Aspects of the Earthquake Damage Scenario Project of Quito, Ecuador* 121

M. Erdik *Developing a Comprehensive Earthquake Disaster Masterplan for Istanbul* ... 125

S. Balassanian and A. Manukian *Seismic Risk on the Territory of the City of Yerevan, Armenia* 167

R. Yadav, P. Singh, A. Dixit, and R. Sharpe *Status of Seismic Hazard and Risk Management in Kathmandu Valley, Nepal* 183

F. Kaneko *Earthquake Disaster Countermeasures in Saitama Prefecture, Japan* 199

M. Klyachko *Arguments for Earthquake Hazard Mitigation in the Kamchatka Region* 215

C. Ventura and N. Schuster *Seismic Risk and Hazard Reduction in Vancouver, British Columbia* 221

H. Monzón-Despang and J. Gándara Gaborit
    Earthquake Hazards in the Guatemala City Metropolitan Area 237

A. Castro de Pérez, M. Benitez, and J. Rivera
    Earthquake Hazard in El Salvador 247

J. Alva-Hurtado   *Seismic Safety of the Lima Metropolitan Area* 251

L. Mendes-Victor, C. Oliveira, I. Pais, and P. Teves-Costa
    Earthquake Damage Scenarios in Lisbon for Disaster Preparedness 265

G. del Castillo and F. Vidal   *Granada Facing an Earthquake* 291

P.-Y. Bard and F. Feuillade   *Seismic Exposure and Mitigation Policy in Nice, France* 301

A. Tselentis, A. Karavolas, and C. Christopoulos
    *The City of Patras-W. Greece: A Natural Seismological Laboratory to
    Perform Seismic Scenario Practices* 315

Index 327

Disclaimer:

Neither the sponsoring nor supporting agencies assume responsibility for the accuracy of the information presented in this book or for the opinions expressed herein. The material presented in this publication should not be used or relied upon for any specific application without competent examination and verification of its accuracy, suitability, and applicability by qualified professionals. Users of information from this publication assume all liability arising from such use.

PREFACE

Urban seismic risk is growing worldwide and is, increasingly, a problem of developing countries. In 1950, one in four of the people living in the world's fifty largest cities was earthquake-threatened, while in the year 2000, about one in two will be. Further, of those people living in earthquake-threatened cities in 1950, about two in three were located in developing countries, while in the year 2000, about nine in ten will be. Unless urban seismic safety is improved, particularly in developing countries, future earthquakes will have ever more disastrous social and economic consequences.

In July 1992, an international meeting was organized with the purpose of examining one means of improving worldwide urban safety. Entitled "Uses of Earthquake Damage Scenarios for Cities of the 21st Century," this meeting was held in conjunction with the Tenth World Conference of Earthquake Engineering, in Madrid, Spain. An earthquake damage scenario (EDS) is a description of the consequences to an urban area of a large, but expectable earthquake on the critical facilities of that area. In Californian and Japanese cities, EDSes have been used for several decades, mainly for the needs of emergency response officials. The Madrid meeting examined uses of this technique for other purposes and in other, less developed countries. As a result of this meeting, it appeared that EDSes had significant potential to improve urban seismic safety worldwide.

Capitalizing on the ideas developed in the Madrid meeting, a pilot project was launched in the fall of 1992: the Quito Earthquake Risk Management Project. The purpose was to develop an EDS for the capital of Ecuador, in order to reduce the consequences of future earthquakes to that city. The unique features of that project included: work performed in large part by local specialists, but with the help of international experts; review and approval of the technical methods and findings by an impartial, international committee of experts; and oversight and implementation of recommendations by a board consisting of the political and business leaders of Quito, chaired by the Mayor. The technical phase of this project was scheduled to be complete in mid-1993.

The fall of 1993 seemed, therefore, to be a good time for representatives from earthquake-threatened cities around the world to gather and determine if the approaches used in the Quito project might be useful elsewhere. Further, it seemed that such a meeting would be an excellent opportunity to compare the hazard mitigation strategies and earthquake recovery experiences of these cities. Earthquake-threatened cities would learn from each other.

For this purpose, a NATO Advanced Research Workshop was organized. In order to accomplish its goals, financial support was sought from several public and private sources. The Kandilli Observatory and Earthquake Research Institute, of the Bogaziçi University in Istanbul, Turkey agreed to be host. Experts in urban seismic safety were invited from more than a dozen cities around the world, from both developing and developed countries. These experts represented a broad range of specialties - from earthquake engineering, seismology, and urban planning, to emergency response, environmental management, and insurance. Elected officials were present as well. All participants shared a commitment to improve international urban earthquake safety. The workshop took place from October 8-11, 1993. Its proceedings comprise this book.

We wish to acknowledge our deep appreciation to all the individuals and organizations that made this workshop possible. First, funding came from NATO, the Office of U.S. Foreign Disaster Assistance of the Agency for International Development, and OYO Corporation. The Kandilli Observatory and Earthquake Research Institute of Boğaziçi University was the site of the workshop. We extend our deep appreciation to the director of the Kandilli Observatory and Earthquake Research Institute, Professor Ahmet Işikara, for his kind hospitality and to the numerous students and staff of the Institute for their help throughout the workshop. Some material used in the workshop, as well as transportation and lunch during the field trip, were generously provided by the Şeker Insurance Company. We wish to thank in particular the President of that company, Mr. Yurdal Sert. The design of the workshop benefitted from conversations with J. Gunnar Trumbull, and, later, Christina Hwang. Formative ideas came from discussions with the authors of the papers in Part I: Shirley Mattingly, Bruce Bolt, Liam Finn, Chris Rojahn, and George Mader. They not only wrote stimulating papers, but helped conceive the overall structure of the workshop. Finally, we are grateful to Sarah Wyss, of GeoHazards International, for assistance in assembling these proceedings, and to Dr. Maxx Dilley, of OFDA, and David Hollister, of the Asian Disaster Prevention Center of the Asian Institute of Technology, for helping us identify and contact some of the workshop participants.

M. Erdik   B.E. Tucker

OVERVIEW

This volume contains twenty-one papers covering guidelines for developing earthquake damage scenarios (EDSes), the Quito Earthquake Risk Management Project, and the state of urban seismic safety. These papers are not only related, but actually complementary. This overview is meant to aid the reader in seeing the connection between the papers and in selecting the portions of most interest.

1. Introduction

In "Some Remarks Concerning Worldwide Urban Earthquake Hazard and Earthquake Hazard Mitigation," B. Tucker, J. Trumbull and S. Wyss intend to establish the importance of the papers that follow. They address two issues: the greatest urban earthquake risk (as measured in lives, if not capital) is faced, increasingly, by developing nations; and, the resources devoted to urban earthquake hazard mitigation are, increasingly, concentrated on the needs of industrialized nations. The authors first summarize evidence for this disparity between where earthquake mitigation resources have been spent and where earthquake risk will be located. Next they discuss the prospects for reducing it, using traditional means. And finally, recognizing that new means must be found, they propose two alternative approaches. Examples of these two approaches are presented in the remainder of this volume: documentation of the growing urban earthquake risk in developing countries, and development of methods to transfer earthquake risk mitigation techniques from cities in developed countries to cities in developing countries.

2. Guidelines for Earthquake Damage Scenarios

After the introduction, the first five papers elaborate on the basic components of and present guidelines for developing earthquake damage scenarios. Their authors served as key technical advisors on the Quito Earthquake Risk Management Project. In their papers, these experts draw on experience from their entire professional careers. Each author places emphasis on his/her specialty: Shirley Mattingly, on urban emergency response; Bruce Bolt, on seismology; W.D. Liam Finn, on geotechnical engineering; Christopher Rojahn, on structural engineering; and George Mader, on urban planning. They do not, however, restrict their comments to their specialties. Thus, there is a stimulating overlap among the technical approaches discussed by Finn, Bolt and Rojahn, and between the strategies discussed by Mattingly and Mader.

The term "guidelines" has been used in reference to these papers, but it will be apparent reading these papers that they do not, usually, prescribe step-by-step recipes. Procedures and approaches are described, but often there are differences between the authors' recommendations. The diversity in the needs and capabilities of earthquake-threatened cities means that no single approach applies to all. Approaches that worked in some cities will not necessarily work in others. For this reason no single set of guidelines, representing a consensus of thought, was sought.

These papers are meant, rather, to stimulate anyone—scientist, engineer, government official, business leader, international aid provider—considering a project to develop an EDS for a city.

The papers list some of the questions that should be asked, suggest some methods for answering them, describe some advantages and disadvantages of different approaches, and provide ample references to orient the reader.

S. Mattingly, in "Socioeconomic and Political Considerations in Developing Earthquake Damage Scenarios in Urban Areas," discusses the context which will define the parameters of the EDS and explains how to set up an EDS project so that it will result in positive changes in the subject city. She recommends analyzing who are the key players, getting them involved from the beginning of the project, and devising a means to make changes in the future.

B. Bolt, in "Seismological Information Necessary for Beneficial Earthquake Risk Reduction," covers the basic seismological considerations and then delves into more fundamental issues. Particularly interesting is his discussion of acceptable risk. Many countries have placed emphasis on life safety rather than economic loss. He points out the difficulty of managing jointly life safety and property damage. Also, recent earthquakes have shown the limitations of emphasizing life safety as the sole criterion to drive rehabilitation. For example, the incapacitation of the Bay Bridge, the loss of electrical power, and the damage to Stanford University as a result of the 1989 Loma Prieta earthquake taught us that a slightly larger earthquake might have produced severe social consequences, beyond human casualties. EDSes for urban areas must, therefore, address the functioning of key institutions (e.g. hospitals, banks, public utilities) as well as life safety. Another particularly interesting concept introduced by Bolt is that a political entity (e.g. city, state, or nation) should seek uniform safety within its boundaries. He introduces the use of benefit-to-cost ratios in managing public seismic risk reduction programs. Further, he suggests that our goal should be to reduce the risks of earthquakes to levels comparable with those of more familiar threats in our societies. He recommends that if seismological evidence indicates that strong ground shaking is likely to occur in a given metropolitan area in the next 10-20 years, a program of risk reduction should be undertaken at once.

In "Geotechnical Aspects of the Estimation and Mitigation of Earthquake Risk," W.D. Finn describes the procedures for characterizing the impact of geotechnical factors on the hazards and risks imposed by earthquakes. By way of introduction he describes some examples of seismic risk assessment in Indonesia and Ecuador. His paper then presents a thorough summary of methods for zoning ground motion levels, liquefaction, and landslide potential. The paper encompasses an extensive list of references and can serve as a state-of-the-art discussion on the incorporation of geotechnical factors in earthquake hazard assessment.

C. Rojahn, in "Estimation of Earthquake Damage to Buildings and Other Structures in Large Urban Areas," provides an overview of the factors that affect earthquake structural damage potential, and of the methods used to estimated earthquake-induced structural damage. He comes the closest to providing a step-by-step approach in the section describing how to develop structural damage estimates for scenario earthquakes. The required resources for such estimates and the available potential supporting information are also identified.

Finally, G. Mader, in "Creating the Scenario and Drafting Earthquake Hazard Reduction Initiatives," goes beyond procedures for calculating earthquake damage and loss to the purpose of such calculations. He attempts to bring closure to preceding papers by describing the purpose of an EDS (or, in his terms, an "exercise"), the contents of an EDS, the different possible levels of technical information used, and the ingredients of successful EDSes. He focuses on the purpose of the scenario: namely, to determine the most appropriate types of measures for reducing earthquake damage. He stresses that a great deal can be accomplished with a small amount of expensive technical work, which can be particularly important in developing countries. He illustrates this by presenting two approaches, representing very different degrees of technical input.

3. Quito Earthquake Risk Management Project

The volume includes two papers written by people who have recently used the guidelines presented to develop an EDS for the city of Quito, Ecuador. As mentioned before, at the time these papers were prepared, only the technical phase of the Quito project had been completed, making a thorough evaluation of the project impossible.

J. Fernandez, J. Valverde, H. Yepes, and J.-L. Chatelain, in "The Quito Project: Technical Aspects and Lessons," give a preliminary report on the Quito Earthquake Risk Management Project. They describe the need for this project and its organization. The technical aspects are summarized, including the seismic intensity distribution maps, the damage calculations, the consequences on the city, and the recommended mitigation activities. Perhaps most interesting to the participants in the workshop were the lessons learned. In the authors' opinion, several things seemed to work well. The project served as a catalyst to get different scientific and engineering groups together. For the first time in the city's history, an earthquake safety project attracted the support of the Mayor. Finally, the project results had credibility because of the involvement of the diverse disciplines and international reviewers. Nevertheless, several aspects of the project needed improvement. Communication was difficult, taking place among several different countries and in several different languages. More of the technical work should be done locally. The expectations that each involved party placed on the other parties should have been more clearly defined at the outset. The need to translate the technical final report into a popular publication to raise awareness among the public should not be overlooked.

T. Abdo, in "Governmental Aspects of the Earthquake Damage Scenario Project of Quito, Ecuador," describes in understated but riveting terms the challenges the project presented the city officials. Most basically, the project makes public and explicit some pressing, interconnected problems. First, the city has the normal problems of any rapidly growing urban area, including increasing population density, growth into geologically hazardous areas, lifelines with little redundancy, and many critical emergency services concentrated in a few areas of the city. Second, the city's seismic hazard is significant. Third, there is little public awareness of the city's risk. Fourth, while the federal government, not the city, has the legal responsibility to manage disasters, the city officials feel that the city is not prepared for a major earthquake and must, therefore, take some responsibility to reduce this risk. Abdo continues by mentioning the political risks of awakening the conscience of the public to earthquake hazards, and concludes by noting that once a city starts such a project, it must be ready to finish it.

4. Reports from Earthquake-Threatened Cities

In the remainder of the volume, individuals from fourteen other earthquake-threatened cities, in developed and developing countries, attempt to summarize the earthquake safety of their cities and risk mitigation steps to be taken. The purpose of these presentations is to compare these cities' needs and approaches to reducing seismic risk, in the context of their particular social, economic, political, and technical settings. These papers are important also in assessing how well the guidelines outlined by the first five papers might work in other cities, keeping in mind the lessons learned from the Quito project. They are written in a common format in order to facilitate comparisons of the cities' different strategies and experiences. These reports have been ordered according to their geographic location relative to Istanbul, the site of the workshop where these papers were presented. We start with Istanbul itself, then move to the first participating city to the east, Yerevan, Armenia, and continue eastward around the world.

In "Developing a Comprehensive Earthquake Disaster Master Plan for Istanbul," M. Erdik presents a thorough description of the earthquake disaster master plan that will be constructed for Istanbul. This paper will serve as an important reference, although probably many other cities will not have the resources to create such a complete plan. While the responsibility of the Governor of Istanbul for commissioning the development of this plan should not be forgotten, the plan may have been largely impelled by the estimate that a plausible earthquake would cause in excess of 20,000 deaths and about 30 billion US$ in physical losses. Erdik's summary of the effects of the Erzincan earthquake provides firsthand insights into post-earthquake crises and rehabilitation efforts. Another value of this paper is its list of examples of preparedness, emergency response and post-earthquake activities. This will be a good reference for anyone trying to imagine possible types of actions.

S. Balassanian and A. Manukian, in "Seismic Risk on the Territory of the City of Yerevan, Armenia," summarize the long history of disastrous earthquakes in their city. In their opinion, the seismic risk of Yerevan is now higher than ever before. They report the purposeful downgrading of seismic hazard during the Soviet regime, resulting in catastrophic losses in the 1989 earthquake. Under the Armenian Republic, many reforms were made following the earthquake, including the formation of the National Survey of Seismic Protection.

A. Dixit, P.L. Singh, R. Yadav, and R.D. Sharpe discuss the state of urban seismic safety in the Kathmandu Valley, in "Status of Seismic Hazard and Risk Management in Kathmandu Valley, Nepal." They conclude that not only is earthquake hazard significant, but the risk of disaster is extreme because of poor building design and construction, as well as the lack of preparedness. Kathmandu's needs are complicated by the rather low economic and technical ability to reduce risk and the mixed blessing of near total reliance on international aid. In order to be effective, this aid must eventually be coupled with local commitment. Nepal is now making many of the required steps for improving seismic safety, including, most recently, the introduction of a new building code. The authors feel that an EDS would complement this building code project.

In "Earthquake Disaster Countermeasures in Saitama Prefecture, Japan," F. Kaneko describes one of the most comprehensive EDSes ever developed. It calculates human casualties to five significant digits, that is, to .001%. It considers such diverse social effects as deaths, injuries, homeless, refugees, houses without water, and houses without electricity. In addition, it estimates the economic losses resulting from damage to buildings, water systems, sewers, gas and electric power systems, roads, bridges, and railways. This project was commissioned by the government and is supported by widespread public awareness of Japan's seismic risk. Deadly earthquakes are in the public's memory, and damaging earthquakes are frequent. The technical results of this EDS were summarized in a pamphlet and distributed to school children. In addition, the methodology used was transferred to a personal computer, in order to allow the city to estimate damage on its own for other natural and social conditions, including different earthquake magnitudes and epicenters, as well as seasons of the year, times of day, and wind velocity.

In "Arguments for Earthquake Hazard Mitigation in the Kamchatka Region," M. Klyachko describes the seismic safety conditions of this major city on Kamchatka. In addition to the usual seismic hazards, this city has a significant risk of fires and spills of fuel, as well as the difficulties associated with a severe climate and extremely isolated geographical location. The author has headed various programs to improve the city's safety, including ranking buildings in need of retrofitting. However, emergency response capabilities must still be improved, by training personnel, adding communication equipment, and modernizing medical facilities. There is no long-term preparedness program. Impeding all efforts to correct this situation is a low public awareness level of seismic risk or personal measures to be taken during an earthquake.

C. Ventura and N. Schuster, in "Seismic Risk and Hazard Reduction in Vancouver, British Columbia," describe the seismic safety state of the city of Vancouver. Very large earthquakes occur, but only rarely. Public officials and the population at large are reportedly aware of this situation. Vancouver is one of the very few cities described in this volume whose seismic safety is increasing with time. While the occurrence of a large earthquake tomorrow could be devastating, the level of expected devastation is decreasing as preparations are made, through public education, improvement of emergency response capabilities, and regulation of construction practice.

In "Earthquake Hazards in the Guatemala City Metropolitan Area," J.L. Gandara Gaborit and H. Monzon-Despang describe their city's seismic setting. Clearly the need for seismic risk reduction is great. The city constitutes about half of the country's total urban population and contains about four-fifths of all non-agricultural investment, about two-thirds of its industry, and about half of total formal employment. The city is located near several faults capable of generating damaging earthquakes. In the face of this threat, however, little has been done. The city's urban planning department has no master plan of development, no disaster mitigation planning, and is capable of monitoring less than half of the formal construction. The public perception is that earthquakes occur only in the past and the distant future, creating an environment in which preparation is postponed. A significant amount of research has been conducted on disaster mitigation, but the results of that work are not readily available, being for the most part in the form of internal reports. The authors recommend that all available earthquake-related information should be collected, organized and made available in one disaster-related center.

A. Castro de Perez, M.L. Benitez, and J.L. Rivera, in "Earthquake Hazard in El Salvador," describe the programs of this country. Some of the social and economic consequences of that country's last devastating earthquake, in 1986, are detailed, as well as the emergency response organization that was created to cope with such disasters. San Salvador is distinguished in terms of high seismic risk simply because of its proximity to a fault that is capable of generating large earthquakes.

J. Alva, in "Seismic Safety of the Lima Metropolitan Area," summarizes this city's seismic vulnerability. About one third of the country's population resides in this capital city. Its seismic vulnerability has been increasing dramatically in recent years, due to an increase in population, deterioration of existing buildings, and construction of new buildings on poor land. A characteristic of Lima's seismic hazard is that damage from past earthquakes has been highly localized. Based on this record, and knowledge of where population density is greatest and where the most vulnerable buildings are located, maps of future expected damage have been prepared.

L. Mendes-Victor, C. Oliveira, I. Pais, and P. Teves-Costa, in "Earthquake Damage Scenarios in Lisbon for Disaster Preparedness," present the result of their city's remarkable EDS. As in the case of the Saitama EDS, Lisbon's is extremely detailed and complete. It has been developed over more than a decade and is the product of experts in such disciplines as seismology, civil engineering, demographics, and municipal civil defense. The consequences of two different earthquake sources were estimated for different types of buildings, and were expressed in terms of the total cost of damage and the total population affected. The fluctuation of the population as a function of time of day is being accounted for. This EDS is being used by the civil defense to make an emergency response plan.

In "Granada Facing an Earthquake," G. del Castillo and F. Vidal outline urban seismic safety in the area of greatest seismic activity of Spain. Granada's risk is high not only because of its proximity to earthquakes but also because of the city's soil conditions and topography. Large site amplification effects have been observed in the past. While there are several groups responsible for civil defense, and they have a seismic emergency plan, there is a need to involve urban planning into earthquake risk mitigation, because future growth is expected to take place in an area of poor soil conditions. An EDS might help to motivate officials to take such development into consideration.

P.Y. Bard and F. Feuillade, in "Seismic Exposure and Mitigation Policy in Nice, France," describe a unique combination of geologic and social conditions. Nice is different from other cities considered in this volume because its seismicity may be among the lowest, its total population among the smallest, and its economy among the highest. Nevertheless its problems are not simple. The city contains rather unfavorable geotechnical conditions; the city center is located on young, unconsolidated alluvium. There is considerable capital, in terms of services, communication and transportation at risk. Although the community has the economic and technical ability to reduce its risk, it is difficult to get this problem into public consciousness. An EDS is currently being considered for Nice, by national and local government officials in order to improve mitigation policies. This paper contains an interesting discussion of the development and enforcement of France's building codes, as well as the state of earthquake awareness in France.

C. Christopoulos, A. Karavolas, and G.-A. Tselentis summarize the seismic risk of their city in "The City of Patras -W. Greece: A Natural Seismological Laboratory to Perform Seismic Scenario Practices." Patras is located in the most seismically-active region of Europe, and has been destroyed many times in past from earthquakes. It contains hazardous soil conditions and active regional faults, including one running directly through most densely populated part of city. Some of the construction is poor, as evidenced by the damage suffered by 45% of the city's buildings due to a recent event of magnitude 5.5. The city is expecting to develop an EDS in the near future.

5. Questionnaires on Urban Seismic Safety

In addition to preparing papers for the workshop, the participating city representatives completed a questionnaire which covered factors influencing their city's seismic safety: demographics, seismicity, emergency response capabilities, construction practices, public awareness, and research. These revealed some differences, including the level of seismic activity and the distance to faults having destructive earthquakes, as well as the size and growth rates of the city, the economy, the level of indigenous scientific research and engineering, and the political system. We attempted to quantify and compare urban seismic vulnerability. One measure of earthquake risk is the size of the threatened population, divided by the distance from the threatening fault. The cities participating in our survey differ in this measure by about 3 orders of magnitude, from a low of about .002 million people per kilometer for Nice and Kamchatka, to a high of about 2 million people per kilometer for San Salvador.

The causes of high seismic vulnerability are numerous. One is overcrowding, as a result of natural births and rural-to-urban migration. The questionnaires showed population density to be quite variable, even within a city, but often the density was several thousand per square kilometer. This overcrowding leads to expansion into geologically hazardous areas, for example, steep slopes and loose, water-saturated ground. Often, inadequate or nonexistent land-use planning aggravates this problem.

Inadequate construction practices are another cause of high vulnerability. Kathmandu, in fact, lacks a building code, although one is in preparation. Among those cities that have a building code, there is a large difference in the percentage of the city's buildings to which the code applies: many cities report a high percentage (greater than 80%), but Lima and Lisbon reported less than 25%. Perhaps more important, ultimately, than the existence and application of a building code, is the prevalence of so-called "informal" (extralegal) construction. Such construction is common.

Another cause of vulnerability is inadequate infrastructure and emergency response capability. Infrastructure systems are often old, poorly-maintained and overtaxed. There is a lack of sufficient emergency response agencies. Often emergency response agencies are not well-equipped, organized and trained for the special demands of earthquakes. A common need, in the public and private sectors alike, is up-to-date, tested emergency response plans. Large differences in the per capita number of emergency response officials exist: the per capita number of fire department personnel varies by a factor of about 60 from city to city, while the per capita number of police personnel varies by only about a factor of 6. Another measure of a city's state of emergency preparedness is the frequency of meetings among police and fire department leaders. By this measure, there is considerable variation: the average frequency is once a month, but in one city (Yerevan) there is one such meeting every week, and in another city (Kathmandu) such meetings occur only once a year.

We attempted to measure the frequency of news stories about earthquake hazards and the perception of the level of awareness. Most cities have reports about earthquake hazards in the news media about once every month. In Kathmandu, Nice and Vancouver, however, such reports occur only about once a year. In Yerevan, they purportedly never appear. In Patras, they appear weekly, while Saitama prefecture has them almost daily. Our questionnaire revealed no correlation between the frequency of reports in the news media and the level of awareness among the business leaders, government officials and general public. For example, in Yerevan, where local news media never report on earthquake hazards, and in Nice and Vancouver, where news media make such reports only once a year, there is a fairly high level of awareness. This is obviously a complex issue, since it would be difficult to ascertain whether the media coverage on earthquake risk generates public awareness or vice versa.

6. Conclusions

6.1 EARTHQUAKE DAMAGE SCENARIOS

Several ingredients are key to a successful EDS. First, a team effort is required, involving scientists, engineers, planners, government officials, and businessmen. The earlier all these elements are involved the better. Next, only the minimum necessary amount of technical information should be developed. What is appropriate varies greatly according to resources available and the ultimate use of the results. Certainly it is possible to accomplish many public policy goals (e.g. identifying a need for a comprehensive, long-term seismic reduction program and the highest priorities of that program) with limited expense on technical issues.

An EDS can be used for many purposes. As described by George Mader, emergency response specialists have participated in exercises whereby damage caused by earthquakes was described in a compressed time period and the participants of the exercise were required to make emergency response decisions. These procedures test the emergency response system and allow it to be improved. Recently, EDSes have also been used to help people concerned with reconstruction after an earthquake and preparation before the earthquake. In such cases, EDSes are used to make recommendations that will lead to fewer injuries and deaths and less property damage. EDSes can also be used, as in the Quito Earthquake Risk Management Project, to target research needs and public policy priorities, increase earthquake hazard awareness among city leaders and the public, and summarize scattered information concerning the probability of damaging earthquakes and their effects. An EDS performed jointly by locals and international experts (rather than by one individual), achieves an enhanced level of objectivity in assessing risks and making recommendations.

There are several possibilities for future research concerning EDSes. EDSes should include estimates of indirect losses, since for some societies these can be comparable with direct losses. There should be more attention on how best to express the likelihood of the chosen earthquake. Rigorous estimates of earthquake probability are rare, particularly where geological and seismological information has only recently been gathered. Even in those cases (e.g. California), where scientists have devoted considerable effort to calculating probabilities, surprises are common. Regardless of the probability of the scenario earthquake, the scenario has significance beyond the particular earthquake considered. The difficulty in estimating a particular earthquake's probability of occurrence should not interfere with the easier task of identifying the structure and importance of a city's earthquake risk management program. Finally, research is needed on how best to express the uncertainty of damage and loss estimates in an EDS, intended for use in influencing public policy. Even if the location, magnitude and focal mechanism of an earthquake were perfectly well known, estimates of its consequences in terms of human casualties and economic losses are uncertain. Effective use of EDSes will require knowing how to communicate those uncertainties to the user.

While EDSes have been useful to some of the cities (e.g. Saitama, Lisbon and Quito) in improving the seismic safety, they are by no means the only solution. Vancouver, for example, identified its seismic vulnerabilities and raised the awareness of the public and elected officials without an EDS.

6.2 REPORTS ON URBAN SEISMIC SAFETY

Finding means to reduce the high vulnerabilities of urban areas in developing countries is difficult. Resources are scarce for preparation and mitigation activities, even in the wealthier communities. It is much easier to find funding for relief actions. Another common difficulty is the low awareness among the general populace and city leaders of their community's risk.

The most striking observation from these reports is not the high seismic vulnerability of these cities or the difficulty of finding solutions, but the increase in most cities of seismic vulnerability with time. Only a few communities, such as Vancouver and Saitama prefecture, have long-term, comprehensive programs to reduce their seismic risk. Most cities have no risk reduction program. In order to be effective, such programs must reduce risk faster than the natural forces—such as urban growth—are increasing it. It appears that in Vancouver and Saitama, the risk is decreasing with time. Yet for some cities, it appears that risk is higher now than ever before, and increasing.

6.3 GENERAL REMARKS

While this compendium of the state of urban seismic safety is not comprehensive, it is instructive. Perhaps future efforts should be made to compile periodically some standardized summary of the state of urban seismic safety worldwide. There are natural advantages to comparing cities' seismic vulnerabilities, as opposed to nations'.

A good measure of the state of a city's seismic safety is whether safety increases or decreases with time. Such a measure could be applied around the world, regardless of economy, political system, or geological setting. While a philosophically attractive goal may be to achieve single (low) life safety ratio (in the terms introduced in Bolt's paper) everywhere around the world, it may be more practical in the short term simply to seek risk reduction. In any case, this would be an improvement over the present situation, where the seismic risk of many cities is increasing.

Communities at risk should develop a program to decrease their city's vulnerability with time. Such a program would entail: building public awareness, disseminating information, and training personnel; developing effective mechanisms for the enforcement of land-use planning and building regulations; and retrofitting urban facilities essential to the operation of the socioeconomic system after earthquake disasters.

The inevitability of earthquakes and the rapid population growth in earthquake-prone urban centers worldwide make preparedness and risk reduction imperative. Preparation of an EDS is one important method of earthquake risk management.

SOME REMARKS CONCERNING WORLDWIDE URBAN EARTHQUAKE HAZARD AND EARTHQUAKE HAZARD MITIGATION[1]

BRIAN E. TUCKER
J. GUNNAR TRUMBULL
SARAH J. WYSS
GeoHazards International
123 Townsend Street, Suite 655
San Francisco, CA 94107

1. Introduction

By the year 2000, approximately 3 billion people -- nearly one-half of the world's population -- will live in urban areas. Of the fifty largest cities, half will have populations over 10 million, and half will lie within 200 kilometers of faults known to produce earthquakes of magnitude 7 or greater. Death tolls from recent urban earthquakes have been large: the 1976 Tangshan earthquake in China reportedly killed 250,000 people; the 1990 earthquake in Tabbas, Iran killed 40,000; the 1991 earthquake in Spitak, Armenia killed 25,000. The rapid growth of the world's cities will make such disasters more deadly and more frequent.

These facts are not new. It has long been recognized that the consequences of natural hazards, including earthquakes, to urban areas are increasing. This trend has been used as an argument for renewed efforts in natural hazard reduction, such as creating the United Nations' International Decade for Natural Disaster Reduction.

This paper addresses two related, but less commonly recognized problems: the greatest urban earthquake risk (as measured in lives, if not capital) is faced, increasingly, by developing nations; and, the resources devoted to urban earthquake hazard mitigation are, increasingly, concentrated on the needs of industrialized nations. First we summarize evidence for the disparity between where earthquake mitigation resources have been spent and where earthquake risk will be located. Next we discuss the prospects for reducing the risk, using traditional means. Finally, recognizing that new means must be found, we propose two alternative approaches. Excellent examples of these two approaches are presented in the remainder of this volume.

2. A Comparison of the Earthquake Risk in Developing and Industrialized Countries

Figure 1 summarizes the trend in worldwide urban earthquake threat, with emphasis on the threat to cities in developing countries (Jones, 1992).

---

[1] This paper has been adapted from "Trends in Urban Earthquake Hazard and Earthquake Hazard Mitigation in Developing and Industrialized Countries," a paper by the same authors, published in the Proceedings of the IDNDR Aichi/Nagoya Conference, Disaster Management in Metropolitan Areas for the 21st Century, November 1-4, 1993.

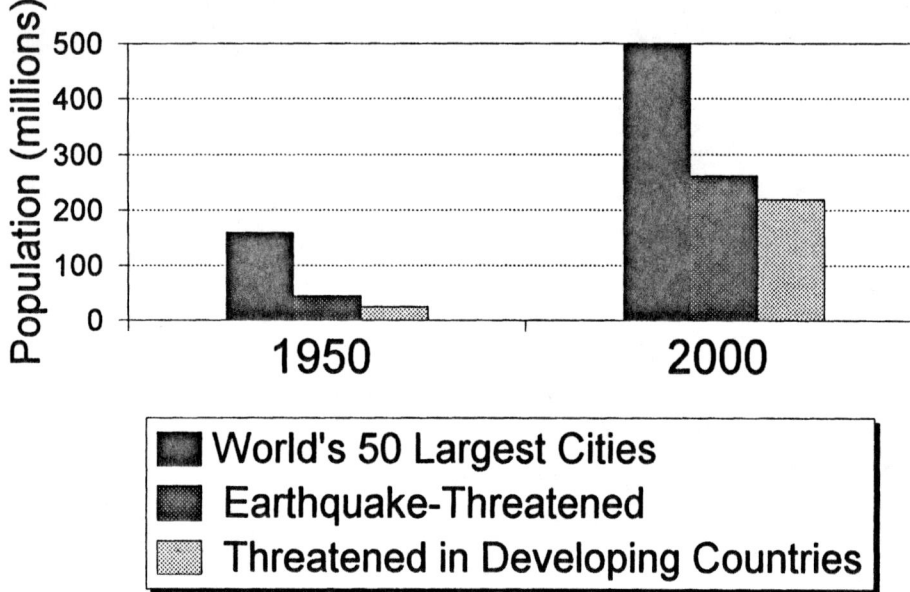

Figure 1. From Jones, 1992, including data from R. Bilham.

In order to make observations about worldwide urban earthquake threat in developing and industrialized countries, it is necessary to discuss terminology. To characterize a city as being from either an industrialized or a developing country is clearly imprecise. While a city may belong to a developing country, the city itself may be highly developed. For example, the level of development of Bangkok is much different than that of the rest of Thailand. It may be difficult to draw convincing conclusions based on classifications of cities according to the level of development of the countries in which they are located. Another complication is variation in a city's classification with time. For example, if Leningrad was considered industrialized in 1950, will St. Petersburg be in 2000? If Istanbul was considered developing in 1950, will it be in 2000?

We also need a definition for earthquake threat. In Figure 1, a city is considered "earthquake-threatened" if it is located within 200 kilometers of a fault capable of generating a magnitude 7 earthquake. (This definition and the corresponding classification of cities comes from R. Bilham.) Attempting to characterize a city's earthquake hazard in these terms is clearly subject to personal judgment, with the possibility of error and misinterpretation. No account of the probability of earthquakes is taken, nor of the importance of smaller events, nor of the frequency of larger events. In addition, the September 30, 1993 Maharashtra, India earthquake and the January 17, 1994, Northridge, California earthquake remind us that we have not identified all faults capable of generating destructive earthquakes.

Finally, Figure 1 makes an implicit definition of "urban," by considering only the world's fifty largest cities. When drawing conclusions from this figure about trends in worldwide urban earthquake hazard, one should keep in mind that the population of the world's fifty largest cities represents only about 1/5 of the world's urban population in 1950 and only about 1/6 in 2000.

Figure 1 shows that in 1950, about 1 in 4 of the people living in the world's fifty largest cities was "earthquake-threatened," while in the year 2000, about 1 in every 2 will be. Further, of those people living in earthquake-threatened cities in 1950, about 2 in 3 were located in developing countries, while in the year 2000, about 9 in every 10 will be. Such observations lead to the conclusion that urban earthquake threat is growing worldwide, and this threat is increasingly a problem of developing countries.

2.1 EFFECT ON LIVES

Records of earthquake-caused deaths during this century support the projection that, over time, earthquakes will increasingly affect populations of developing countries more than those of industrialized nations. Figure 2 shows that in the period 1900-1949, the ratio of number of earthquake-caused deaths in developing countries to the number in developed countries is about 3 to 1, while over the period 1950-1988, that ratio increased to more than 9 to 1 (OFDA, 1990). Clearly, caution must be used in comparing the number of deaths in different geographical regions during different time periods because of normal statistical variations. The differences in population size of the different regions obviously influenced the number of earthquake-caused deaths. Despite these difficulties, however, it is clear from Figure 2 that more people die in developing countries from earthquakes than in industrialized countries.

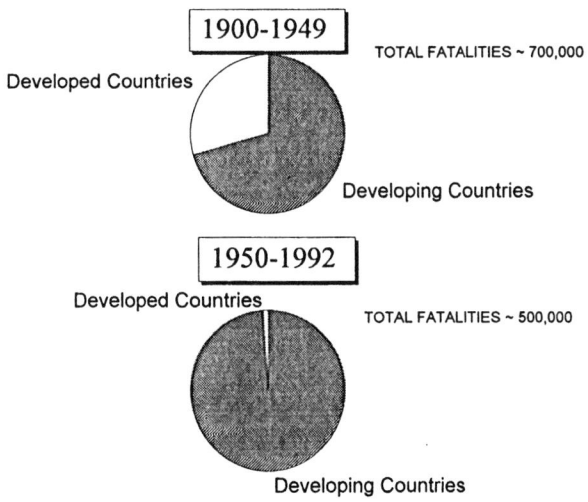

Figure 2. From OFDA, 1990

This generalization is in accord with the consequences of the 1988 Spitak earthquake in Armenia and the 1989 Loma Prieta earthquake in California: although the two earthquakes were of similar size and affected comparably-sized populations, the Armenian event killed 25,000 people, while the Californian earthquake killed 63.

The disparity between the number of earthquake-caused deaths in developing countries and those in developed countries could be explained by many factors, including differences in population density and distribution, differences in earthquake frequency and size, or differences in construction design and practice. It would be difficult to isolate the contribution of each of these variables. We have attempted, however, to remove the effect of variations in earthquake frequency by comparing the average number of fatalities per earthquake, during the two halves of this century, for developing and developed countries. We used the OFDA data set that is summarized in Figure 2. In the 1900-1949 period, the average numbers of fatalities per earthquake in developing and developed countries were effectively indistinguishable, each about 12,000. Over the 1950-1992 period, the average number of fatalities per earthquake in developing countries remained about 12,000, but the figure for developed countries was reduced by a factor of ten, to about 1,200. An improvement in construction quality could have caused such a disparity. Again, caution must be used when attempting to deduce trends involving any process as random as earthquake occurrence.

2.2 EFFECT ON ECONOMIES

The fiscal impact of earthquakes is also growing, particularly in developing countries. Figure 3 shows that the economic cost of earthquakes, worldwide, has increased four-fold in the past 30 years (Smolka, 1993).

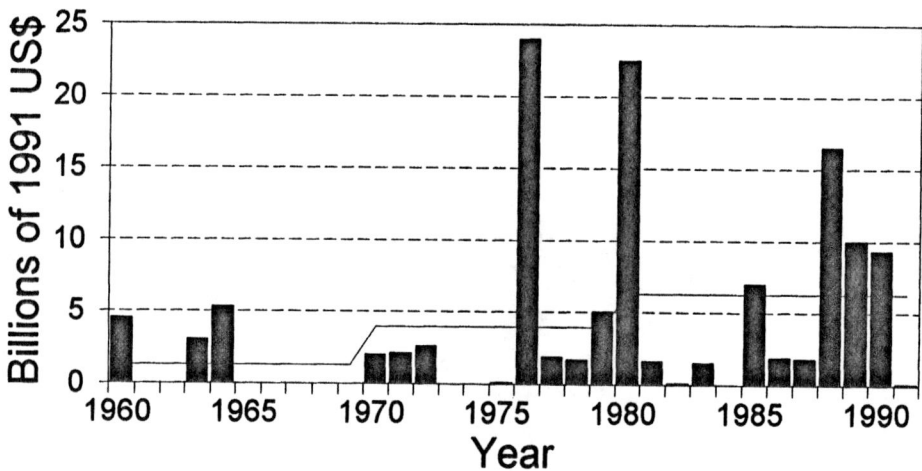

Figure 3. From Smolka, 1993

Earthquake-caused fiscal losses represent a greater percentage of developing countries' gross national products (GNP) than do such losses in developed countries. Data from Guatemala, El Salvador, Nicaragua, on the one hand, and Italy and the US, on the other hand, serve as examples. Guatemala's earthquake of 1976 caused US$1.1 billion worth of damage, which represented 18%

of the country's GNP. El Salvador's 1986 earthquake caused US$1.5 billion worth of damage, which represented 31% of the country's GNP. The 1972 Managua earthquake cost US$5 billion and represented 40% of Nicaragua's GNP. In contrast, Italy's 1980 earthquake caused US$10 billion worth of damage, yet this comparatively high figure represented only about 7% of the GNP; similarly, the 1987 Loma Prieta earthquake in the US caused US$8 billion worth of damage, which represented just 0.2% of the country's GNP and 6% of the Gross Regional Product of the San Francisco Bay area. The greatest current estimate of the damage due to the January 17, 1994 Northridge (Los Angeles) Earthquake is $30 billion, which constitutes less than 1% of the US's GNP and about 8% of the Gross Regional Product of the Greater Los Angeles area (i.e. the Los Angeles, Orange, Santa Barbara, Riverside, Ventura and Imperial counties).

3. A Comparison of Earthquake Hazard Mitigation Efforts in Developing and Industrialized Countries

Although earthquakes pose a great mortal and financial threat to the developing world, the majority of the world's resources devoted to earthquake hazard mitigation are focused on the US, Western Europe and Japan. Coburn and Spence (1992) estimate that only 2% of earthquake mitigation research and development is spent on developing nations, while the rest is spent on industrialized nations.

We have attempted one measure of the relative earthquake hazard mitigation effort in the developing and industrialized worlds: the level of participation from different countries in international professional conferences related to earthquake engineering. Another measure could be the topics of papers delivered in these conferences; for example, one could compare the number of papers concerning high-rise steel construction with the number concerning adobe buildings. Clearly, these measures are incomplete and subject to error; they must be tested and improved. In the meantime, they may provide some useful initial comparisons.

Figure 4 plots the attendance at the first ten World Conferences of Earthquake Engineering held between 1956 and 1992, classifying all participants as coming from either developing or developed countries. At these meetings, there was an average of four times as many participants from developed countries as from developing ones. Perhaps more significant, there is a trend of increasing participation of developed countries compared with developing ones. Only at one of these conferences, held in Chile in 1969, did participants from developing countries outnumber those from developed countries. Such data suggest that, over this time period, the level of effort focused on mitigating the effects of earthquakes in developing countries has remained rather constant, while the level of effort focused on the problems of developed nations has increased.

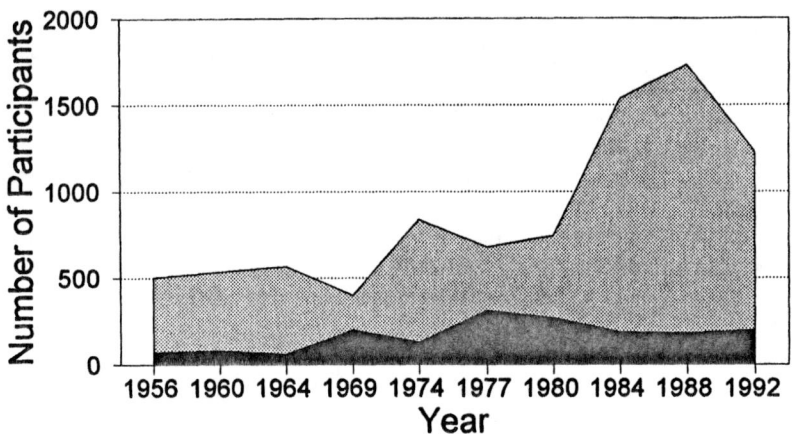

Figure 4

4. The Disparity between Where Earthquake Hazards Occur and Where Earthquake Hazard Mitigation Efforts are Focused

The information summarized above suggests that, in the 20th century, the greatest need (at least in terms of loss of life) for earthquake hazard mitigation has been in developing countries, while the focus of the world's earthquake hazard mitigation efforts has been on the needs of industrialized nations. The data also suggest that if current trends persist, this will be even more characteristic of the next century.

One unusual and striking measure of the disparity between earthquake threat and earthquake mitigation effort in developing and developed nations is presented in Figure 5. There the ratio of number of attendees at World Conference of Earthquake Engineering per earthquake-caused death is plotted for developing and industrialized nations. For developing countries, the ratio has remained approximately constant over the last five decades, while for industrialized nations, this ratio has increased 400-fold. The trends on this figure are more significant than the numbers themselves. The first section of this paper argued that, while urban earthquake hazard is increasing for developing countries, the world's resources devoted to earthquake hazard mitigation are increasingly devoted to developed nations. Figure 5 suggests that, over the last five decades, this disparity has been growing.

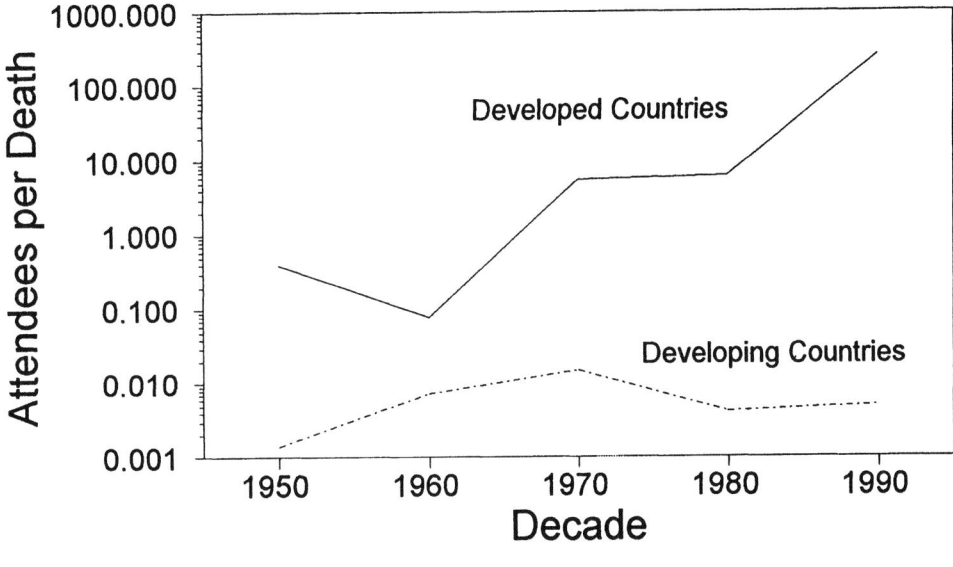

Figure 5

5. Prospects for Change

Consideration of the traditional sources of support for earthquake hazard mitigation in developing countries leads to the conclusion that the disparity described above will not be reduced in the near future and, in fact, may continue to increase. One traditional source of support has been the United Nations. Even a casual reader of today's current events knows of the increasing demands on the United Nations and of the number of its member states that are significantly in arrears in their dues. A reduction in the UN's support for natural hazard mitigation must be expected. A reduction has already occurred in one particularly relevant part of the UN, the United Nations Office of the Disaster Relief Coordinator; UNDRO has shifted emphasis from natural to manmade disasters (Tomblin, 1993).

Another traditional source of support for earthquake hazard mitigation effort has been the national foreign disaster agencies, such as Great Britain's Overseas Development Agency, Japan's Japanese International Cooperation Agency, and the US's Agency for International Development. The fundamental priority of these agencies has always been, understandably and necessarily, domestic. They seek to achieve national goals with international aid. These goals follow shifts in world and domestic politics, and thus assistance has not always been applied where the need was greatest, nor sustained long enough to realize lasting benefits. While these agencies continue to provide natural hazard assistance, the end of the Cold War and changes in the global economy have raised questions in some countries about whether the goals and budgets of these agencies should be modified.

The effectiveness of national foreign disaster agencies in meeting the needs of developing countries has been debated. For example, Brown and Sarewitz (1992) describe why, in their opinion, the technical assistance provided by the US has not served to foster an independent science and technology capability of the most needy developing nations. They argue that US development philosophy focuses on supplying technology and expertise to developing nations, rather than on contributing to indigenous capability. Further, this aid has been targeted at just a few countries. According to them, about 80% of all the funding dedicated to science and technology agreements between the US and developing countries has gone to just four nations: Egypt, India, Pakistan and China. These authors point out that, although numerous federal agencies provide support for nonfederal US scientists and engineers to conduct research in developing countries, the agenda for this type of research usually reflects the priorities of the US, which may or may not be the same as those of the host nation.

While Brown and Sarewitz consider only the US, and discuss science and engineering in general, their observations may apply in particular to the foreign aid policy of other industrialized nations and to earthquake hazard science and engineering. If their analysis can be thus extended, one might conclude that, even in the unlikely event that the national foreign aid agencies continue their work at the level that existed during the Cold War, they will not be able to meet all of the rapidly increasing needs of developing countries for earthquake hazard reduction.

6. What Can Be Done?

We have attempted to demonstrate that there is a growing problem and that traditional solutions do not appear adequate. To summarize, the problem is that the greatest urban earthquake risk is increasingly faced by developing nations, whereas the resources devoted to urban earthquake hazard mitigation are increasingly concentrated on the needs of industrialized nations; the traditional solution to this problem is to rely on national and international aid programs. We would like to propose seeking solutions through two different activities.

The first activity is to research and document this problem, and then increase the awareness of the problem among the world's public and among the leaders of earthquake-threatened cities. If the public in both developing and developed countries were aware of this disparity between need and resources, perhaps they would help reduce it. To the extent which this disparity is unknown or viewed as natural, it will be tolerated. Of course, to some degree the disparity is natural: most earthquake hazard mitigation effort is paid by the direct beneficiaries of that effort, and, therefore, most resources are available to solve the problems of the industrialized nations. At some point, however, the disparity may appear unacceptably large.

Documentation of this disparity could also be directed towards the leaders of earthquake-threatened cities in developing countries. If they knew the risk their citizens were facing, and, as important, that the traditional sources of post-earthquake aid were overtaxed and would be unavailable at former levels following an earthquake, they might look for solutions. Not only are the international and national foreign aid agencies less able to provide post-earthquake assistance, but re-insurance companies will not be offering insurance at the same levels as before. After recent catastrophic losses, re-insurance premiums have greatly increased in cost. Local insurance companies may no longer be able to afford adequate coverage (Smolka, 1993). In 1983, Los Angeles business and industry leaders formed a council on emergency planning and preparedness in response to an announcement by that city's Mayor that the public and private sectors would have to rely on themselves for three days following a disastrous earthquake. Similar education of government leaders of earthquake-threatened cities might result in a similar efforts of self-reliance. These are possible benefits of education.

The second activity that we would like to propose is to increase technology transfer from earthquake-threatened developed cities to earthquake-threatened developing cities. "Technology transfer" is an old—perhaps even hackneyed—concept. The new aspect of our proposal is to not rely exclusively on international and national agencies to drive this transfer. Of course, any transfer that derives from such sources is welcomed, but based on the thoughts expressed above, it appears unreasonable to think that the needs of developing cities will be met entirely from such sources. We think that technology transfer fostered by such agents as cities themselves, international trade and professional societies, and nonprofit foundations might be productive. There are already sister city programs; one possible modification of this might be the adoption of an earthquake-threatened developing city by an earthquake-threatened developed city. Most basically, we believe that the most productive technology transfer of earthquake mitigation techniques will come between cities and professionals, not nations. Of course, international and national organizations could sponsor city-city and profession-profession exchanges.

7. Conclusions

If people, regardless of where they live, are eventually to be subject to the same seismic risk—the underlying principle of some hazard mitigation programs—more attention needs to be directed to the earthquake hazards of developing countries. Less ambitiously, if human and economic loss due to earthquakes in developing nations is not to increase, both absolutely and in comparison to earthquake losses in industrialized nations, new resources must be found and applied to earthquake hazards in developing countries. However, traditional methods either to increase resources devoted to the needs of developing countries or to shift resources from developed to developing countries may not work, given the current slow world economy, the rapidly-shifting global political alliances, the multitude of demands on local governments in developing countries, and the understandable priority that the wealthiest countries place on reducing the risk on their people and capital.

Two alternative approaches were proposed. The first involves an international effort to educate the public and to sensitize the leaders of earthquake-threatened cities in developing countries of the increasing risk of these cities and their diminishing prospects for foreign aid, either from governmental or private sources. Such an effort might eventually generate the needed support. The reports in this volume from earthquake-threatened cities are an example of this kind of effort.

The second promising approach is an international effort relying on city, professional and nonprofit agencies, to develop methods of efficiently transferring the technological advances pioneered in developed cities to developing cities. The first five papers in this volume, which suggest guidelines on how to perform earthquake damage scenarios and the following papers, which concern the Quito project, exemplify such international efforts at technology transfer.

References

Bilham, R. (1992), personal communication

Brown Jr., G.E. and D.R. Sarewitz (1992), Fiscal Alchemy: Transforming Debt into Research", Issues in Science and Technology, Volume VIII, Number I, pp 1-7.

Coburn, A. and R. Spence (1992), Earthquake Protection, published by John Wiley & Sons, pp.355.

Jones, B. G. (1992), personal communication.

OFDA (1990), Disaster History: Significant Data on Major Disasters Worldwide, 1900-Present, prepared for the Office of U.S. Foreign Disaster Assistance, Agency for International Development, by Lagat-Anderson.

Smolka, A., Munich Re-Insurance Company (1993), personal communication.

Tomblin, J. (1993), personal communication.

SOCIOECONOMIC AND POLITICAL CONSIDERATIONS IN DEVELOPING EARTHQUAKE DAMAGE SCENARIOS IN URBAN AREAS

SHIRLEY MATTINGLY
Director of Emergency Management
City of Los Angeles
200 North Main Street, Room 300
Los Angeles, California 90012

1. Introduction

In some areas of the world, like California, we think of ourselves as living in earthquake country, but damaging earthquakes can also occur from Kamchatka to Santiago de Chile. Some of the world's better known active faults, such as the San Andreas which bisects California, have been extensively studied, and the risks they present are relatively well understood. However, many other faults, undiscovered or unstudied, present unknown or ill-defined hazards to millions of unsuspecting urban dwellers. If cities and their inhabitants are to be protected from the devastation earthquakes can cause, this hazard must be understood and the risk reduced. The earthquake damage scenario (EDS) is a key tool to achieve both these ends.

This paper discusses the context which will define the parameters of the earthquake damage scenario study and loss reduction plan. It encompasses the human and environmental factors, the social, political and economic power structure and forces at work within the community and beyond. This is the "soft science" to which all the "hard science" which follows will be related, and this "soft science" is absolutely critical to the success of the project. If you go wrong here, you will expend vast amounts of effort and resources doing work which will not reduce the vulnerability of your community, which will continue to go about its business largely ignorant of or ignoring the risks.

To avoid this pitfall, this paper defines a critical, four-step process of analysis and actions. It is recommended that this process be followed by anyone considering initiating or taking part in the development of an urban earthquake damage scenario, whether you are international agencies or advisors, such as GeoHazards International, deciding where to pursue your next study, or the local residents of an urban area at some degree of risk, deciding whether or not to follow Quito's example in undertaking this kind of project. The key considerations and elements in the decisions about whether to do it and how to go about it will be discussed. The four-step process focusses on: (1) Analyzing your environment and likelihood of success; (2) Bringing together the essential elements; (3) Convening the key players and giving them a stake in the process; and (4) Institutionalizing your effort by creating a mechanism to facilitate change and keep seismic hazard reduction on the public policy agenda.

2. Our Working Premise

An underlying assumption forms part of the foundation of the Quito earthquake damage scenario project: earthquake-prone cities have much to gain from sharing earthquake reduction strategies and experiences. Really big earthquakes occur infrequently in any one city. They are so unusual that it is extremely difficult to get urban decision-makers to focus on the threat they pose and to dedicate scarce resources and time to mitigating their effects. Gaining and holding their attention, at the same time they are engulfed by other urban problems such as crime, housing shortages, and ever-increasing demands on health care delivery systems and other basic public services, require that they recognize the seriousness of the seismic hazard.

The earthquake threat becomes meaningful when people relate to the experience of others and personalize the risk, imagining what it would be like if it happened to their city. Short of suffering an earthquake oneself, the best ways to become sensitized to a quake's devastating effects are twofold: listening to the real life experiences of others, and defining the earthquake's impacts on one's own environment, through an earthquake damage scenario.

There is much to learn from the real life experience of other cities in terms of their risk reduction efforts, what has succeeded and what has not, as well as their actual earthquake events. For instance, Los Angeles has learned much from Tokyo and Mexico City by studying what those cities have accomplished and applying it in Los Angeles. The city of Quito has gleaned useful information through the Quito Project from the wide experiences of the international representatives on the Project's advisory boards.

The world's cities at risk of earthquake, regardless of differences in social, economic and political structures, share affinities. They all face other problems and needs considered more pressing by the public and public officials alike. Even in terms of the natural and manmade hazards they face, earthquake is only one of many. For instance, the release of hazardous materials is a danger virtually everywhere. Hazards related to wind and water threaten many cities, and severe landslides and flooding may be storm- or earthquake-induced.

In large earthquake-prone cities, we know we face special seismic risks due to the vast numbers of people and buildings and amount of investment at risk. Response to and recovery from a quake are especially complex processes because of the interdependencies in the socio-economic systems and the physical infrastructure and lifelines. In many cities, the need to preserve historic and cultural monuments vulnerable to earthquake is a growing concern. Complex human factors such as increasing unemployment and homelessness also complicate the issues faced by earthquake hazard reduction planners.

The process of preparing for and creating an earthquake damage scenario or EDS is intended to be suitable and applicable anywhere in the world, because it recognizes both diversity and commonality among earthquake hazard prone cities and it is based on an analysis of local conditions.

The Quito EDS Project, from its inception, built on experiences gained in diverse cities in California and Japan. The State of California's Department of Conservation developed earthquake damage scenarios for the major urban areas of Los Angeles and the San Francisco Bay, working in concert with utilities and local officials to ensure accurate portrayal of the impacts and understanding of the significance of the studies. Before the study of the Southern San Andreas Fault was released in 1982, the people of Los Angeles knew for generations that they lived with seismic risk, but their understanding of that risk was limited. The scenarios helped convince a variety of audiences--elected officials, bureaucrats, private businesses, and the public—that they would face horrendous problems when a major quake hits and that they needed to reduce their vulnerability. The scenarios resulted directly in a variety of actions by the municipal government, with the cooperation of the local private sector and the research community, to increase awareness and understanding of the threat and to reduce the vulnerability of the community.

Los Angeles provides one example of a city where earthquake damage scenarios have been used effectively. To learn from this experience, however, and help us apply it to other localities, we should consider some of the factors which predisposed Los Angeles to utilize rather than shelve the studies. When the studies were being developed, Los Angeles benefitted from:

(1) An existing city organization involved in mitigation and emergency planning, and this organization needed and wanted the EDS information;

(2) Willing political leaders who were receptive to the information and possessing the aptitude to use it in tackling difficult problems;

(3) A concern for the citizenry and good of the community which was strong enough to stand up to the conflicting interests of building owners and developers, at least part of the time;

(4) Inside advocates for the hazard reduction initiatives, i.e. city officials and employees who supported the projects;

(5) Close consultation between the people preparing the scenarios and these city supporters; and

(6) Good timing. The "political climate" was right.

Lessons gleaned from this experience have helped guide the Quito Project in development of the earthquake damage scenario for Quito.

3. Step One: Analyzing Your Environment

Before any city begins an EDS project, an analytical process should be undertaken to assist in assessing your likelihood of success. A series of questions should be addressed in the following approximate order.

3.1. WHY DO IT?

If your city is earthquake prone, why should you choose to do this rather than some other project to reduce vulnerability? How will this project fit in with, compete with, or stimulate other activities to reduce vulnerability to earthquakes or other hazards? These questions explore the inherent value of earthquake damage scenarios, which the experiences of California and Japan have shown to be very high and the current experience of Quito shows to have great potential. However, while we believe that scenarios are invaluable tools, developing one may not be the best project for your city right now. To answer these questions, you need to consider what your other options or alternatives might be, and what other considerations should be taken into account. What would make it worth it to go to so much trouble and expense?

The answer lies in defining your specific goal, your purpose in undertaking the project. Your goal could be to protect people, investments and economic livelihoods, and/or to correct hazardous conditions, and/or to avoid creating more hazardous conditions in your community. Presumably, the overriding concern will be life safety; however, you also need to consider both short term and long term economics. We are talking about making an investment decision. Do you want to invest time and effort in this earthquake damage scenario project? How would its potential impact, its potential for causing change, measure against its cost and against other possible projects or efforts toward the same goals? You will find tradeoffs and political considerations to be dealt with in making such decisions.

3.2 WHAT ARE THE POLITICAL CONSIDERATIONS?

Experience in California and other parts of the United States indicates it is often very difficult to get elected leaders to look beyond their term in office. Investments with short term paybacks are much more attractive than those with long term paybacks. Programs to increase earthquake safety may not save lives until decades in the future. Because of this, you need to consider the variety of possible motivators to gain the support of elected leaders. For instance, is there what is called "political mileage" for a community leader or elected official in supporting such a project? Are there votes to gain or lose? Could spearheading the project open new avenues for establishing contacts which could be useful to achieve other goals or to further other interests?

3.3 HOW VIABLE IS THE PROJECT?

Realistically, how viable will this kind of project be in your environment? Will you be able to attract supporters, people or organizations willing to provide financial support, technical support, and leadership? What kinds of obstacles will you encounter and how hard will they be to overcome? Typical obstacles include apathy or denial on the part of those whose support you need and lack of sufficient funding or expertise to do the kind of job you would like, with the level of accuracy you desire.

How much time and energy do you have to dedicate to what could be a protracted struggle? In Los Angeles, it took one destructive quake and ten years of work to get the City Council to pass an ordinance requiring the seismic strengthening of 8,000 unreinforced masonry buildings which were known to be an imminent collapse hazard in case of earthquake. Persistence and resilience are useful attributes when pursuing seismic hazard reduction.

3.4 WHAT ARE THE CONSTRAINTS?

Generalizing from experiences to date, it is clear that the utility and success of EDS study projects will be constrained:

> (1) If there is a lack of receptivity to the information or willingness on the part of decision makers to take on the challenge;

> (2) If the study does not reflect the language and values of the community it describes;

> (3) If there is inadequate cooperation and consultation between the people preparing the

scenarios and the potential users; and

(4) If there is no structure or organization to take it and use it; that is, you must build in some assurance that someone will emerge to carry it on.

3.5 WHAT ARE THE POTENTIAL OUTCOMES?

As part of the analysis preceeding the project, you should evaluate and seriously consider what can you expect its potential outcomes to be. You should not, of course, expect attitudes and practices to change overnight. The experience of the City of Los Angeles and its unreinforced masonry ordinance has already been cited: after the problem was identified and the solution defined, it took ten years to overcome the obstacles and enact the program and another twelve years to come close to completing the work.

On the other hand, do not underestimate the project's potential benefits, even beyond the immediate goal of reducing the earthquake hazard. For instance, by bringing together diverse community leaders to jointly oversee the project, you may create a cooperative working relationship which can carry over to other, unrelated projects. Use of the decision making model described below can help you decide how likely you are to effect change through the use of an EDS. In the end, scenarios provide a tool which does no good unless it is used.

3.6 WHAT IS THE LIKELIHOOD OF SUCCESS?

Whatever the stated goal of your EDS project, in order for it to make an impact, it must change people's behavior. Your goal may be "to reduce future earthquake impacts; to make mitigation and preparedness a part of how daily business is conducted" in your community. To achieve a change in attitude or awareness and behavior requires action in the form of an active decision-making process. Your likelihood of success with the project will relate strongly to how well you have planned in advance and how well the necessary elements described below are brought together. These key elements for the project are a clear perception of the problem, a viable solution, the key players including advocates, and an opportunity to bring them together.

4. Step Two: Bringing Together the Elements for Success

Four separate elements or streams must converge at the same time in order for decisions or policy to be made:
- Problems
- Solutions
- Participants, and
- Opportunities for choice.

Whether or not something is a problem depends on the observer. In the case of the seismic hazard problem, in your city there may be a low level of awareness that the problem exists. Of course, one of the advantages of conducting an EDS study is that it can play a big role in increasing the realization and comprehension of the problem. Even though there may be some level of awareness of the hazard, many people may feel that money to be spent on the EDS should be spent addressing higher probability problems, something more likely to affect them.

The Garbage Can Theory would lead us to predict that there would have to be sufficient agreement among the decision makers that a problem exists, and there would have to be a proposal that seems to be an acceptable solution to the problem. Solutions to the seismic hazard are known; the EDS study is a first step, to be followed by a seismic hazard reduction plan of actions to reduce the hazards and vulnerabilities defined by the EDS. There are models of such plans and model processes to follow to develop them. Technical solutions to seismic hazard reduction problems also exist: earthquake effects can be predicted and mapped, and structures can be designed and constructed or retrofitted to withstand collapse.

The third decision element is comprised of the participants in the decision making process. Where will the initiative come from? Who should initiate it, and who should lead or push the project? Who would be most effective in your environment: a government insider or official, an academic, or an international group or person such as GeoHazards International? If the answer is you, then ask yourself, who do you know, and who can you access? Who can you engage as the project's advocates and supporters? A number of key people will need to be involved, and fulfilling the need for leadership and advocacy is critical.

The fourth decision element is opportunity. Since destructive earthquakes are infrequent, they are not often an object of public concern. When should you undertake the project? According to the Garbage Can Theory, the right time is when you have a convergence of all four decision elements. What we have often found in pursuing seismic hazard reduction programs is that the opportunity most often arises when a strong earthquake occurs in a populated area. Policy decisions and actions most often occur when public concern is heightened. The earthquake which awakens concern could be your own or somebody else's. It could be half-way around the world or in your backyard, but of course it will have the biggest impact if the public in your city are affected one way or another by the quake. They may relate to it because of media accounts, such as by viewing dramatic footage on television.

In Los Angeles, where there is a large population with roots and family in Mexico, Mexico City's 1985 quakes had a tremendous mental and emotional impact on the public and politicians alike. It was reflected in two ways: first, the Hispanic population clamored to help, and relief flights were rapidly organized to take needed goods to the striken city. The City of Los Angeles organized, incredibly, a trainload of 64 bulldozers and heavy pieces of equipment for debris removal, and sent them to Mexico City. Second, that quake served to remind the residents of L.A. that they themselves were also at risk. The Mexico City quake motivated Los Angeles officials and politicians to provide funding for a new program to develop local capability for urban search and rescue.

The 1989 Loma Prieta earthquake is another example of an event spurring action not only in the affected areas but elsewhere as well. Hundreds of miles away in Los Angeles it resulted in the acceleration of the program to reinforce hazardous unreinforced masonry buildings. It also encouraged the electorate to vote, months later, to increase their own taxes in order to provide funding to reinforce City-owned buildings and bridges at a cost of $396 million dollars. Similar programs were approved in several other cities.

In the state capitol California legislators introduced over 300 pieces of legislation directed at reducing the impacts of future earthquakes and approved many new programs, including retrofitting of existing vulnerable bridges statewide and establishing a satellite communications system to link state and local emergency response agencies.

Based on experiences like this, it is clear that there is a window of opportunity for advocating seismic safety policies following earthquakes. However, one must be ready to take advantage of them, because the window may not stay open for long, and new events will spawn new concerns and priorities. Further, in the absence of awareness-heightening earthquakes, you may need to orchestrate or generate an opportunity to raise interest in initiating the EDS project.

5. Step Three: Identify and Convene the Key Players

Determining who will oversee and participate in the study is a critical step in the strategy for success. Social science research in the U.S. on implementation of vulnerability reduction measures has suggested that such measures can succeed if they respond to the values of the stakeholders, the interested parties in the community. These people should be involved as early as possible in the process.

5.1 IDENTIFY THE KEY PARTICIPANTS

To identify the EDS team, you must know and eventually penetrate the formal and informal power structure in your city. Think about how things work, who influences whom, and who has input into what kinds of decisions. What kinds of intergovernmental relationships exist between city and provincial and state and national governments, and how do these relationships figure into government policy decisions?

It also would be useful to look at who makes investment decisions. To what extent are investment decisions regulated by government and what level of government? To what extent are these decisions influenced or affected by insurers and financial institutions? These questions should help you define the key people whom you want to involve in the project. You might also want to consider what role public opinion plays in public policy decisions, and to what extent public opinion is influenced by the media. You may decide the media has a role due to its ability to influence the public.

Who do you expect to use the scenarios? Who are your clients? In identifying multiple clients, keep in mind that they need to be involved from the beginning, and who your key clients and what their interests are will dictate many of the characteristics of the study. In Los Angeles, the EDSes have had three major uses:

> (1) A planning tool providing better data for government decision making. The damage pattern assumptions formed a realistic basis for tabletop and functional exercises to test emergency response and recovery plans. Information about the expected damages to lifelines systems motivated city officials to form new planning partnerships with utilities, the business community, and other governmental agencies.

> (2) A marketing tool to promote mitigation and preparedness measures. The scenarios made the potential impacts real and convincing to many audiences, and they have been used as the basis for public awareness campaigns. They have also motivated public dedication of resources to reduce vulnerability.

(3) A basis for further studies. For instance, the scenarios led to an urban planning consultant's study entitled Pre-Earthquake Planning for Post-Earthquake Rebuilding which delineated damage which would be caused to private structures throughout various sectors of the city by several scenario quakes. This study further defined the city's problems and led to solutions, resulting directly in the development of a blueprint for recovery and reconstruction from a major quake.

The uses for which you design your EDS study may be similar to or different from these, but the extent to which you can identify likely uses and users prior to beginning the project will impact positively on your likelihood of success.

5.2 CONVENE THE KEY PLAYERS

Actual participation and involvement by the right people is critical to success. After you have gone through the process of identifying the key players, you have to get them around the table. The technique used in Los Angeles and subsequently in Quito was to enlist the Mayor to convene the key players, and this worked well. In your environment, however, it could be another well known or influential individual or you could use another strategy altogether, based on your knowledge of what or who could bring the right people together.

Who is convened should reflect the interests of the community and the ability to influence or create change. In Quito, the group which was convened to provide project oversight and institutional input was comprised of about fifteen individuals, representative of a broad spectrum of interests, painstakingly selected by the local project organizers in consultation with OYO Pacific, now GeoHazards International. This group was identified as the Conseil d'Affaires Sociales et Economiques (CASE). In addition to the CASE's initial two-day meeting, they have convened a second time formally and a third time informally through participation in a two day workshop which has proven to be a critical step in the scenario development process.

The workshop, to be more fully described in another paper, represents the critical point in the CASE members' involvement, in that it was the point at which they took ownership of the scenario and integrated it with their own reality and perceptions. In the evaluations of a similar workshop in Los Angeles, participants responded extremely positively to the experience of identifying the impacts of the damage scenario themselves.

6. Step Four: Institutionalize

To make a lasting impact, you must institutionalize your effort by creating a mechanism(s) to facilitate change and place and keep seismic hazard reduction on the public policy agenda.

Whom you select as your project's steering committee or advisory board, how you organize them, and who leads the effort will impact on this step, how you institutionalize. You need a complete strategy going in, of course,z with some built in flexibility to take advantage of unfolding events or unanticipated power relationships. In Quito, all levels of government were included in the CASE, as well as representatives of the utilities, financial institutions, engineering and construction interests, and the insurance industry. The Mayor is chairman of the CASE, and this sets the framework for the organization which will emerge to carry on the work set in motion by the EDS study. With the CASE reporting to the Mayor, it will fall to him to take the initiative in institutionalizing the effort. The leadership of the advisory panel you create to oversee the study will strongly influence how your city will approach institutionalization.

There is no single model of how to institutionalize a seismic hazard reduction effort. In the United States, the State of California's Seismic Safety Commission has served as a model for similar panels in other states. Under agreement with the U.S. Federal Emergency Management Agency, the Seismic Safety Commission has developed a manual to assist other states; this useful manual, **Creating a Seismic Safety Advisory Board: A Guide to Earthquake Risk Management**, can be obtained from the Commission at 1900 K Street, Suite 100, Sacramento, CA 95814, USA.

In addition to California's successful state government-sponsored Commission, several other models have been used for different purposes, to meet varying needs. In Los Angeles, what has worked to accomplish one objective proved inadequate to deal with another, so other avenues were explored and other mechanisms instituted. For instance, the City of Los Angeles has created three separate panels to promote hazard reduction and preparedness. First, the Mayor and City Council created by ordinance an organization within the City government structure to be responsible for preparing for, responding to, and recovering from all types of emergencies. It includes all local governmental agencies and relies on the active participation of the Mayor and top City officials. It promotes teamwork, mutual planning, training, exercising, and public awareness, and does a fine job within its jurisdiction. However, it soon was realized that it was also necessary for business and industry to engage in emergency planning, preparedness, and efforts to reduce seismic and other hazards.

A cooperative, voluntary private sector--public sector partnership was formed to engage and provide a linkage with the business sector. This panel, the Business and Industry Council for Emergency Planning and Preparedness, or BICEPP, shares emergency plans, provides training, and forms the basis for providing mutual aid during an emergency to its members.

Both of the above organizations are led by boards of directors and both are active, through monthly meetings and other activities, in promoting hazard reduction. Nevertheless, it was recognized that neither could actually fulfill the role of a seismic safety commission or board, so the City's Mayor created in 1993 a Blue Ribbon Panel on Seismic Hazard Reduction. This panel most closely parallels the Quito CASE and the State of California's Seismic Safety Commission. It includes representatives from the business, scientific and engineering communities, utilities, government, building owners and insurers. It is charged with evaluating new scientific data regarding seismic risk and seeking new incentives and proposals for risk reduction which will be consistent with the city's current economic and political realities. It will propose and promote a comprehensive approach for mitigating the hazards of a plausible major earthquake in Los Angeles. Staff support for the panel is provided through the City's Building and Safety Department.

What does it take to establish such an advisory board, panel or commission? Essential ingredients include: leadership and sponsorship from the mayor or other strong leader; respected, influential, and interested membership; a clear and well-conceived charge or goal; and staff and necessary financial and information resources. The board's major efforts should be directed toward the development and implementation of a seismic hazard reduction plan for the city or region.

Any city contemplating the initiation of an earthquake damage scenario project should think through and plan for this entire process, including institutionalization, which the EDS will empower the city to undertake.

SEISMOLOGICAL INFORMATION NECESSARY FOR BENEFICIAL EARTHQUAKE RISK REDUCTION

BRUCE A. BOLT
*Departments of Geology and Geophysics
and Civil Engineering
University of California
Berkeley, California 94720*

1. Introduction

Seismic hazards are classified into four categories: (1) ground shaking, (2) fault rupture, (3) tsunamis and seiches, and (4) secondary hazards (i.e., avalanches, land and mud slides, differential ground settlement and soil liquefaction, floods from dam and levee failures, and fires). Any city or developed urban area may be threatened by these hazards, depending on the seismogenic conditions of the region (see Figure 1). In this paper, I discuss the way that geological and seismological studies can lay a basis for crucial decision-making on the appropriate level of preparation for the earthquake threat. The discussion gives a critique of not only the type of decision analysis necessary but also the practical results that may be expected from detailed seismological studies.

The reduction of earthquake hazards in urban areas requires a team effort between geologists, seismologists, geotechnical engineers, civil engineers, urban planners, and government officials. It must be stressed that a seismologist alone cannot provide the necessary judgment to make informed assessments of the ground motions which should be considered in a comprehensive study of the earthquake hazard. This has been recognized by the definition of what is called the "Safety Evaluation Earthquake" (SEE) as a preferred scenario earthquake for seismic hazard mitigation planning and for engineering retrofit and adoption of building codes. The SEE differs from a "Maximum Credible Earthquake" (MCE) that might be considered for the area, in that it takes into account the types of structures which are vulnerable to the various earthquake hazards in the area, as well as the risk that is acceptable to the community in light of other social considerations. This paper gives some background to these important matters.

2. Choosing Scenario Earthquakes

Almost every large earthquake has features worthy of close study in the faulted zones, in the damaged areas, and in the instrumental records of the strong ground shaking. Since 1985, three cases stand out.

The first case is the 1985 Mexico earthquake, of surface wave magnitude M_S 8.1, which relative to its great energy did little damage near its source along the west coast of central Mexico. The tragic twist was the major enhancement of seismic waves in limited parts of Mexico City at a distance of 420 km from the source. Although local building codes had already incorporated a factor to allow for soft soil layers, the amplification and duration of the ground shaking in parts of

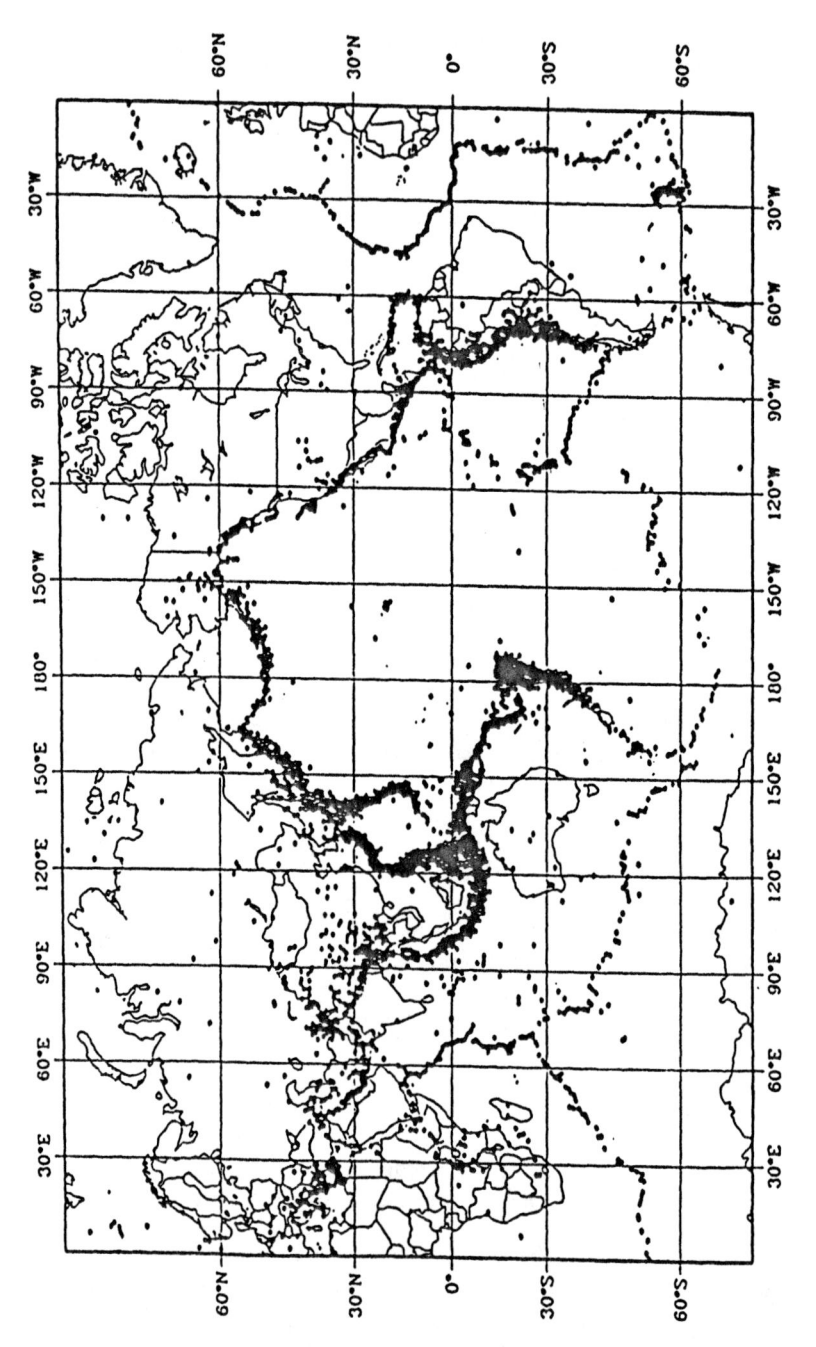

Figure 1. Map of global shallow seismicity 1963-88, M>5, depth<70 km
(courtesy USGS)

Mexico City were greater than had been expected; as a result, a number of 10- to 24-story reinforced concrete structures collapsed and 8000 persons died (Rosenblueth, 1988).

The Armenian earthquake on December 1988, M_S 6.9, resulted in 25,000 deaths and the displacement of half a million people. It caused economic losses that reached $16 billion, with particularly high damage in the cities of Spitak, Leninakan and Kirovakan. Although Armenia has its share of competent seismologists and engineers, clearly the presence of such experts was not sufficient for earthquake risk reduction to prevail against the contending societal, economic, and political pressures in that country. The result was an unconscionable amount of fatalities and economic dislocation. A great many buildings were not seismically resistant, even though the area has been known from antiquity to be geologically unstable (Bolt, 1993).

The Loma Prieta, California, earthquake of 17 October 1989, M_S 7.1, focused U.S. public attention on earthquake safety more than any other case in recent decades. There was major damage in the cities and urban areas of the San Francisco Bay region, including Oakland, Santa Cruz and San Francisco (Benuska, 1990; Housner, 1990). Its conjunction with a World Series baseball event explains why many people were—uncharacteristically—safely in their timber-frame homes at 5:04 p.m. to watch TV coverage rather than in normal freeway commuter traffic or in congested areas of critical danger. Because the eyes of the nation were fixed on San Francisco, the scenario of an actual damaging earthquake was played out visually before a wide audience. Afterwards, public reaction was strong that significant resources must be expended to attain greater earthquake safety. Political action by the Governor and Legislature was immediate, with new state laws and programs.

Effective steps for damage reduction in future earthquakes in urban areas depends strongly on understanding their properties. As stated in the 1906 objectives of the Seismological Society of America, "It is possible to insure ourselves against (earthquake) damage by proper studies of their geographical distribution, activities and effects on buildings." In this respect, certain fundamental seismological lessons were learned from studies of the three earthquakes above, and it is useful to consider these briefly.

In recent years, the concept of a "characteristic earthquake" on a particular fault has been formulated. If faulting processes repeat, there is a hope that the prediction of future earthquake behavior can be reduced to a known set of basic earthquake types. Confidence in modeling future earthquake sources is highly relevant to vulnerability reduction of urban areas because response analyses of critical engineered structures and lifelines such as viaducts and utility systems require the definition of an earthquake source that is reliably established by geological and seismological analysis. Modern quantitative analyses of the necessary quality were pioneered by the extremely low-risk requirements for operation of nuclear power reactors in earthquake regions. An example of such analysis, stimulated by the Loma Prieta earthquake experience, is the study in the San Francisco Bay area of earthquake sources and likely ground motions needed for engineering retrofit and design of the Golden Gate Bridge, the Bay Bridge (a span of which fell on 17 October 1989), and the Bay Area Rapid Transit railroad (BART) extensions.

The first key question in strong ground motion specifications concerns the relevant maximum earthquake; that is, what is the realistic maximum limit to the fault length that will rupture? In attempting to estimate the probability of rupture lengths, geological evidence such as the identification of fault segments that are separated by such mechanical barriers as changes in fault strike or lateral fault offsets, and seismological evidence, such as historical hypocentral patterns, are used. The second problem is the estimation of the appropriate attenuation factors for the seismic waves from source to site (Joyner and Boore, 1988). The third major question is the effect of surficial alluvial and soil layers on the incoming seismic waves.

Although significant progress has been made in understanding the tectonic, geological and geophysical basis of earthquakes, case histories such as Mexico City, Armenia, and Loma Prieta show that drawing conclusions about future ground motions to be used in the design of structures is not a routine exercise. In practice, the uncertainties involved are allowed for by the application of safety factors. Consequently, larger-than-necessary ground motions, given a realistic lifetime of structures, may be the result. Such conservative judgments would generally be applauded if there was not also a need, on the grounds of other pressures and societal demands, for limits on construction costs.

3. Prediction of Ground Shaking Intensity

Industrialized countries with earthquake hazards are now in the third era of seismic risk. The first era was characterized by rapid, almost uncontrolled industrial and urban growth. Seismological knowledge was rudimentary and instruments to measure ground shaking were primitive. The modern development of California, for example, from about the middle of the last century, was punctuated, but hardly affected, by major earthquakes in 1857 in southern California, and in 1865, 1868, and 1906 in northern California.

The years after 1920 mark the initiation of a second era of seismic mitigation. The devastation and 143,000 deaths in Tokyo from the shaking and great fire in the 1923 Kwanto earthquake stimulated many earthquake studies. Heavy damage to the public schools in the Long Beach, California earthquake of 1933 led to political intervention and the passage by the state legislature of the Field Act, which set strict construction standards for public schools. During the subsequent 50 years, a band of dedicated pioneers of many nationalities has studied earthquake problems and applied quantitative techniques to strengthen structures against earthquake forces. In the 1960s, large research expenditures resulted from the strict regulatory requirements for earthquake-resistant design of nuclear power plants in a number of seismically-prone countries. This activity brought to bear advanced technical thinking on risk reduction by earthquake scientists and engineers, and a great amount of basic geological, seismological, and engineering knowledge was thereby transferred into much wider practice to critical structures such as large dams, hospitals, utilities and high-rise buildings through building codes, city ordinances and so on.

This second era of earthquake-damage mitigation ended in the 1970s. For example, the 1971 San Fernando earthquake in California, M_S 6.5, sharply illustrated the diversity of damage in a densely-populated urban area. It stimulated the establishment in 1974 of the California Seismic Safety Commission with the task to develop broad policy throughout the state. In China, the tragic losses of over 250,000 people in the 1976 Tangshan earthquake were clearly seen as the result of the collapse of non-seismic-resistant buildings.

The hallmark of the present or third era is the ability to achieve more quantitative and cost-effective efforts in risk reduction in a broad urban setting, with particular attention to mega-cities and densely populated industrial complexes. Contemporary damaging earthquakes such as those mentioned, as well as Coalinga, California in 1983, Philippines, 1990 and Erzincan, Turkey, 1992, stimulated this aspect of hazard-reduction efforts. The balance between risk reduction and general benefits has become a central issue (Panel, 1989). Part of the debate hinges on the effectiveness of past research and its application and justification for future efforts. On the seismological side, one critical advance in the last two decades that can be thoroughly justified as cost-effective is the successful instrumental recording of strong ground shaking in earthquakes in California, Japan, Italy, Turkey, India, China, Mexico, Chile and elsewhere. From these field records, the strengths, durations, and frequencies of the large seismic motions near the source of the waves can be directly measured. These ingredients are essential for prediction of the shaking of the ground at specified places in future earthquakes.

In brief, the availability of strong motion recording now makes it possible, given a specified active fault source, for seismologists to predict numerically the radiated seismic waves. By computer calculations, these motions can then be transferred through the complex of rock structures between the fault and the specified site (Joyner and Boore, 1988), and there through any soil column, to yield the expected ground shaking. This ability means that, in principle, overlays or templates of the expected seismic intensities can be produced for any seismically vulnerable urban development such as Manila, Tokyo, Patras, Mexico City, and San Salvador. Of course, adequate data on local soil properties, topography and so on must be available. Such maps are basic to development planning in urban areas, to the assessment of the vulnerability of older structures, and to the design of earthquake-resistant new structures.

In practice, the full benefit of quantitative intensity mapping sometimes requires substantial one-time costs because subsurface properties are often unmeasured. It is known that strong ground shaking varies dramatically from one earthquake region to another. In the 1989 Loma Prieta earthquake, Oakland and the San Francisco peninsula were 70 km distant from the shaking source yet severe damage occurred in pockets, leaving most of the Bay Area scared but unscathed. Buildings on the campus of the University of California at Berkeley, built on rock, were subjected to a maximum horizontal ground acceleration of one tenth of gravity ($0.1g$), and the strongest motion lasted for only a second or two. There was no significant structural damage. At the same distance from the fault rupture, part of the Cypress I-880 viaduct in Oakland, built on soft soils (Bay mud), failed (Housner, 1990). At that site, ground motions were about three times the acceleration and five times the duration of those at the nearby campus. The significant differences in the strength of shaking on soil and rock in the Loma Prieta earthquake were not a surprise technically. They had been described only a few years before in soil engineering amplification studies of the 1985 Mexico City damage patterns.

When the various sources of variability in earthquake shaking are taken into account in earthquake scenarios for an urban region, the reliability of intensity maps for application of engineering codes for mitigation of large future earthquakes can be made quite adequate. It is essential also to compare and incorporate in detail the effects of past great earthquakes in the region of study. As in the 1989 Loma Prieta earthquake, it is known, for example, that the intensity in the great 1906 San Francisco earthquake also varied markedly. In some places in the Sacramento Valley high intensities were recorded, but in Sacramento the felt intensity of shaking was low.

Such historical damage reports are not always straightforward to interpret because intensity is a strong function of the frequency of the waves and the duration of the shaking. Special professional training and experience is required. Unfortunately, the earthquake intensity maps produced just after the 1906 San Francisco earthquake did not specify in sufficient detail the wave frequency of significant shaking that might be expected in the future. Although improvement in intensity mapping techniques has occurred in recent years, the pattern of the Loma Prieta earthquake made it clear that, for maximum benefit in decision-making on rehabilitation and new construction, maps are needed that allow for earthquake source variability and the effect of soils. The same conclusion is true for seismically-vulnerable urban regions outside of California.

4. The Assessments of Earthquake Occurence and Related Risk

All statements of risk contain, either explicitly or implicitly, elements of probability (Keeney et al., 1986). Probability statements on the chance of earthquakes or earthquake vulnerability, both for engineering design and for policy decisions, are becoming common.

Numerical statements of such odds are sometimes difficult to interpret unless they are compared with odds for other hazards. Thus, in California the risk of death per year to an individual from a motor vehicle accident is about 1 in 4,000; from earthquakes in the most exposed Californian metropolitan areas, the risk is perhaps 1 in 50,000. But much more is involved in interpreting risk than these simple propositions. The *individual* risk clearly varies with individual situations. There is also a collective or *societal* risk. Such refinements in conceptual models have been much discussed in recent years and with the improved modern geological databases, allow probability risk statements to be made that are helpful to engineers, planners and government officials.

One illustration is the assessment of the odds of large earthquakes along segments of the San Andreas and other major active faults in California (Working Group, 1990). After the Loma Prieta earthquake, pre-earthquake probability evaluations of this type were given much publicity. The combination of different bodies of seismological observations had indicated that there was a better than 1 in 2 chance of a major magnitude earthquake occurring in the San Francisco Bay Area in the next 20 years. A more-specific study had given the chance of a 6.5 to 7 magnitude earthquake along a 30-km-long segment of the San Andreas fault in the southern Santa Cruz Mountains as 30% in 30 years. This value was higher than for the adjacent San Francisco Peninsula segment of the San Andreas fault to the north and led to the impression that the earthquake was predicted.

Care is needed in formulating such statistical statements. Among the explanations required are: What is the *range* of earthquake size involved rather than the specification of a particular magnitude? And are such statements predictions at all or only summary accounts of past events? Public acceptance depends on making only well-defined simple statements, in which the extent of reliability is clearly expressed.

Probability models have now been used to prepare ground shaking hazard maps for many countries and regions, for example, the United States, New Zealand, Japan, Italy, Canada, China, and many others. These maps give the expectation of exceedance in a given time (such as 500 years) of seismic intensity parameters, such as ground accelerations (Committee, 1989). In computing the expectation of these parameters, the older concept of discrete hazard zones, drawn mainly on the basis of the historical seismicity, has been largely abandoned and replaced by the contours expressing rate of occurrence of earthquakes of various magnitudes weighted by geological evidence of active fault systems. To be effective, such maps must, of course, be incorporated, at least broadly, in building code provisions, with the explicit understanding that a balance of risk is implied between the odds of larger shaking and the high cost of overdesign.

5. Decisions on Acceptable Risk

Even when probability models are worked out in appropriate ways and are clearly explained, there still remains the difficulty of lack of agreement on the major goals of hazard abatement. Unquestionably, the trend in recent years in many earthquake-prone countries has been to maximize life safety rather than economic loss. For example, the Uniform Building Code in its

1988 and earlier editions specifically states that, "The purpose of this code is to provide minimum standards to safeguard life or limb, health, property and public welfare while regulating and controlling design and construction" (UBC, 1988). The practical problem, of course, is how to manage joint treatment of life safety and property damage. Not only may there be incompatibilities, but, when minimal standards apply, damage to structures can be significant even though casualty loss is low.

The tested effectiveness of modern building codes has indicated that, in general, older structures present the greatest risk. The trade-off between life safety and reconstruction costs is well illustrated by recent studies of the seismic resistance of state-owned buildings in California. It is estimated that over $20 billion worth of state properties are involved, and much of this property is vulnerable to damage. The California Seismic Safety Commission, after testing the proposed hazard evaluation methodology on 40 state-owned buildings, recommended that priorities for upgrades of state-owned structures should be based on a benefit-cost ratio (BCR), defined as the number of lives saved per reconstruction dollar. As a consequence, structural engineers were retained to provide a prioritized list of state-owned structures based on the BCR method. Such a list was essential to obtain cost estimates so that the state government could fund a realistic schedule of upgrading.

The estimation of the BCR measure of benefit depends on the evidence that certain classes of construction perform poorly in earthquakes whereas other classes resist the shaking. The capacity can be quantified through a life safety ratio (LSR), which is a factor calculated from the expected number of fatalities per 10,000 occupants before reconstruction, given the class of the structure and the appropriate shaking intensity for the specified seismic zone. Thus, from experience, unreinforced masonry buildings have been allocated a particular LSR value and reinforced concrete structures another.

The computational equation is

$$BCR = \frac{(LSR)(ECO)(SCF) - (LSRG)(ECO^*)}{10,000\ (RC)},$$

where ECO (equivalent continuous occupancy) is the average number of persons occupying the building in some appropriate period such as each day of the year, SCF (seismicity correction factor - a point where seismology enters the assessment) depends on the earthquake occurrence rate in the zone, and RC (reconstruction cost) is the cost to rehabilitate this class of structure in order to reduce the hazard to the prespecified life-safety goal (LSRG). The single asterisk denotes the value after reconstruction.

BCR ratings deserve wider national and international use. In California, they have been used to allocate funds for more detailed engineering studies and, when combined with additional engineering evaluation, to set priorities for reconstruction. After the Loma Prieta earthquake reemphasized the grave danger of collapse in earthquakes of certain classes of structures, such as unreinforced masonry buildings, in certain locations, it was realized that the rating list for the state-owned buildings required reexamination. Because the BCR values are highly sensitive to the LSR ratings, several modifications have been recently suggested that give weight not only to structural material but also to details of the structural system.

In conjunction with the BCR method with its emphasis on life safety, a separate system of structural seismic performance ratings based only on engineering judgment and intensity maps has also been developed for use on California buildings. Each building is judged as "good," "fair," "poor," or "very poor." For example, a "poor" structure would be expected to suffer significant structural and nonstructural damage leading to appreciable life hazard. These ratings are dependent on structural engineering considerations and not on the occupancy of particular buildings.

Past difficulties in achieving more widespread earthquake safety suggest that emphasis on a life safety criterion to drive rehabilitation also deserves critical reexamination (Fischoff et al., 1981). One of the lessons after the 1989 Loma Prieta shaking was the seismic fragility of many crucial facilities in modern urban and industrial society. Failure of "lifelines" — electrical power, water, sewage, communication, and transportation — can prostrate the economy. The severance of the San Francisco Bay Bridge on 17 October 1989 and a widespread power failure in San Francisco, 70 km from the seismic source, prove this point. The same problem has long concerned authorities in Japan. The soaring real estate values in Tokyo continue to encourage the filling of coastal land tracts, and these have become heavily-populated industrial and commercial zones. It is estimated that a magnitude 7.9 earthquake, similar to the Great Kwanto 1923 earthquake, could produce liquefaction and hence disrupt lifelines over 26.5 square miles of reclaimed land along the city's major waterways. Such economic loss has evidently been regarded as acceptable, perhaps because of lack of widespread public debate.

An illustrative case comes from the serious damage that occurred to unreinforced structures on the campus of Stanford University in the Loma Prieta earthquake. The costs of damage rehabilitation are estimated to exceed $160 million. There is little doubt that the damage would have been significantly more severe at the Stanford campus, given the types of unreinforced structures at risk, if the seismic source had been closer or of longer duration. In such a case the institution and its complex research facilities would have been seriously diminished as a center of higher education for years. The lesson is that, in decision-making on risk reduction, the failure to allow for the functioning of key institutions such as hospitals, banks and utilities, as well as life safety, can have the gravest consequences. Earthquake damage scenarios for urban areas must address this aspect of earthquake hazard mitigation.

A remark needs to be made about earthquake insurance as an important component in reducing earthquake risk (The Earthquake Project, 1989). The greatest overall benefit from earthquake insurance accrues when there is a link between the availability of low-cost insurance and a requirement to upgrade the seismic resistance of the structure. In the case of homes, inspections at the time of purchase should establish premium levels according to the degree of risk inherent in the dwelling and its location (Roth and Sam, 1988). This assessment, again, depends upon reliable seismic hazard maps. Governments need to examine the trade-offs between disaster insurance and disaster relief programs in order to optimize the advantages of insurance mechanisms. An important side benefit could be the widespread reduction of risk, not by government regulation, but by market incentives contained in graduated insurance premiums.

6. Seismic Hazard Estimation Procedures

First, we need to have geological input. This starts with considerations of the geotectonic conditions of the region. This section of the study must recognize the important differences between (a) seismogenic zones which are at the margins of the tectonic plates along subduction collision zones, such as Japan, Taiwan, Central and South America, Greece, and Italy; (b) earthquakes which occur along more passive margins, such as the San Andreas fault in California and the Anatolian fault in Turkey; (c) intraplate seismicity, such as central China (e.g., Tangshan), the midwest of the United States (e.g., New Madrid tectonic zone) and the Rhine Graben of Europe; and (d) continent-to-continent collision, such as India and Nepal. Urban areas away from the active plate margins are subject to intraplate earthquakes which in some ways are more difficult to assess, but are exemplified by a great deal of the seismic hazard to many large cities worldwide.

The next step is to consider the active faults of the region and the type of displacements occurring on them, as well as their age. This stage requires the development of structural geological maps when these are not already available. Maps of the structural geology around the urban area are also of great importance. Attention has to be given to the occurrence of bedrock scarps, widespread differential erosion and offsets in overlying deposits. Assessments are needed of probable length, continuity, and type of movement on local faults. Geophysical work is often useful here, including magnetic and gravity measurements across the area. Maps of landslides, major settlements or inundation maps for flooding potential are also required.

Next is the seismological input. The first step is to document the earthquake history of the region by looking at historical earthquake catalogues and studying all sources of felt reports. The resulting statistics should show the locations, magnitudes and ground motion intensities for each historical earthquake where this is available, and the whole information must be transferred to appropriately scaled regional maps. The historical record must be complemented by the (usually much shorter) instrumental record of earthquake locations in the vicinity, often out to distances of some hundreds of kilometers from the urban area in question, together with the earthquake magnitudes. From these data, recurrence curves of the frequency of regional earthquakes can be constructed using earthquakes even down to small magnitudes. Such curves are the basis of probability risk estimates for both comparative purposes and decision making on urgency of and response to the hazard.

At this stage, the seismologist must rely on reviews by engineers knowledgeable of local building practices, of the available historic records of ground shaking damage, and of intensity information in the urban area, often at a time well before the present development and construction. After joint consultations, an estimation of the variation in the seismic intensity and a prediction of the likely peak ground accelerations in the SEEs can be made.

The geological and seismological evidence assembled in the above sections can then be used to predict the appropriate scenario earthquake. This is usually given in rather broad terms, involving the moment magnitude of the largest earthquakes expected in an appropriate interval of time (say 500 years) and the effect of distance from the source to the urban area. It is then necessary to interpret the way that the ground shaking from such a source or sources would be distributed throughout the urban development. The prediction aims at stating which earthquakes would give the most severe ground shaking throughout the area, but in a way that is most helpful to engineers. Thus, highrise buildings are mostly affected by long- period seismic waves while small, weakly-constructed buildings are mainly affected by high frequency shaking of the ground. There may be specific faults on which surface rupture occurs which should also be listed together with the likely mechanisms of the rupture on them (strike-slip, thrust and so on).

Sometimes an urban area or city region is fortunate enough to have already an instrumental record of ground shaking produced by a regional earthquake. When this is the case, the above general estimates can be partly validated against the actual record and the key building response parameters related to the recorded amplitudes (such as peak acceleration or ground displacement) and duration of strong ground motion as well as the response spectrum (see Figure 2). Such an accelerogram should have been obtained on rock within the urban area or on soil deposits. In the latter case, the effect of the soil layers on the accelerogram can be judged by comparison with synthetic records obtained by adjustments made using theoretical models of seismic waves through soil.

7. Procedures in the Quito Earthquake Project

The general considerations discussed earlier and the specific procedures summarized in the last section were applied in the Quito Earthquake Project carried out in Ecuador in 1992-93. The illustration is an important one because Quito is located in a region where a considerable amount of geological and seismological observations have been made, and where a reasonably long historical record exists of earthquake effects on the city as it developed from colonial times. On the other hand, the breadth and depth of these studies was only moderate in many respects. There have not been the resources to conduct exhaustive and extensive tectonic studies, nor has the geological mapping of Ecuador been extended as thoroughly as that in, say, Italy, Japan or California. In Quito itself, detailed information on the effect of the topographical relief and the particular soil conditions of the urban area were not available and simple estimation procedures needed to be employed.

In this case, a seismology technical advisory group recommended that the risk study required three seismic intensity distribution maps. These should be based, in turn, on three distinct, hypothetical earthquakes: (1) a 6.5 earthquake source rupture on the Quito fault which runs below the city. It was thought that although this was not the most likely source of the shaking, it was significant because of its proximity; (2) a 7.3 earthquake on the Asedro fault, 80 km north of Quito. Regional earthquakes usually 80-120 km from the Quito metropolitan area have been a repeated source of earthquake damage for the past three centuries. This fault was thought to be the probable source of earthquakes in 1587 and 1868, and possibly a 1987 earthquake, all of which inflicted significant damage within the city. This source was considered representative of a maximum regional earthquake that might occur; and (3) a 7.7 earthquake along the subduction zone on the Pacific coast to the west of Quito. A known seismic gap on that plate margin makes a significant earthquake there likely in the near future.

The next step was to select appropriate attenuation laws from published studies in other seismic regions because insufficient indigenous instrumental measurements are available. The subduction zone scenario earthquake required special consideration on attenuation. Finally, the single strong motion accelerogram recording of the 1987 earthquake in Quito was used to guide the synthesis of likely time histories and response spectra.

8. Present Seismological Abilities

In seismology there has been an undoubted improvement in providing earthquake intensity scenarios in recent decades, mainly as a result of the understanding gained from the many recordings of strong ground motions from various earthquake source types and sizes, and for various site conditions. New techniques of ground shaking simulation by computer modelling have been introduced and have been validated to a limited extent.

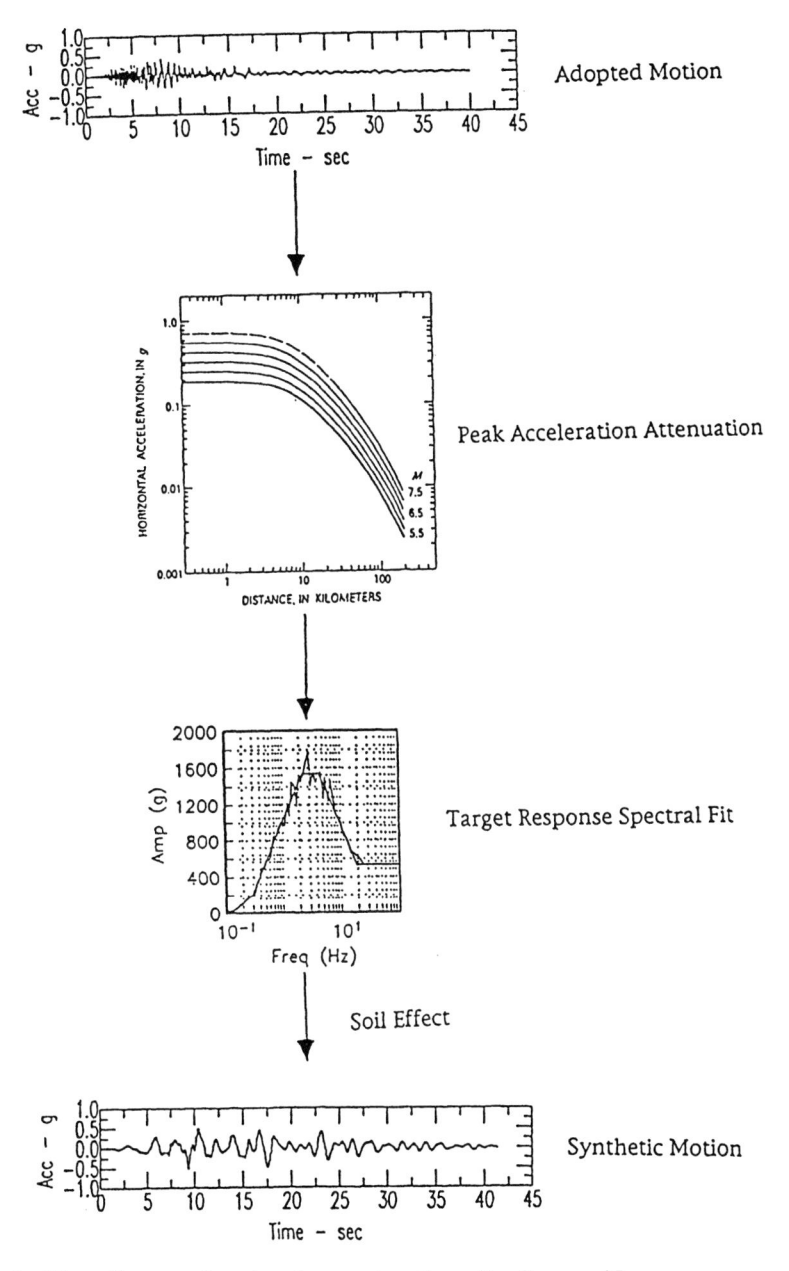

Figure 2. Flow diagram showing the construction of a site-specific response spectrum for a scenario Safety Evaluation Earthquake. Recorded ground motion in a similar earthquake elsewhere is modified for attenuation and spectral compatibility and then for soil effects.

Increased support by legislatures in a number of countries has occurred in the past decade for earthquake safety efforts, particularly in those countries that have a high earthquake risk. The nongovernmental effort and support has also improved in recent years, with successes by individual groups and cities in obtaining significant new funds from industrial and other private sources, including international and national construction companies. Yet such sources will always remain relatively limited and government programs must provide the focus and impetus for a cohesive and effective program for seismic risk reduction in provinces, cities and counties.

In the mustering of broad political support, the cost-effective aspects of earthquake risk reduction provide the main argument for action to achieve the ultimate safety goal. The benefits of mitigation are obvious, but history shows that earthquake risk reduction is characterized by bursts of activity and political support after earthquakes have already caused death and injury and wrecked the economy. The decay curves for such beneficial reactions unfortunately have a life of only a few years before public effort recedes.

The present era of earthquake safety programs coincides with the International Decade of Natural Disaster Reduction (IDNDR). This initiative has been agreed to by the United Nations as a major effort to reduce, by the year 2000, the risk from earthquakes, volcanoes, floods, and other natural hazards. Specialist technical organizations such as GeoHazards International can help developing countries with knowledge, experienced analysts and education to achieve the IDNDR goal. The interaction is, of course, not a one-way street. A well-planned program of hazard analysis, risk assessment and emergency preparation can leave long-lasting local benefits and is transferable to other urban centers.

Can the promise of an era of minimal earthquake risk be delivered in a decade? The most difficult problem is finding the capital, against competing economic demands, for retrofit of vulnerable buildings and lifelines. Earthquakes such as the 1989 Loma Prieta earthquake in California make it quite clear, however, that major seismicity near metropolitan areas will have serious economic effects, not only regionally, but nationally. After such earthquakes, industries and institutions will not be able to operate effectively for a considerable time, reducing the living standards of the whole country. If the seismological evidence indicates that strong ground shaking in a metropolitan area is likely in the next 10 or 20 years, a program of risk reduction should be undertaken at once. Despite the remaining uncertainties in predicting scenario earthquakes and technical gaps in earthquake engineering, there are really no insurmountable reasons why earthquake risks to both the individual and society cannot be reduced during the International Decade to levels comparable with those of more familiar threats.

References

Benuska, L., *Loma Prieta Earthquake, Reconnaissance Report*, Earthquake Spectra, 6, suppl., 1-448, 1990.

Bolt, B.A., *Earthquakes*, W.H. Freeman, New York, 1993.

Committee on Seismic Risk, *Earthq. Spectra*, 5, 675, 1989.

The Earthquake Project, *Catastrophic Earthquakes: The Need to Insure Against Economic Disaster*, National Committee on Property Insurance, Washington, DC, 1989.

Fischoff, B. et al., Acceptable Risk, Cambridge Univ. Press, New York, 1981.

Housner, G.W., (Chairman), *Competing Against Time*, Report to Governor George Deukmejian from the Governor's Board of Enquiry, California Office of Planning and Research, North Highlands, 1990.

Joyner, W.B. and Boore, D.M., in *Earthquake Engineering and Soil Dynamics II -Recent Advances in Ground-Motion Evaluation*, Geotechnical Special Publication 20, American Society of Civil Engineers, New York, pp. 43-102, 1988.

Keeney, R.L. and von Winterfeldt, D., *Risk Anal.*, **6**, 417, 1986.

Panel on Earthquake Loss Estimation Methodology, *Estimating Losses from Future Earthquakes*, National Research Council, Washington, DC, pp. 1-82, 1989.

Rosenblueth, E., *The 1985 Mexico Earthquake*, Earthquake Spectra, **4**, No. 3, 1-65, 1988.

Roth, R.J. and Sam, S.O., *California Earthquake Zoning and Probable Maximum Loss Evaluation Program,* California Department of Insurance, Los Angeles, 1988.

Uniform Building Code, International Conference of Building Officials, Whittier, CA, 1988.

Working Group on California Earthquake Probabilities, *U.S. Geol. Surv. Circ. 1053*, pp. 1-51, 1990.

GEOTECHNICAL ASPECTS OF THE ESTIMATION AND MITIGATION OF EARTHQUAKE RISK

W.D. LIAM FINN
Department of Civil Engineering
University of British Columbia
2324 Main Mall
Vancouver, B.C. Canada
V6T 1Z4

1. Introduction

This paper describes the procedures for characterizing the impact of geotechnical factors on the hazards and risks imposed by earthquakes. Definitions of the concept of risk and its constituent parts proposed by UNDRO (United Nations Disaster Relief Organization) (Varnes, 1984; van Essche, 1986), have been adopted here. A modified form of the definitions is given here specifically for earthquake risk.

1.1 HAZARD, VULNERABILITY AND RISK

- *Seismic hazard* (H) means the probability of occurrence within a specified period of time and within a given area of a potentially damaging earthquake or level of ground shaking.

- *Vulnerability* (V) means the degree of loss to a given element or set of elements at risk resulting from the occurrence of a specific seismic hazard. It is expressed on a scale from 0 (no damage) to 1 (total loss).

- *Specific risk* (R_s) means the expected degree of loss due to the specified seismic hazard. It may be expressed as the product of H times V.

- *Elements at risk* (E) means the number of elements at risk in the category for which vulnerability is being considered. It can mean population, number of buildings or any other collection of entities for which loss can be calculated.

- *Total risk* (R_t) means the expected number of lives lost, persons injured, damage to property, or disruption of economic activity due to the occurrence of the specified seismic hazard, and is therefore the product of the specific risk (R_s) times the elements at risk (E). Thus:

$$R_t = (E)(R_s) = (E)(H)(V) \qquad (1)$$

These terms are useful in focusing attention on the major elements in earthquake risk. Different professional skills are needed to deal with the different elements. Seismologists define the basic seismic hazard distribution maps for motions on rock or stiff sites. Geotechnical engineers redefine the basic seismic hazard distribution for the effects of local soil conditions and play a role also in evaluating the vulnerability of lifelines which are primarily affected by ground deformations. Structural engineers play the major role in developing vulnerability functions for various classes of buildings.

2. Examples of Risk Assessment

Two examples of seismic risk evaluation in developing countries carried out by multidisciplinary international groups will be described briefly in order to provide an integrated view of the process of risk assessment. The global view of the process presented by the examples makes it easier to understand the role of geotechnical hazard zonation in the final estimation of risk to be described later.

The examples also demonstrate the effective improvements in risk estimation that can be achieved despite limited resources and a shortage of instrumental data on seismic response in the study areas. The process of organizing what information is available for estimating seismic risk also provides valuable information for planning risk-reduction measures. Data and resources are usually scarce in undeveloped countries, and it is encouraging to realize that significant contributions can be made to earthquake risk mitigation by supplementing whatever data is available by judicious selection from the worldwide database.

2.1 NORTH SULAWESI PROVINCE, INDONESIA

After the April 1990, North Sulawesi earthquake (M_s = 7.3), an assessment of seismic risk in North Sulawesi Province was conducted by the Indonesian Geological Research & Development Center and the US Geological Survey under the auspices of the International Decade for Natural Disaster Reduction (Thenhaus et al., 1993). The principle components of this study were: (1) mapping the ground motion hazard in the region; (2) developing sources of information for the estimation of economic loss; and (3) demonstration of the techniques for estimating loss at a particular site in the region.

The data for earthquake occurrence were compiled from the US National Earthquake Information Service (NEIS) database. From this data, and tectonic and geological knowledge of the area, seismic source zones were developed for probabilistic analysis. Available ground motion recordings for Indonesia were insufficient for establishing attenuation laws for scaling ground motions for earthquake magnitude and distance from the site. The attenuation relation for Japan, developed by Fukushima and Tanaka (1990), was selected for use on the basis that the attenuation characteristics for the two Western Pacific Island arc settings would be similar. This attenuation law was developed for the Japanese subduction zone and incorporates terms for different site conditions, ranging from rock to soft soils. Using standard procedures (Bender and Perkins, 1987) for developing ground motion estimates at various probabilities of exceedance, Thenhaus et al. (1993) developed horizontal ground acceleration estimates for rock and soft soils with a 10% chance of exceedance in 50 years. This is the hazard level adopted for use in building codes in Canada and the United States. These ground motion patterns are shown in Figure 1. It is clear that the ground motions on the soft soils are almost twice as large as those on bedrock at similar locations from the source zones.

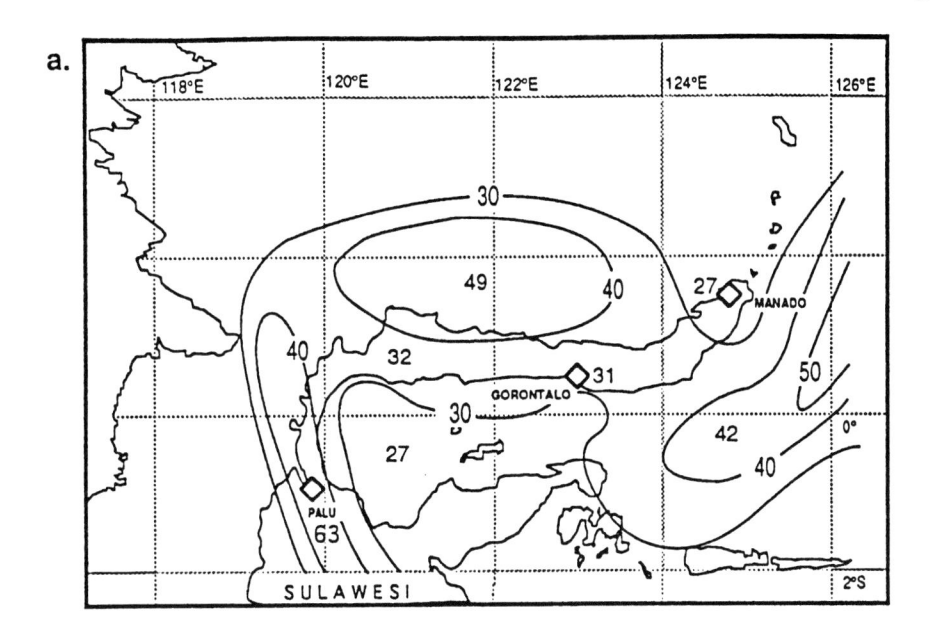

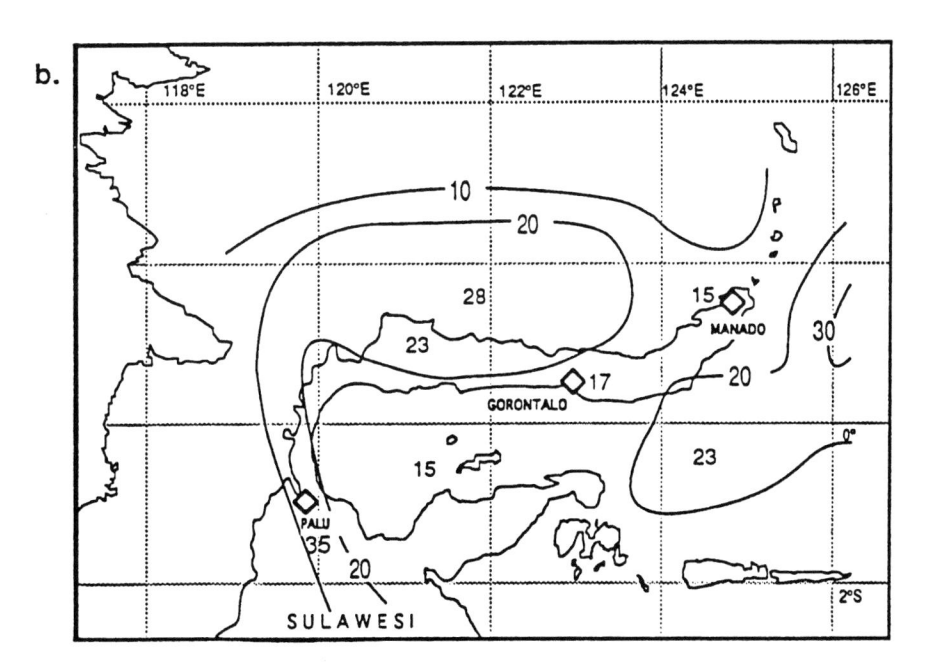

Fig. 1. Map of North Sulawesi showing peak horizontal acceleration values in (a) soft soil, and (b) in rock having a 90 percent probability of not being exceeded in 50 years. Contours are in percent of gravity (after Thenhaus et al., 1993).

The effect of local soil conditions is clearly shown for the town of Gorontalo by the Modified Mercalli Intensities derived from a study of the damage to masonry buildings during the 1990 earthquake. These intensities are shown in Figure 2, superimposed on the depths of surface soils. The greatest damage specified by MMI-VIII occurs over the deepest surficial deposits, and damage decreases with decreasing thickness of soil.

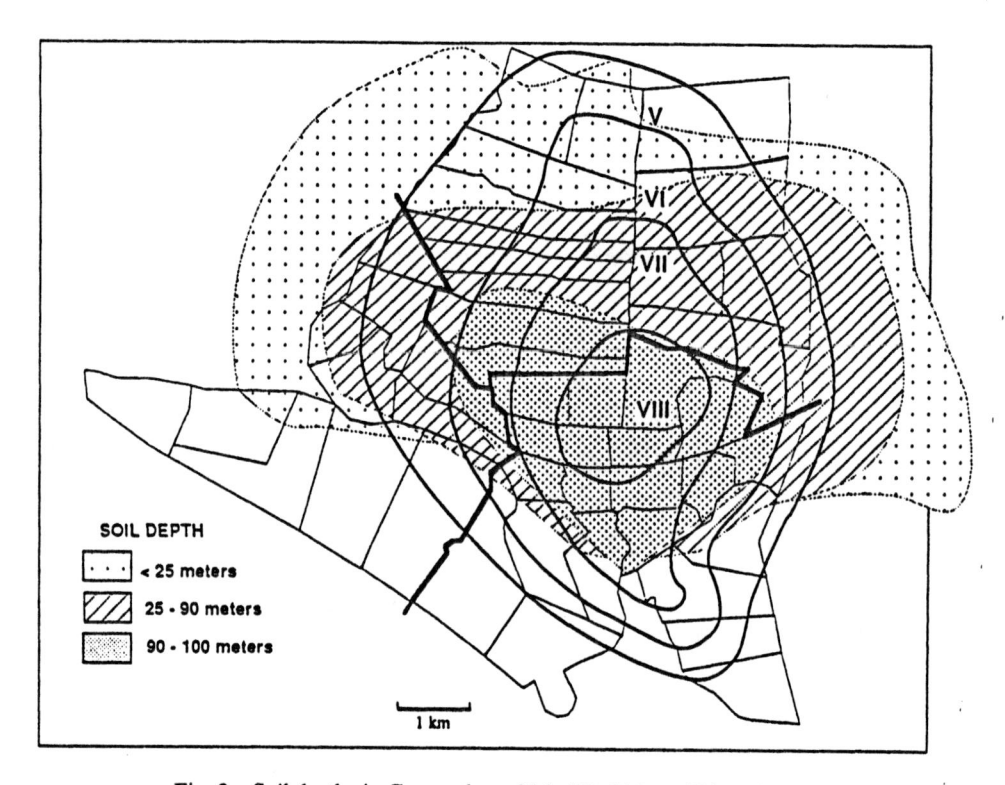

Fig. 2. Soil depths in Gorontalo and Modified Mercalli isoseismals of the 18 April 1990 earthquake (after Thenhaus et al., 1993).

The MMI corresponding to the computed distribution of peak horizontal ground accelerations, a_{ph}, was estimated using the equation by Trifunac and Brady (1975)

$$MMI = (\log a_{ph} - 0.14) / 0.30 \qquad (2)$$

This equation is based on 374 peak horizontal acceleration components with MMI's between III and X on sites varying from hard granitic rock to thick quaternary alluvium. Therefore, it applies to sites of some undefined degree of softness. It is likely to overestimate the intensity at hard sites and underestimate intensity at deep sites. The MMI computed by Eq. (2) were modified on the basis of the MMI estimated from damage during the 1990 earthquake.

The expected damage to masonry housing in Gorontalo was estimated using vulnerability curves which express the percentage loss as a function of MMI. A vulnerability curve for masonry structures in Gorontalo was developed on the basis of observed damage during the 1990 earthquake. This curve is shown in Figure 3 together with vulnerability curves for different types of construction, including masonry construction in the United States. It is interesting to note that the vulnerability curve for masonry construction in Gorontalo is much lower than the US curve. Thenhaus et al. (1993) suggests that the better performance of the Gorontalo structures is due to the fact that the roof is supported independently of the walls, whereas in US construction the walls are load-bearing. This example shows how important it is to calibrate whatever procedures are being used by local data. From the percentage loss, the estimated number of buildings and the local value of housing, an estimate of the economic loss due to damage to masonry buildings, can be estimated. Loss estimates for Gorontalo during the 1990 earthquake are given in Table 1 (Thenhaus et al. (1993)).

Table 1. Estimation of Economic Loss to Masonry Dwellings in Gorontalo from the 18 April 1990 Earthquake (M_s=7.3)

Intensity[1]	Value[2]	Vulnerability[3]	Loss[4]
5	$2,666,315	0.00	0
6	$3,014,736	0.075	$ 226,105
7	$2,824,210	0.15	$ 423,631
8	$1,826,315	0.3	$ 548,842
		TOTAL	$1,198,578

[1] intensity level observed in the 18 April 1990 earthquake from Figure 2.
[2] estimated value of masonry housing contained in each MMI contour.
[3] estimated vulnerability obtained from Figure 3.
[4] estimated losses corresponding to each intensity level and total estimated loss.

When the procedures had been calibrated using the data from 1990 earthquake, earthquake risk studies were then conducted for a large subduction earthquake M=8.5.

This example identifies the major steps in assessing seismic risk and some of the procedures used to compensate for lack of local seismic and site response data. The study also showed the significant impact that local soil conditions can have on the level of seismic hazard and on the percentage of damaged buildings.

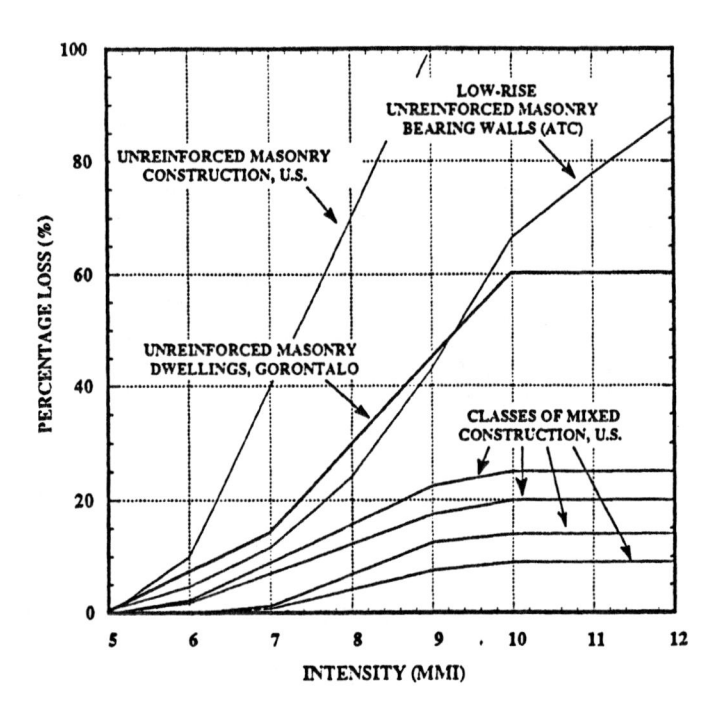

Fig. 3. Comparison of the estimated vulnerability of masonry housing in Gorontalo (heavy line) with vulnerability of various construction types in the United States (light lines). Vulnerability relationships for US data are from Algermissen (1989). Vulnerability relationship for low-rise, unreinforced masonry, bearing walls is from the Applied Technology Council (1985) for California construction (after Thenhaus et al., 1993)..

2.2 QUITO, ECUADOR

The Quito project was initiated with the primary aim of reducing the social and economic consequences of future earthquakes in Quito. The strategy adopted was to estimate the effects of the occurrence of a few representative earthquakes and then devise practical countermeasures to reduce these effects.

The study was managed by GeoHazards International of San Francisco, California. Direct funding and some professional help for this project came from OYO Corporation in Japan. The major part of the work, in establishing the characteristics of the local building stock and establishing local soil conditions, was carried out by professors and students at the Escuela

Politecnica Nacional. Limited funding was supplied for the support of the students. The seismic characteristics needed for the study were developed by the Escuela Politécnica Nacional, the Institut Français de Recherche Scientifique pour le Développment en Coopération (ORSTOM) who are conducting long-term studies of seismicity in Ecuador. An international advisory committee, of which the writer was a member, provided general advice and occasional reviews of progress. A major technical objective of the study was to establish a distribution of Modified Mercalli intensity over the metropolitan area from which damage could be assessed using appropriate vulnerability curves.

The procedure followed was to divide the metropolitan terrain into 19 zones based on topography, geological characteristics and soil conditions. The peak horizontal accelerations on bedrock for each of these zones were estimated for the target earthquakes. For each zone, a dynamic analysis was carried out on a soil column typical of the zone to estimate the local amplification of ground motions. At the site where ground motions were available, the data were used to calibrate the computed seismic response. The original distribution of peak ground accelerations throughout the metropolitan area was revised to take the amplification factors into account and then converted to MMI. These MMI were then calibrated using the damage statistics from the 1987 earthquake and any perceived effects of ground slope.

Damage ratios were established for various classes of structures using vulnerability functions linking MMI and percent damage. The functions are based on those developed by the Applied Technology Council for use in the United States (ATC-13, 1985) but were calibrated and modified for local and regional conditions. The damage ratios for reinforced concrete frames with shear walls and flat slabs (7 storys) are given in Table 2 as examples of the kinds of results obtained from the study. Similar damage statistics were developed for each of the 21 classes of buildings into which the total building stock was divided. Damage estimations for critical facilities and lifelines are still under development.

Table 2. Loss Statistics on Reinforced Concrete 7-Story Buildings in Quito.

MMI Intensity Damage Ratio	VI	VII	VIII	IX	X
0.00	2.5	0	0	0	0
0.50	59	8.6	0	0	0
5.00	38.5	89.2	66.4	11.7	0.4
20.00	0	2.2	33.6	83.9	56.9
45.00	0	0	0	4.4	42.7
80.00	0	0	0	0	0
100.00	0	0	0	0	0
Total (%)	100 (2.22)	100 (4.94)	100 (10.04)	100 (19.35)	100 (30.62)

This project is in the process of being completed and final reports are not yet available. Therefore, specific conclusions cannot be drawn yet about many aspects of the study. But it is a very cogent example of how useful estimates of seismic risk can be developed with limited resources guided by state-of-the-art professional advice. The results of the study are expected to significantly advance the opportunities for earthquake risk mitigation by land use planning, improvement in design and construction procedures, and the planning of post-disaster responses that can cope with the expected damage scenario.

In both the Sulawesi and Quito examples, the important parameter for estimating damage was Modified Mercalli Intensity, MMI. Because most damage in past earthquakes has been correlated with MMI, this will remain the prime parameter for estimating damage in the foreseeable future. Reliable estimates of damage depend on the MMI and the availability of suitable vulnerability curves for the region at risk. Therefore, the core activity in any earthquake risk study is the development of the distribution of MMI and the establishment of appropriate vulnerability curves.

There are three aspects to establishing the MMI. They are: (1) basic probabilistic seismic hazard maps for rock or firm ground conditions; (2) modification of this map to reflect the properties of surficial deposits, particularly the impact of soft soils; and (3) the conversion of the hazard map to a map of seismic intensity distribution (SID). Determining the effects of foundation soils in modifying the seismic hazard and the seismic intensity is the primary role of the geotechnical engineer. A major task for structural engineers is the adaptation of vulnerability curves from one set of environmental conditions to another, in order to reflect the different type, quality and control of construction.

The examples did not consider directly landsliding or liquefaction. The effects of these cannot be incorporated directly into the seismic hazard map based on ground motion parameters. Their effects on damage potential must be investigated separately. The total estimate of damage will be the sum of the damage due to the distribution of seismic intensity and the damage due to liquefaction and landslides.

3. Zoning for Ground Motion Levels

3.1 FUNDAMENTAL ASPECTS OF SITE CHARACTERIZATION

The foundation for effective microzonation of seismic hazard is a critical understanding of how local soil conditions affect ground motions. Although surface topography and buried geological structures affect site response, it is often impractical to take these into account, and most sites are assumed to be horizontal layers on an elastic half-space or, in some instances, on a rigid base. A useful insight into some aspects of site response can be achieved by considering a very simple site model.

Consider a simple uniform site consisting of an elastic layer of thickness, H, mass density ρ_1 and constant shear wave velocity, V_{s1}, resting on an elastic half-space with mass density ρ_2, and shear wave velocity, V_{s2}. The natural periods of the surface layer are given by $T_n = 4H/(2n-1)V_{s1}$. The impedance ratio of the site is defined by the ratio $\kappa = \rho_1 V_{s1}/\rho_2 V_{s2}$. The seismic response of the site to input motions depends on the impedance ratio, the natural periods of the site, T_n, and the relationship between T_n and the frequency content of the input motions. The impedance ratio can be included in site response analyses but is too awkward to incorporate directly in general methods of zonation.

The period, T_n, is a very important variable. Sites with the same period, T_n, will all have the same response to the same input motion. Constant period can be ensured by maintaining a constant H/V_{s1} ratio while allowing the layer thickness, H, and the shear wave velocity to vary.

Therefore, neither H nor V_{s1} alone would be a reliable index for the response of these sites to the same input.

Sites with constant velocity, V_{s1}, and variable thickness, H, will have different responses to the same input motion. However, now there is a one-to-one correspondence between the layer thickness, H, and the site period, T_n. Therefore there will be a strong correlation between response and layer thickness and for these sites layer thickness will be an effective parameter for microzonation. Layer thickness plays an important role in the modification of seismic input for code design.

Similarly, for sites with constant layer thickness, H, and different periods, the shear wave velocity, V_{s1}, would be an effective parameter for microzonation. Previous ground response studies in California (Borcherdt, 1970; Borcherdt and Gibbs, 1976; Borcherdt et al., 1975, 1978) have shown that for sites underlain by comparable thicknesses of various geologic units, the average amplification of ground motions increases with decreasing average shear wave velocity. This tends to confirm the key role played by V_{s1}, other things being approximately equal. Borcherdt (1990) and Borcherdt et al. (1991) have developed zonation techniques for soil amplification effects based on the average shear wave velocity in the upper 30 m of the ground, although they do not restrict this to cases of constant thickness.

The examples considered above involve constant input to sites of varying characteristics. Now consider a uniform elastic site subjected to varying input corresponding to motions from different earthquakes. Since the layer is elastic the amplification factors do not depend on the amplitude of the input motions, only on their frequency content. For simplicity assume that the input initially excites peak response at the fundamental period of the site. If the frequency content of the input motion is shifted to shorter periods, then the response at the fundamental period will decrease and, as the frequency shift continues, greater amplification of the motions may occur at one of the higher natural frequencies. Analyses of the response of the old lakebed sites in Mexico City indicate that the period of peak response was associated with the properties and local thicknesses of the surface clay layers rather than the fundamental period of the entire sedimentary deposit (Finn and Nichols, 1988).

These results indicate that layer thickness or the average shear wave velocity over some depth cannot always be reliable quantitative indicators of site response, but field data suggests that in many practical cases they give useful quantitative information on the relative response of different sites.

How layer thickness and velocity have been used to characterize sites in practice will now be reviewed.

3.2 CHARACTERIZATION OF EFFECTS OF LOCAL SOIL CONDITIONS ON GROUND MOTIONS

Damage patterns in Mexico City after the 1985 Michoacan earthquake demonstrated conclusively the significant effects of local site conditions on seismic response of the ground. Peak accelerations of incoming motions in rock, generally less than 0.04 g were amplified about 5 times on the clay soils of the old lakebed with devastating effects for structures with periods close to site periods (Finn and Nichols, 1988). In the 1989 Loma Prieta earthquake, major damage occurred on soft soil sites in the San Francisco-Oakland region where the spectral accelerations were amplified 2-4 times over adjacent rock sites (Housner, 1990). Clearly seismic risk evaluation should incorporate the effects of local soil conditions. How to do this effectively is the central problem of effective microzonation for both hazard and risk.

3.2.1 *Site Factors*. The parameters, such as peak horizontal ground acceleration or peak ground velocity which define seismic hazard at a given probability of exceedance are estimated from the activity of seismogenic sources and the attenuation of the parameters with distance from the source. The expected values from the attenuation laws are related to rock or firm soil conditions. Therefore, the effects of local soil conditions are not included in this estimate. These are usually taken into account in building codes by the introduction of a foundation factor, F, or a site coefficient, S, into the formula for computing the seismic shear force in a building.

Codes in the United States such as the Uniform Building Code (UBC, 1989), use four site categories, S, related to broad categories of soil conditions. The four categories, S, shown in Table 3 range from 1.0 to 2.0. These factors are based on results of site-specific analyses, recorded ground motions and assessments of damage to buildings on various soil profiles. The proper factor to use in a given case is based on the best available knowledge about the site under consideration and engineering judgement.

Note the key role played by the depth of the soil profile in discriminating between the different factors for site amplification. This is clearly a process for microzonation that relies on the thickness of the surface layer.

Hensolt and Brabb (1990) have published a microzonation map of San Mateo County, California, showing the distribution of the UBC soil factors. This map shows the relative hazard at various sites in terms of the factor S. If this kind of map is overlaid on a basic seismic hazard map for firm ground conditions, a revised hazard map can be drawn that reflects in a significant way the effects of local soil conditions. Such a map is feasible in most metropolitan areas as basic soil data is available from construction records. This suggests a very basic elementary way for zoning metropolitan areas for the effects of local soil conditions.

Table 3. Site Coefficients

Type	Description	S Factor
S_1	A soil profile with either: a) A rock-like material characterized by a shear wave velocity greater than 2,500 ft/sec or by other suitable means of classification; or b) Stiff or dense soil condition where the soil depth is less than 200 ft.	1.0
S_2	A soil profile with dense or stiff soil conditions, where the soil depth exceeds 200 ft.	1.2
S_3	A soil profile 40 ft or more in depth and containing more than 20 ft of soft to medium-stiff clay but not more than 40 ft of soft clay.	1.5
S_4	A soil profile containing more than 40 ft of soft clay.	2.0

Microzonation based on broad and distinctly different soil categories has the advantage that rather distinct patterns of ground response are associated with each type. Normalized response spectra typical of response on the different types of soil represented by the foundation factors are shown in Figure 4.

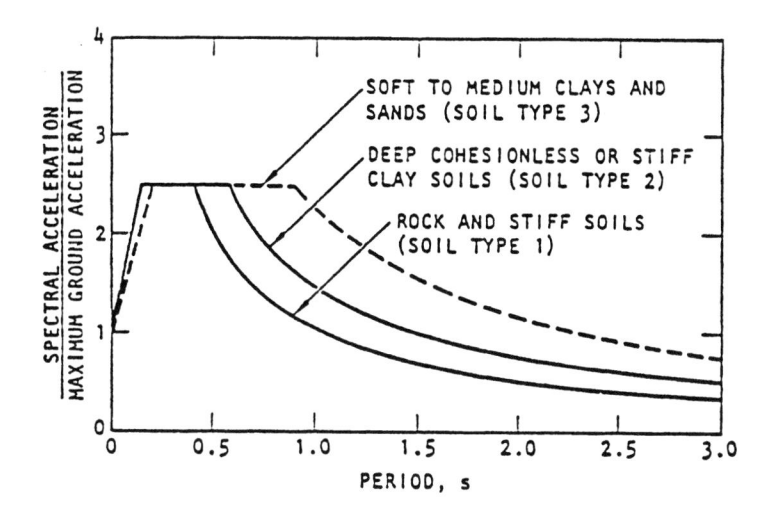

Fig. 4. Normalized Spectral Curves Suitable for use in Building Code (after Seed and Idriss, 1983).

3.2.2 *Site Characterization Using Ground Motion Parameters.* The response of soil sites to weak ground motions is essentially elastic and therefore controlled by the site periods. As discussed earlier over the great variety of geological units, neither sediment thickness nor shear wave velocity alone can be reliable indices of site period and hence of site amplification effects. However, at many sites (e.g. Treasure Island Site), the surface motions are modified significantly by the near-surface sediments within depths of less than 20-30 m. The stiffness characteristics of these sediments play a large role in controlling motion amplification. Therefore, the average shear wave velocity of near surface sediments, being a direct measure of the stiffness, may be a reasonably effective parameter for defining microseismic zones. Strong motions have sufficient intensity that nonlinear effects may become important enough to invalidate conclusions about site response based on weak motion data.

Recent correlations by Borcherdt et al. (1991), based on all available data show a strong correlation between shear wave velocity and the average horizontal spectral amplifications from recorded weak motion data,

$$\text{AHSA} = 701/V_S \tag{3}$$

where V_{s1} is the local shear wave velocity in m/s averaged over the top 30 m of the sediments.

Of major interest for microzonation studies is how well the response patterns based on these weak motions are applicable to the strong motions recorded during the Loma Prieta earthquake in the same area.

The average horizontal spectral amplifications from recorded strong motions is

$$AHSA = 598/V_s \tag{4}$$

Borcherdt et al. (1975) established a relationship between average horizontal spectral amplification (AHSA) determined from weak motion recordings and intensity increments for different formations inferred from to the 1906 San Francisco earthquake, given by

$$dI = 0.27 + 2.70 \log(AHSA) \tag{5}$$

This provides a rough method for modifying intensity I for local soil conditions.

3.3 ANALYSING THE EFFECTS OF LOCAL SOIL CONDITIONS

The effects of local soil conditions on waves propagating from bedrock to the surface are usually evaluated by 1-D shear beam analysis on the assumption that the site can be modelled as a layered half-space. The analyses are capable of modelling nonlinear effects and of identifying the more important characteristics of the surface motions; the resonant period of the site, the lengthening of the period with increasing intensity of shaking, and the amplification or deamplification of motions at various frequencies (Schnabel et al., 1972; Finn et al., 1978; Lee and Finn, 1978). These effects have been very clearly identified in ground motions recorded during earthquakes. Analyses of this kind were used in the Quito project.

The state of knowledge on amplification and deamplification of ground motions at soft soil sites in terms of peak acceleration has been summarized recently by Idriss (1990) based on data from the Mexico City and Loma Prieta earthquakes, and on 1-D response analyses using the shake program SHAKE (Schnabel et al., 1972). The results are shown in Figure 5. The rock motions developed by seismic hazard analysis can be modified in accordance with the median relationship as an approximation to the actual hazard at soft sites. Note that the variation in amplification factor decreases with increasing intensity of rock motions and eventually at an intensity level of about 0.4g the motions begins to attenuate.

The decrease in amplification with increased level of input motion reflects the nonlinear behaviour of soft soil sites during strong shaking. The difference in response of soft sites to low and high levels of input motion is clearly evident in the differences in amplification factors noted between aftershock and mainshock motions from the 1989 Loma Prieta earthquake at the Treasure Island site in San Francisco Bay.

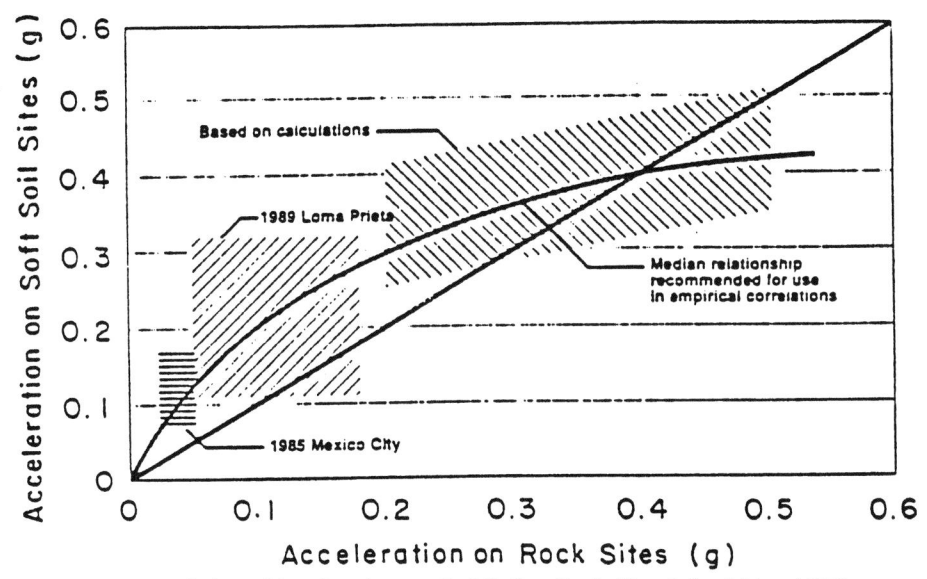

Fig. 5. Variation of Accelerations on Soft Soil vs Rock Sites (after Idriss, 1990).

The amplification factors for surface motions recorded at the Treasure Island Site during the Loma Prieta earthquake of 1989 relative to the rock motions at adjacent Yerba Buena Island are shown in Figure 6. The solid line shows the variation in the NS spectral ratio for the first 5 seconds of the shear wave in the main shock in the period before liquefaction took place at the site. The shaded area is the 95% confidence region for the NS spectral ratios of 7 aftershocks (Jarpe et al., 1989). The amplification factors are drastically reduced in the strong motion phase, although still 2 or greater over a wide frequency band of engineering interest. The peak acceleration at the surface of 0.16 g shows an amplification of about 3. The reduction in amplification with increased shaking is due to the nonlinear stress-strain response of the soil.

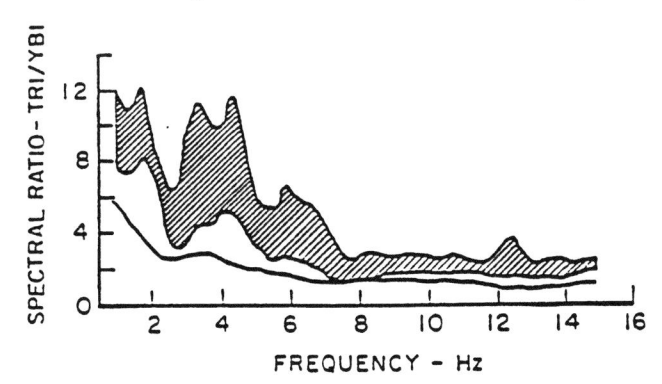

Fig. 6. Amplification Factors for Strong and Weak Motions at Treasure Island Site (after Jarpe et al., 1989).

3.4 EFFECTS OF TOPOGRAPHY

Aki (1988) used the simple structure of a triangular wedge (Fig. 7a) to illustrate the effects of topography. This structure may be used to model approximately ridge-valley topography as shown in Figure 7b by Faccioli (1991). An exact solution exists for the wedge for SH waves propagating normal to the ridge and polarized parallel to the ridge axis. Displacement amplification at the vertex is $2/v$ where the ridge angle is $v\pi (0 < v < 2)$. In Figure 7b, the amplification of the crest relative to the base is v_1/v_2. Thus the simple solution provides a rough estimate of the relative amplification at the crest of the ridge compared to the valley.

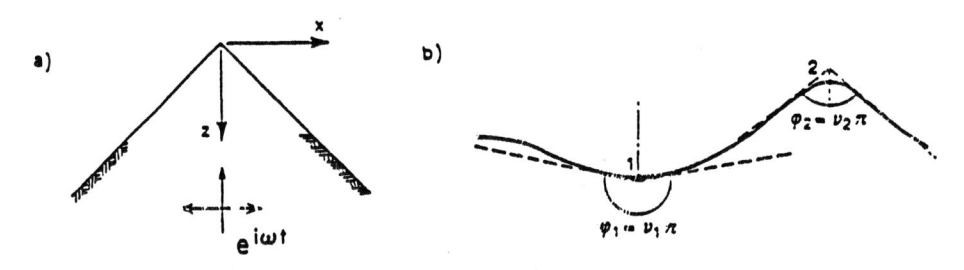

Fig. 7 (a) Approximating a Ridge Formation by a Triangular Wedge, (b) Infinite Wedge Excited by Plane SH Waves (after Faccioli, 1991).

A case history illustrating the variation in amplification over a ridge structure is provided by data from the Matsuzaki array in Japan (PWRI, 1986). The mean values and standard error bars of peak accelerations normalized to the crest acceleration for five earthquakes are plotted in Figure 8 (Jibson, 1987) as a function of elevation. The range in peak accelerations for the five earthquakes is rather limited, ranging from a low of a few gals at station 5 to a maximum of about 100 gals at the crest. The amplification of the crest relative to the base is about 2.5. The amplification factor increases rapidly as the crest of the ridge is approached.

Amplification of motions at the crest of a ridge relative to the base is also supported by damage patterns during the 1980 Friuli earthquakes in Italy (Brambati et al., 1980) and in the Chilean earthquake of 1985 (Celebi, 1987,1991).

The new French seismic code (Jalil, 1992) includes a factor for amplifying the motions on ridges higher than 10 m. The factor τ ranges between 1.0 and 1.4, depending on the geometry of the ridge. The field data suggests that motions on the ridge should be increased by about 10-20%. However, corrections to hazard maps for the effects of topography is very much a matter of judgment at the present time.

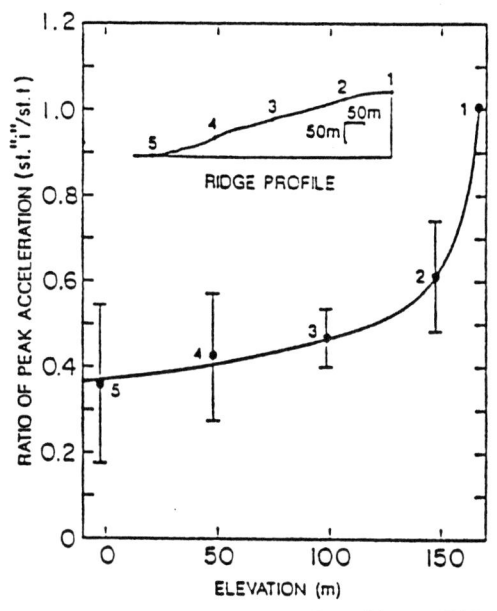

Fig. 8. Relative Distribution of Peak Accelerations Along a Ridge From Matsuzaki Array in Japan (after Jibson, 1987).

3.5 SUMMARY OF PROCEDURES FOR INCORPORATING SOIL CONDITIONS INTO SEISMIC HAZARD EVALUATION FOR GROUND MOTIONS

1. Map the distribution of soil types in the UBC (1989) seismic code and assign the corresponding site coefficients as amplification factors to the motions determined by the basic seismic hazard analysis for rock or firm ground.

2. Characterize the relative amplification potential of the different site conditions by mapping the distribution of average shear wave velocity in the top 30 m of sediment. The average horizontal spectral amplification for weak motions is then given by

   For weak motions: $AHSA = 700/V_s$

   For strong motions: $AHSA = 600/V_s$

   The increments in intensity for different formations with AHSA can be estimated by

   $$dI = 0.27 + 2.70 \log(AHSA)$$

The shear wave velocities can be directly measured by various geophysical procedures or estimated by correlation with penetration resistances, either standard penetration, or cone penetration.

3. For sites classified as soft, approximate amplification factors can be derived from the mean curve relating acceleration in rock to acceleration in soft ground developed by Idriss (1990) and given in Figure 5 above.

4. Ground motions on ridges may be increased by 10-20% before estimating the Modified Mercalli Intensities for damage estimation. However, topographic effects on seismic hazard are often ignored in practice at present.

5. Dynamic nonlinear analysis of representative columns of soil for different zones of the city, established on the basis of geological topographical and geotechnical criteria, may be used to establish amplification factors for ground motions. This procedure proved particularly useful in the seismic risk studies in Quito where there was a shortage of basic data from previous earthquakes.

4. Zoning for Liquefaction

One of the more significant factors leading to ground failure during earthquakes is the liquefaction of loose- to medium-dense sands below the water table.

The mechanics of liquefaction are now well understood, and the potential for occurrence can be estimated with a reasonable degree of confidence at a specific site. During shaking, the sand tends to compact. The water in the pores cannot escape quickly enough, at least in the finer sands, to accommodate instantaneously the compaction. Therefore, stresses are thrown on the water which increase the porewater pressure and reduce the effective or intergranular stresses between the sand particles. Sand, a frictional material, depends on the effective stresses between the grains to mobilize shear strength and resistance to displacement. Therefore the increasing porewater pressure leads to loss of strength and stiffness.

Attention was focused on this problem for the first time as a result of the widespread ground failures during the 1964 earthquake in Niigata, Japan (Kawasumi, 1968). Most of the damage in Niigata attributable directly to the earthquake was associated with liquefaction, and such damage has been a significant factor in most major earthquakes since then. The most recent example in North America is the extensive damage caused by liquefaction in the Marina District in San Francisco during the Loma Prieta earthquake of 1989 (Seed et al., 1990).

Liquefaction results in severe damage characterized by foundation failures, slope failures, and damage to lifelines such as highways, rail lines, water, gas and oil distribution systems, and power systems, due to lateral spreading of the liquefied soils. Therefore, it has become standard practice to develop liquefaction potential maps to identify the critical areas. Mapping of liquefaction hazard began in the 1970's and has developed rapidly since that time. Youd (1991) has detailed the worldwide activity, especially in Japan and the United States for mapping liquefaction potential.

Liquefaction potential depends on the susceptibility of the soil deposit to liquefaction and the duration and intensity of expected ground shaking . Susceptibility depends primarily on sediment type, gradation, density, and depth of the water table.

For susceptible soils to have the opportunity to liquefy, they must be subjected to the necessary intensity and duration of shaking. Therefore, mapping for liquefaction potential is based on susceptibility maps showing the location and extent of susceptible materials and opportunity maps which display expected levels of ground shaking. When one map is superimposed on the other, the locations of liquefaction potential, in the event of an earthquake, can be determined.

4.1 SUSCEPTIBILITY MAPS - LEVEL 1

Susceptibility maps, at the lowest level of certainty, can be developed by characterizing in geomorphologic terms the types of deposits that have liquefied or not liquefied in past earthquakes. The first such comprehensive typology was developed by Youd and Perkins (1978) and is shown in Table 4.

A similar table, based on geomorphologic units, and applicable to ground motions of MMI-VIII or JMA Intensity V, has been presented by Wakamatsu (1992) in Table 5.

Table 4. Susceptibility of Sedimentary Deposits to Liquefaction During Strong Shaking (Youd and Perkins, 1978)

Types of Deposit	General Distribution of Cohesionless Sediments in Deposits	Likelihood that Cohesionless Sediments, When Saturated, Would be Susceptible to Liquefy (by Age of Deposit)			
		< 500 yrs	Holocene	Pleistocene	Pre-Pleistocene
Continental Deposits					
River Channel	locally variable	very high	high	low	very low
Flood plain	locally variable	high	moderate	low	very low
Alluvial fan and plain	widespread	moderate	low	low	very low
Marine terraces and plains	widespread	---	low	very low	very low
Delta and fan-delta	widespread	high	moderate	low	very
Lacustrine and playa	variable	high	moderate	low	very low
Colluvium	variable	high	moderate	low	very low
Talus	widespread	low	low	very low	very low
Dunes	widespread	high	moderate	low	very low
Loess	variable	high	high	high	very low
Glacial till	variable	low	low	very low	very low
Tuff	rare	low	low	very low	very low
Tephra	widespread	high	high	?	?
Residual soils	rare	low	low	very low	very low
Sebka	locally variable	high	moderate	low	very low
Coastal Zone					
Delta	widespread	very high	high	low	very low
Estuarine	locally variable	high	moderate	low	very low
Beach: High wave energy	widespread	moderate	low	very low	very low
Beach: Low wave energy	widespread	high	moderate	low	very low
Lagoonal	locally variable	high	moderate	low	very low
Fore shore	locally variable	high	moderate	low	very low
Artificial					
Uncompacted fill	variable	very high	---	---	---
Compacted fill	variable	low	---	---	---

Table 5. Susceptibility of Detailed Geomorphologic Units to Liquefaction Subjected to Ground Motion of the JMA Intensity V or MMI VIII (Wakamatsu, 1992)

Geomorphologic Conditions		Liquefaction Potential
Classification	Specific Conditions	
Valley Plain	Valley plain consisted of gravel or cobble	Liquefaction not likely
	Valley plain consisted of sandy soil	Liquefaction possible
Alluvial Fan	Vertical gradient is more than 0.5%	Liquefaction not likely
	Vertical gradient is less than 0.5%	Liquefaction possible
Natural Levee	Top of natural levee	Liquefaction possible
	Edge of natural levee	Liquefaction likely
Back Marsh		Liquefaction possible
Abandoned River Channel		Liquefaction likely
Former pond		Liquefaction likely
Mash/Swamp		Liquefaction possible
Dry River Bed	Dry river bed consisting of gravel	Liquefaction not likely
	Dry river bed consisting of sandy soil	Liquefaction likely
Delta		Liquefaction possible
Bar	Sand bar	Liquefaction possible
	Gravel bar	Liquefaction not likely
Sand Dune	Top of dune	Liquefaction not likely
	Lower slope of dune	Liquefaction likely
Beach	Beach	Liquefaction not likely
	Artificial beach	Liquefaction likely
Interlevee Lowland		Liquefaction likely
Reclaimed Land by Drainage		Liquefaction likely
Spring		Liquefaction likely
Fill	Fill on boundary zone between sand and lowland	Liquefaction likely
	Fill adjourning cliff	Liquefaction likely
	Fill on marsh or swamp	Liquefaction likely
	Fill on reclaimed land by drainage	Liquefaction likely
	Other type fill	Liquefaction possible

Youd and Perkins (1978) used these susceptibility and opportunity criteria to produce demonstration maps for the San Fernando Valley. The susceptibility map is shown in Figure 9 and the opportunity map is shown in Figure 10. In constructing the susceptibility map, Youd and

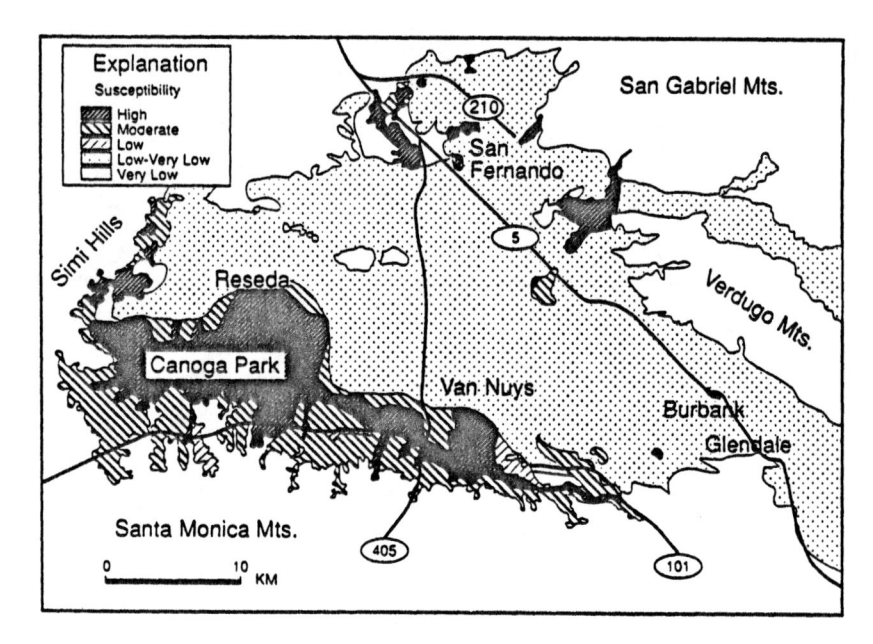

Fig. 9. Zonation map for the San Fernando Valley, California showing probable susceptibility to liquefaction (modified from Youd et al., 1978).

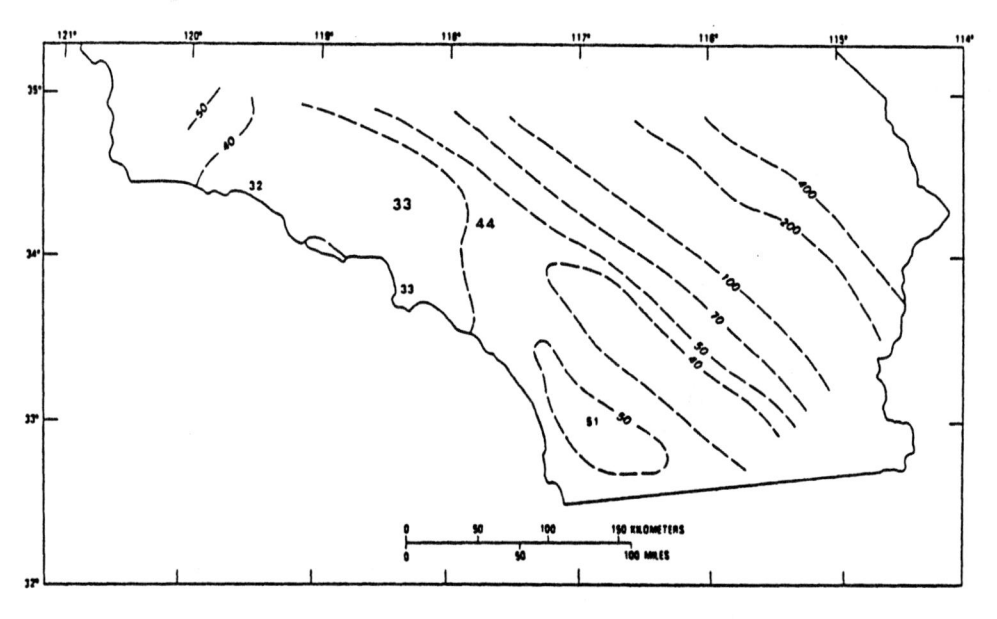

Fig. 10. Return period of liquefaction opportunity for southern California (after Tinsley et al., 1985).

Perkins (1978) selected geologic criteria from in Table 4 and criteria on the effects of the depth of groundwater from Youd and Perkins (1978). The operational criteria given in Table 6 are representative of those used in the San Fernando Valley study. Table 6 provides a very compact list of criteria that are useful for a preliminary estimate of liquefaction susceptibility.

Table 6. Criteria Used in Compiling Liquefaction Susceptibility Map for the San Fernando Valley, California (modified from Tinsley et al., 1985)

Sedimentary Unit	Depth to Ground Water, in Metres			
	0 - 3	3 - 10	10 - 15	> 15
Holocene: Latest Earlier	Very high to high[1] High	Moderate[2] Moderate	Low Low	Very low Very low
Pleistocene: Late Middle and early	Low Very low	Low Very low	Very low Very low	Very low Very low
Tertiary and pre-Tertiary	Very low	Very low	Very low	Very low

[1] Areas are mapped as having very high susceptibility if fluvial channel and levee deposits are known to be present; sediment deposited in other sedimentary environments is considered to have high susceptibility

[2] Fluvial deposits having high susceptibility occur rarely and are not widely distributed; other sediments are moderately susceptible to liquefaction.

Iwasaki et al. (1982) also developed a set of simplified criteria for developing liquefaction susceptibility maps based on topographic criteria that are in general agreement with the geologic criteria given in Table 4. Iwasaki's criteria are given in Table 7.

Table 7. Microzonation Procedure Based on Topographical Information (modified from Iwasaki et al., 1982).

Rank	Topography	Liquefaction Potential
A	Present river bed, old river bed, swamp, reclaimed land, interdune lowland	Liquefaction likely
B	Fan, natural levee, sand dune, flood plain, beach, other plains	Liquefaction possible
C	Terrace, hill, mountain	Liquefaction not likely

More recently, a liquefaction susceptibility classification based on microtopography has been developed by the National Land Development Technology Center in Japan (NLDTC, 1991). It is shown in Table 8.

The value of this approach has been demonstrated by Wakamatsu (1991). She investigated the topography around Azuma Village and identified the principle types of formations based on Table

8. These are shown in Figure 11. Earthquake damage to wooden houses in this area was determined by interviews with the residents. Damage was assessed in terms of settlement, tilting, and damage to the building.

Table 8. Liquefaction Susceptibility Classification Based on Microtopography (NLDTC, 1991. Adapted from Kuwano et al., 1992).

Liquefaction Vulnerability	Microtopographical Units
High	former river channel or pond, reclaimed land, margin of natural levee, lowland between sand dune, riverbed with fine sand, fill with high water table
Medium	delta, natural levee, back swamp, fan with gentle slope delta, valley plain, reclaimed land by drainage
Low	Sand dune, fan, gravelly riverbed

The extent of damage to the houses due to liquefaction was established in this way and is shown in Figure 12. A comparison of Figures 11 and 12 shows that damages are light for houses on the natural levee and heavy in the former river channels or ponds. This example demonstrates the utility of even simple criteria for zoning for liquefaction to indicate areas of potential damage during an earthquake.

4.2 OPPORTUNITY MAPS - EMPIRICAL

For preliminary planning, it can be assumed that all susceptible soils lying within the liquefiable range of the scenario earthquake will liquefy. The range of liquefaction occurrence for a given earthquake can be established based on historical data. Such a relationship is shown in Figure 13 (after Youd and Wieczorek, 1984). This procedure gives a very generalized picture of the liquefaction potential hazard and does not reflect in any significant way the geotechnical or geologic properties of the materials under consideration.

A number of other criteria are available for assessing the farthest distance to liquefied sites for a given earthquake magnitude. Kuribayashi and Tatsuoka (1975) developed the equation

$$\log R = 0.77 M_J - 3.6 \qquad (6)$$

where M_J is the earthquake magnitude defined according to the Japan Meteorological Agency scale.

Ambraseys (1988) reviewed the data from the previous investigations and added about seventy additional data points. He converted all magnitudes to seismic moment magnitude, M_W, and treated data for shallow and intermediate depth earthquakes separately. The data points for shallow earthquakes were bounded by the equation

$$M_w = -0.31 + 2.65 \times 10^{-8} R_e + 0.99 \log R_e \qquad (7)$$

Fig. 11. Microtopography of the area surrounding Azuma Village (Wakamatsu, 1991).

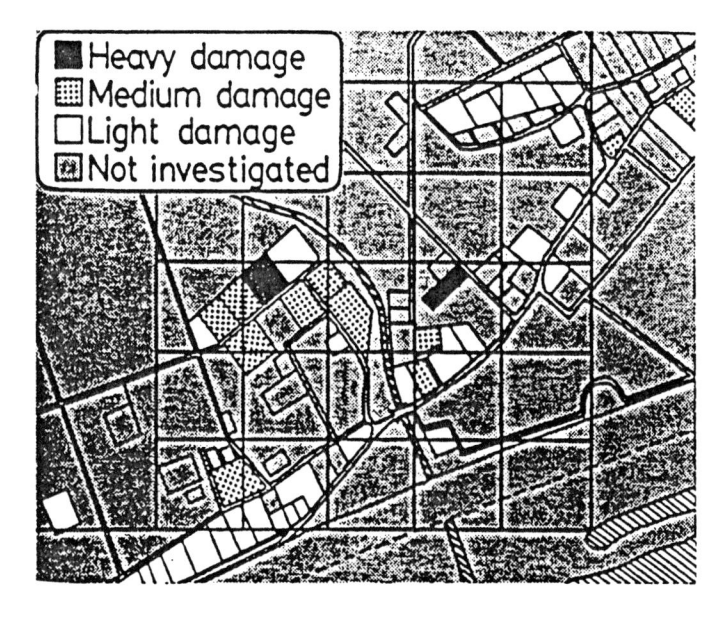

Fig. 12. Damage of houses due to liquefaction (assessed through interviews) (after Kuwano et al., 1992).

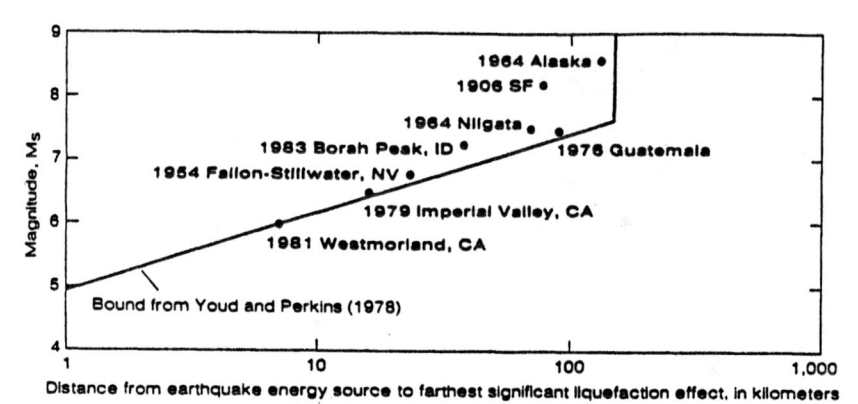

Fig. 13. Earthquake magnitude vs distance to farthest significant liquefaction effect (after Youd and Wieczorek, 1984).

where R_e is the farthest epicentral distance in cm. The data points for intermediate depth earthquakes were shown generally to lie well beyond the bounds for shallow-focus events, indicating that the intermediate depth earthquakes may generate liquefaction at larger distances than the shallow-depth events. Ambraseys' curve is compared with that of Youd in Figure 14.

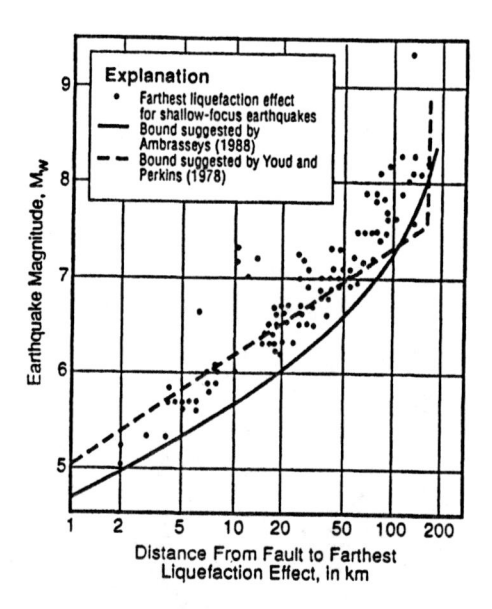

Fig. 14 Relationship between distance from the earthquake energy source to the farthest liquefied site (Ambraseys, 1988) or significant liquefaction effects (Youd and Perkins, 1978) and moment magnitude M_w for shallow earthquakes (modified from Ambraseys, 1988; by Youd, 1991).

4.3 SUSCEPTIBILITY MAPS - LEVEL 2

4.3.1 *Historic Criteria.* The generalized criteria described above are useful for giving an estimate of the potential exposure to liquefaction problems. Such maps can provide a basis for preliminary land use planning and for conducting more detailed geologic and geotechnical investigations.

The first step in improving the quality of the susceptibility map in a seismic region is to map the previous occurrences of liquefaction in the region. This information will clearly delineate the zones of greatest liquefaction potential for the occurrence of similar or greater earthquakes. This mapping can also be used to extrapolate susceptibility to other areas not identified by the generalized criteria discussed earlier.

Mapping the previous occurrences of liquefaction involves extensive field investigations to locate and detect the liquefied areas and to determine their extent. As opposed to Level 1 mapping, this cannot be done as a desk study only.

4.4 OPPORTUNITY MAPS - HAZARD SPECIFIC LEVEL

Opportunity maps based on a peak ground motion parameter such as peak acceleration cannot be used to assess liquefaction potential in the way that the opportunity map based on empirical data could be used. Both duration and the level of peak acceleration are important. Peak acceleration maps can be drawn for each earthquake magnitude in order to take duration into account. However, the interpretation of liquefaction potential would require a susceptibilitiy map that gave the distribution of a measure of acceleration for the specified earthquake magnitude, otherwise the critical values of acceleration cannot be identified. If this level of information is available, a direct evaluation of liquefaction potential can be performed directly without either susceptibility or opportunity maps. Preferably a probabilistic approach should be used.

4.4.1 *Probabilistic Approach.* A probabilistic assessment of liquefaction potential can be established using a program such as PROLIQ (Atkinson et al., 1984) which has been widely used in practice for regional studies. This program computes the probability of liquefaction for a specified liquefaction resistance or the resistance required for a given probabiliyy of liquefaction. The evaluation procedure is based on the simplified method for evaluating liquefaction potential developed by Seed and Idriss (1971). A brief description of this method is found in Appendix 1. The resistance is usually specified by the SPT value, the CPT value, or the normalized shear wave velocity. The results of this study provide a basis for the regional interpretation of whatever field data is available on penetration resistances and shear wave velocities. It is most useful in the urban environment where penetration or shear wave velocity data from foundation explorations are usually available. If a wide distribution of penetration resistances are not available, the reliability of the liquefaction assessment can be improved by running some additional tests and/or by extrapolating, using the geologic and geomorphologic criteria.

4.5 EFFECT OF SURFACE LAYERS

The surface effects of liquefaction of an underlying layer depend on the thickness of the overlying material which is not susceptible to liquefaction. Ishihara (1985) studied this problem and has proposed curves shown in Figure 15 which give the limiting thickness of the surface layer beyond which surface damage is not likely. These curves depend on the intensity of shaking and the thickness of the liquefiable layer. The unliquefiable layer is identified as soil which is not susceptible to liquefaction, even if saturated, or soil that would normally be susceptible to liquefaction but is above the water table or a combination of both.

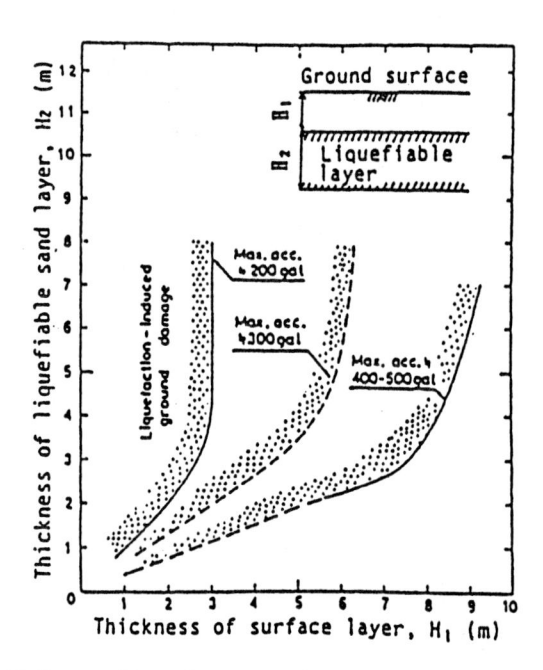

Fig. 15 Proposed boundary curves for site identification of liquefaction-induced damage (Ishihara, 1985).

4.6 GROUND DISPLACEMENTS

When foundation soils liquefy, considerable displacements can occur even on slopes of a few percent. Such movements are especially large near an open face such as the banks of a river or canal. In Niigata during the 1964 earthquake, very large displacements occurred towards the Shimano River (up to 10 metres) and towards the Tusen River (displacements up to 2 m). Hamada et al. (1986), Youd (1980), Youd and Perkins (1978), and Bartlett and Youd (1992) have carried out extensive investigations of ground deformations associated with liquefaction and they have developed equations for estimating these movements. These types of deformations have a major impact on lifelines and on building foundations. Buildings which are not well tied together at the foundation level tend to be pulled apart. On the basis of several post-earthquake studies of damaged and undamaged buildings, Youd (1980) concluded that most buildings can withstand 50-100 mm of differential ground displacement with little damage. On the other hand, few buildings can survive more than 1 m of displacement without major or catastrophic damage. Buildings on pile foundations, unless the piles are designed to resist the forces of the moving soil, can be severely damaged also.

4.6.1 *Hamada Equation.* Hamada et al (1986) related permanent ground displacement, D, to the thickness of the liquefied layer, H, and the greater of the gradients, θ, of the ground surface and the bottom of the liquefied layer expressed in percent (%). The relation was developed by a regression analysis using 60 data sets from the 1983 Nihon-kai-Chubu, the 1964 Niigata, and the 1971 San Fernando earthquakes. The relation is

$$D = 0.75 \, H^{1/2} \, \theta^{1/3} \tag{8}$$

This equation is very heavily weighted by the data from the Japanese earthquakes. Therefore, it probably performs best for estimating deformations on slopes less than 5% and for earthquake magnitudes M = 7.2 to 7.5 causing ground accelerations in the range of 0.2 g to 0.3 g, Although probably somewhat unreliable outside this range, it nevertheless would offer some guidance as to the relative deformation potentials that might be expected in different zones of a metropolitan area.

4.6.2 *Bartlett and Youd (1992).* Bartlett and Youd (1992) have made detailed studies of all readily-available data in Japan and the United States and developed two models for estimating displacements during liquefaction. These equations are for displacements near a free face such as the bank of a river and for ground slope conditions removed from a free face. The equations are

Free face:

$$\mathrm{Log}(D_H + 0.01) = -16.366 + 1.178M$$
$$-0.927 \, \mathrm{Log}R - 0.013R + 0.657 \, \mathrm{Log}W + 0.348 \, \mathrm{Log}T_{15} \tag{9}$$
$$+4.527(100 - F_{15}) - 0.922 \, D50_{15}$$

and

Ground slope conditions:

$$\mathrm{Log}(D_H + 0.01) = -15.787 + 1.178M$$
$$-0.927 \, \mathrm{Log}R - 0.013R + 0.429 \, \mathrm{Log}S + 0.348 \, \mathrm{Log}T_{15} \tag{10}$$
$$+4.527(100 - F_{15}) - 0.922 \, D50_{15}$$

where D_H(m) is the horizontal ground displacement, M (M_W) is earthquake magnitude, R (km) is the distance from the seismic source, W (%) is the ratio of the height of the free face divided by where $D_H(m)$ is the horizontal ground displacement, M (M_w) is earthquake magnitude, R (km) is the distance from the seismic source, W (%) is the ratio of the height of the free face divided by the distance from the free face, S (%) is the gradient of the ground slope, T_{15} (m) is the accumulative thickness of saturated, granular soils in the profile having a modified standard penetration value (i.e., $(N_1)_{60}$) less than or equal to 15, F_{15} isi the average fines content in the T_{15} layer(s), and $D50_{15}$ is the average mean grain size of the T_{15} layer(s). Because log (0) is undefined, 0.01 m was expediently added to all values of D_H prior to fitting these equations. This expediency allowed log (D_H) to be calculated for all zero displacement values that are included in the MLR database.

4.7 MAPPING DAMAGE POTENTIAL FROM GROUND DISPLACEMENTS

Structural damage resulting from permanent ground displacements during earthquakes is a function of the type and amount of differential ground movement. With the exceptions of tectonic fault movements and failures of relatively steep or initially weak slopes, permanent ground deformations are usually associated with liquefaction. Therefore in assessing the severity of expected displacements, Youd and Perkins (1978) developed a parameter S which they called the Liquefaction Severity. S is defined as the measured amount of differential ground displacement in millimeters divided by 25. In the United States this is equivalent to the displacement measured in inches. The S value for areas in Noshiro City with permanent displacements of 1.0 m is S = 40.

Based on their study of the damage resulting from ground displacements during the Alaska and Niigata earthquakes in 1964 and the San Fernando earthquake in 1971, Youd and Perkins (1978) suggested that very little damage would be expected in ordinary buildings for S < 5, but that moderate to severe damage would be likely for $5 \leq S < 20$ and major damage for S > 30.

The S parameter can vary widely even over limited regions. Therefore, while estimates of S are useful for site-specific studies, S is not a useful parameter for mapping expected levels of deformations in a given area with a given probability of occurrence.

To simplify this problem by providing a single number to characterize the general severity of ground failures within a particular locality, Youd and Perkins (1978) introduced the liquefaction severity index LSI which is the general maximum S value in a specific topographic, hydrologic and sedimentologic environment. Typically, LSI is considered to be the maximum S for lateral spreads in geological units such as wide active flood plains, deltas and other areas of gently sloping late Holocene river deposits. Large S values due to localized untypical soil or topographic conditions are excluded. Since LSI is the maximum recorded S, it gives a conservative estimate of displacements and associated damage potential.

Youd and Perkins (1978) have shown that LSI for a given environment which undergoes liquefaction may be related to the moment magnitude of the earthquake, M_w, and the distance, R, from the energy source by

$$\text{Log(LSI)} = -3.49 - 1.86 \log R + 0.98 M_w \tag{11}$$

This equation is based on data obtained in the western United States. Variations in the attenuation of acceleration from the source varies from seismic region to seismic region and the attenuation of LSI may be expected to do likewise. If equations similar to Eq. (11) cannot be derived specifically for a seismic region under study, the results of applying Eq. (11) should be modified empirically for any difference in the attenuation characteristics of the study area compared with the western United States.

5. Zoning for Landslides

Landslides can be a major cause of damage and loss of life during earthquakes. At Haiyun in China in 1920, an earthquake triggered a landslide in loess deposits that killed more than 100,000 people (Li, 1990). In Peru in 1970, a magnitude M=7.7 earthquake generated a rock avalanche at Nevados Huascaran which killed more than 18,000 (Ericksen et al., 1970).

The first step in the mapping of earthquake-induced landslides is to characterize the dominant landslide forms. Keefer and Wilson (1989) put landslide forms into three broad categories as follows.

Category A: Rock and soil falls and slides; rock and soil avalanches. Landslides in this category are generally the most widespread. These landslides have been the most disruptive and travel at speeds from 1×10^{-6} m/hr to 1×10^5 m/hr

Category B: Landslides that move on fairly well-defined shear surfaces and can be analyzed using limited equilibrium methods.

Category C: Primarily landslides that are initiated by liquefaction. Keefer and Wilson (1989) include also lateral spreading of soil and rapid soil flows in this category, but these have been discussed previously under liquefaction.

Keefer (1984) established the relative abundance of the different landslide types (shown in Table 9) on the basis of a sample of 40 historical earthquake data.

Table 9. Relative Abundance of Landslides in Historical Earthquakes (after Keefer, 1984).

Abundance	Landslide Type
Very abundant: > 100,000	Rock falls; disrupted soil slides; rock slides
Abundant: 10,000 to 100,000	Soil lateral spreads; soil slumps; soil block slides; soil avalanches
Moderately common: 1,000 to 10,000	Soil falls; rapid soil flows; rock slumps
Uncommon: 100 to 1,000	Subaqueous landslides; slow earth flows; rock block slides; rock avalanches

Notes: 1) See Keefer (1984) for method of calculating numbers of landslides.
2) Numbers refer to totals for the sample of 40 historical earthquakes.

5.1 MECHANICS OF SLOPE STABILITY

To understand how seismic ground motions can trigger a landslide, one must first look at the factors that control slope stability. The constant generator of movement in the slope is the force of gravity acting down the slope. Under static conditions, this can be augmented by seepage forces caused by water draining from the slope. The resisting force is due to the cohesion and frictional components of the shearing resistance. The frictional resistance depends on the intergranular forces and therefore is affected by changes in the porewater pressures in the slope. Increasing the porewater pressure leads to a reduction in intergranular forces and a reduction in shearing strength. The prime effects of earthquake shaking are to increase the driving forces due to inertia forces acting out from the slope, and to increase the porewater pressures in the case of saturated, loose or medium-dense granular materials. Clearly, slopes which occasionally fail under environmental conditions other than earthquakes, will be at a higher hazard during earthquakes. Therefore, as a first approximation, the seismic susceptibility to landsliding can be mapped by plotting all past landslides.

A major concern is whether earthquake motions will re-activate old landslides. Past experience, most recently in the Loma Prieta earthquake, has demonstrated that old landslides are re-activated but that the resulting displacements are usually small. This is not too surprising since the landslide mass usually comes to equilibrium under smaller driving shear stresses than existed before failure. Some slides in overconsolidated material develop residual strength on the failure surface after significant movement which is substantially less than the initial strengths and such landslides may be particularly vulnerable to the additional forces caused by earthquake shaking.

5.2 SUSCEPTIBILITY MAPS: HIGHER LEVEL

It is usually very expensive to carry out numerous studies of slope stability for the sole purpose of compiling maps showing the susceptibility to sliding. The field studies required to determine the soil properties, the geometry of the slope, and the groundwater conditions usually prohibit extensive studies of this kind. Therefore, although the wherewithal is there to greatly improve reliability of susceptibility maps, the expense makes it prohibitive.

The best use of the susceptibility maps is probably for general land-use planning and especially for controlling the location of critical structures and lifelines.

The small displacements experienced by re-activated landslides suggests that in some landslide terrain it may be feasible to allow development.

5.3 OPPORTUNITY MAPS

The mechanism of landslide development is usually too complicated to develop landslide potential maps in the same way as was done with liquefaction potential. Some attempts have been made to delineate the furthest distance at which landslides have occurred from the epicentres of earthquakes of various magnitudes. Keefer (1984) and Keefer and Wilson (1989) have compared data on landslides during historical earthquakes with magnitude-distance relations for three different levels of Arias (1970) shaking intensity, for the three main categories of landslides. In addition, the area in square kilometres affected by landslides during an earthquake of a given magnitude, have been developed. These correlations are shown in Figure 16. The labelled intensities are exceedence probabilities for landsliding. It is doubtful if much reliance can be put on these figures which involve only the earthquake magnitude. Obviously, from a given earthquake magnitude in a given region, the number of landslides must also depend on the number of slopes at risk, and therefore would depend very much on the geological formations in the area. Nevertheless, the correlations are useful in at least giving some relative assessments of landslide effects from one area to another with similar terrain.

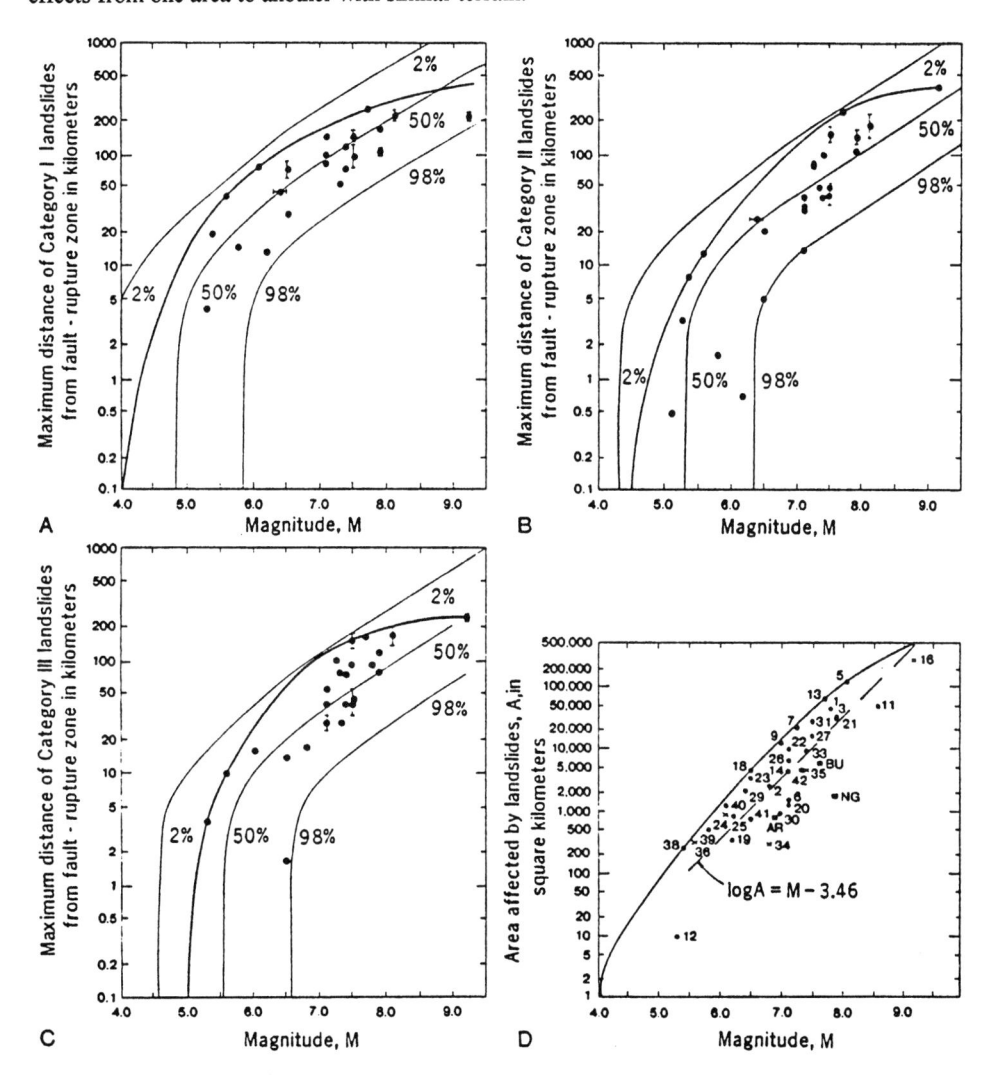

Fig. 16. Comparison of data on landslides in historical earthquakes with magnitude-distance relations for threshold Arias intensities in landslide categories A, B, and C (from Keefer, 1984 a, b and Keefer and Wilson, 1989, compiled by Hansen and Franks, 1990).

5.4 LANDSLIDE DISTRIBUTION AND GEOLOGICAL CONDITIONS

Maps of the distribution of landslides, which show slope angle and the nature of the bedrock, are very useful in defining landslide susceptibility units. Nilson and Brabb (1979) produced relative slope stability maps for the San Francisco Bay region at 125,000 scale using the above method. An early seismic hazard map produced by Rice (1973), identifies five levels of seismic hazard on a base map. The zones are classified according to simple descriptions of bedrock and soil thickness. Areas of potentially high damage are defined as areas underlain by landslide deposits and by thick deposits of colluvium or deeply weathered bedrock on steep slopes. As Nilson and Brabb (1979) point out, their map defines susceptibility to landsliding triggered by other mechanisms such as rainfall and human activities, as well as earthquake loading.

A susceptibility map of the La Honda area was developed by Keefer et al. (1978) based on a published map of unconsolidated Quarternary deposits, a computer-generated map of slope angle, and detailed field observation. Three major classes shown on the map are: (1) potential soil avalanches, falls, slumps, blocks and shallow disintegrating slides in soil or rock. The only selection criteria is that the slope is over 35°; (2) potential lateral spreads and flows indicated by ground underlain by saturated sandy and silty Holocene alluvium; and (3) the distribution of existing slumps and block slides in soil or rock.

Figure 17 shows a susceptibility map by Wieczorek et al. (1985) which uses several procedures to get a composite susceptibility: (1) rock and soil falls with slope angles steeper than 35°; (2) lateral spreads and wet sand flows resulting from liquefaction failure, obtained from a liquefaction susceptibility map. The Wieczorek map recognizes only one model of slope failure, translational failure on an infinite slope to the failure depth of 3 m. The effect of selecting this model is to give the relative susceptibility in a general way between various locations. Three types of geotechnical materials were recognized: (a) well-cemented sandstone and crystalline rocks; (b) unconsolidated and weakly-cemented sandstones; and (c) shales and clays. Average values of shear strength parameters c' and ϕ' were assigned to each group, based on the weakest rock type in each unit. Stability analyses were performed using both dry and saturated conditions to get the static factor of safety (FS). From FS, the critical acceleration to initiate movement is given by,

$$a_c = (FS-1)g \sin\theta \qquad (12)$$

where θ is the slope angle. A number of earthquake strong motion records encompassing both near field and far field sources were used to select a design earthquake for the region. The displacements resulting from these earthquakes are shown in Figure 18. On the basis of Figure 18, the range of critical accelerations that can generate displacements exceeding a critical value such as 50 mm can be determined. This defines the range of accelerations over which damaging landslides will probably occur.

6. Acknowledgements

This study was supported in part by a grant from the National Science and Engineering Council of Canada. The development of the themes in this paper was strongly influenced by three state-of-the-art papers cited in the references: Finn (1991); Hansen and Franks (1991); and Youd (1991).

The permission of the Earthquake Engineering Research Institute to use figures from their referenced material is gratefully acknowledged.

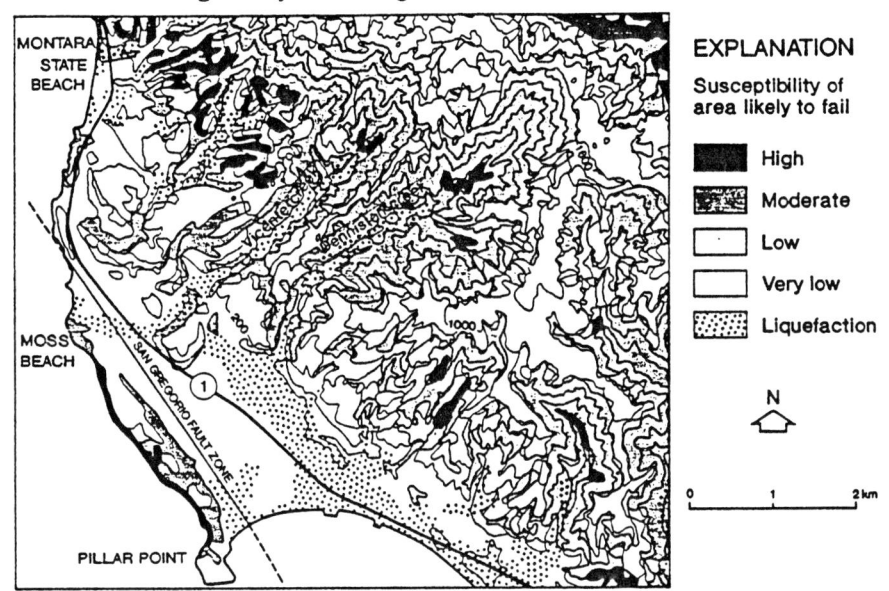

Fig. 17. Maps showing slope susceptibility to slope failure during earthquakes in San Mateo County, California (Wieczorek et al., 1985).

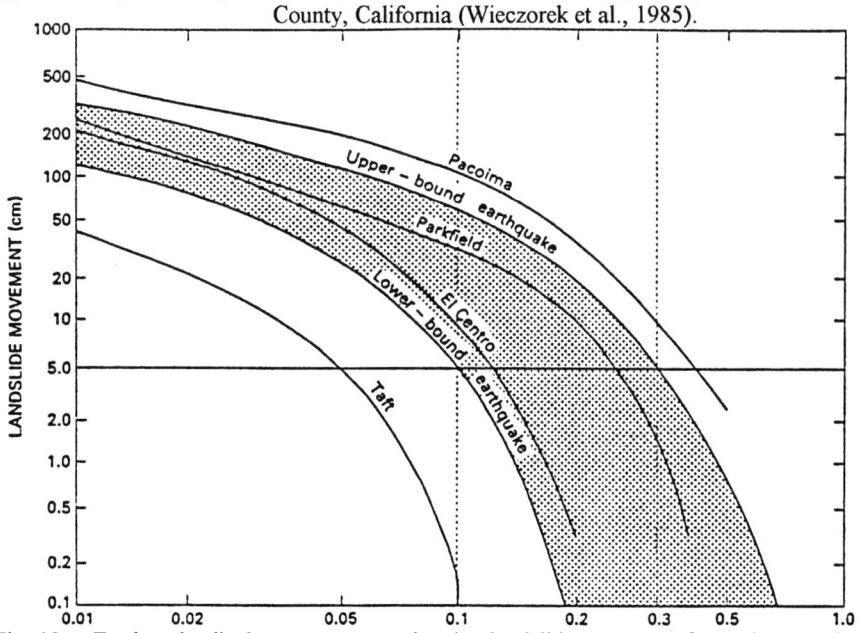

Fig. 18. Earthquake displacement curves showing landslide movement for various levels of critical acceleration for several earthquakes (Wieczorek et al., 1985).

References

Addo, K.O. and Robertson, P.K. 1992. "Shear-wave velocity measurement of soils using Rayleigh waves." Canadian Geotechnical Journal, Vol. 29, No. 4, August, pp. 551-557.

Ambraseys, N.N. 1988. "Engineering seismology." Earthquake Engineering and Structural Dynamics, Vol. 17, pp. 1-105.

Applied Technology Council. 1985. "Earthquake damage evaluation data for California." ATC-13, Federal Emergency Management Agency Contract No. EMW-C-0912, 492 p.

Aki, K. 1988. "Local site effects on strong ground motion." in J. Lawrence Van Thun (Ed.), "Earthquake Engineering and Soil Dynamic (Recent Advances in Ground-Motion Evaluation", ASCE Geotechnical Special Publication No. 20, pp. 103-155.

Algermissen, S.T. 1989. "Techniques and parameters for earthquake risk assessment." Bulletin of the New Zealand National Society of Earthquake Engineering, Vol. 22, pp. 202-219.

Arias, A. 1970. "A measure of earthquake intensity." In R.J. Hansen (ed.) Seismic Design for Nuclear Power Plants, Cambridge Mass., MIT Press, pp. 438-483.

Atkinson, G.M., W.D. Finn, and R. G. Charlwood, "Simple Computation of Liquefaction Probability for Seismic Hazard Applications," *Earthquake Spectra* 1 (1):107-124 (November 1984).

Bartlett, S.F. and Youd, T.L. 1992. "Empirical analysis of horizontal ground displacement generated by liquefaction-induced lateral spreads." Technical report NCEER-92-0021, National Center for Earthquake Engineering Research, State University of New York at Buffalo, August.

Bender, B. and Perkins, D.M. 1987. "SEISRISK III - A computer program for seismic hazard estimation." US Geological Survey Bulletin 1772, 48 p.

Borcherdt, R.D. 1970. "Effects of local geology on ground motion near San Francisco Bay." Bull. Seis. Soc. Am., 60, 29-61.

Borcherdt, R.D., Gibbs, J.F. and Lajoie, K.R. 1975. "Prediction of maximum earthquake intensities in the San Francisco Bay region for large earthquakes on the San Andreas and Hayward faults." U.S. Geol. Surv., Misc. Field Studies Map 709.

Borcherdt, R.D. and Gibbs, J.F. 1976. "Effects of local geology conditions in the San Francisco Bay region on ground motions and the intensities of the 1906 earthquake." Bull. Seis. Soc. Am., 66, 467-500.

Borcherdt, R.D., Gibbs, J.F. and Fumal, T.E. 1978. "Progress on ground motion predictions for the San Francisco Bay region." California, Proc. 2nd International Conf. on Microzonation for Safer Constr. Res. Appl., 1, 241-251, 1978.

Borcherdt, R.D. 1990. "Influence of local geology in the San Francisco Bay region, California, on ground motion generated by the Loma Prieta earthquake of October 17, 1989." Proc. International Symposium on Safety and Urban Life and Facilities, Tokyo Inst. Tech., Tokyo, Japan.

Borcherdt, R.D., Wentworth, C.M., Janssen, A., Fumal, T and Gibbs, J. 1991. "Methodology for predictive GIS mapping of special study zones for strong ground shaking in the San Francisco Bay region, California." To appear in Proceedings, 4th International Conference on Seismic Zonation, Palo Alto, California, August.

Brambati, E. Faccioli, E., Carulli, E., Culchi, F., Onofri, R., Stefanini, R. and Uloigrai, F. 1980. Studio de microzonizzacione sismica dell'are do Tarento (Friuli), Edito da Regione Autonoma Friuli-Venezia, Giulia.

Celebi, M. 1987. "Topographical and geological amplifications determined from strong motion and aftershock records of the 3 March 1985 Chile earthquake." Bulletin of the Seismological Society of America, Vol. 77, No. 4, pp. 1147-1167.

Celebi, M. 1991. "Topographical and geological amplification case studies and engineering implications." Proceedings, International Workshop on Spatial Variation of Earthquake Ground Motion, Elsevier Science Publishers, B.V. Amsterdam.

Ericksen, G.E., Plafker, G. and Concha, J.F. 1970. "Preliminary report on the geologic events associated with the May 31, 1970, Peru earthquake." US Geological Survey Circ., 639, 25 p.

van Essche, L. 1986. "Planning and management for disaster mitigation in urban metropolitan regions: earthquake risk reduction." Proc., Internatinal Seminar on Planning for Crisis Relief, Tokyo, Japan, September.

Faccioli, E. 1991. "Seismic amplification in the presence of geological and topographic irregularities, Proceedings, 2nd International Conference on Recent Advances in Geotechnical Earthquake Engineering and Soil Dynamics, St. Louis, Missouri, Vol. II, pp. 1779-1797.

Finn, W.D. Liam, Lee, W.K. and Martin, G.R. 1978. An effective stress model for liquefaction. Journal of the Geotechnical Engineering Division, ASCE, Vol. 103, No. GT6, pp. 517-533.

Finn, W.D. Liam and Nichols, A.M. 1988. "Seismic response of long period sites: Lessons from the September 19, 1985 Mexican earthquake." Canadian Geotechnical Journal, Feb. 1988, pp. 128-137.

Finn, W.D. Liam, P.K. Robertson and D.J. Woeller. 1990. "Liquefaction Studies in the Fraser Delta," Research Report to Energy, Mines and Resources, Geological Survey of Canada, Vancouver, B.C.

Finn, W.D. Liam. 1991. "Geotechnical aspects of microzonation." Proc., 4th Int. Conf. on Seismic Zonation, EERI, Stanford, California, Vol.1, pp. 199-259.

Finn, W.D. Liam. 1993. "Evaluation of Liquefaction Potential." *Soil Dynamics and Geotechnical Earthquake Engineering*, Balkama, Rotterdam.

Fukushima, Y. and Tanaka, T. 1990. "A new attenuation relation for peak horizontal acceleration of strong earthquake ground motion in Japan." Seismological Soc. of America Bulletin, p. 757-783.

Geli, L., Bard, P.Y. and Jullien, B. 1988. "The effect of topography on earthquake ground motion: a review and new results." Bulletin of the Seismological Society of America, Vol. 78, No. 1, pp. 42-63.

Hamada, M., Yasuda, S., Isoyama, R. and Emoto, K. 1986. "Study of liquefaction induced permanent ground displacements." Association for the Development of Earthquake Prediction, Tokyo, Japan, pp. 1-87, November.

Hansen, A. and Franks, C. A. M. 1991. "Characteristics and mapping of earthquake triggered landslides for seismic zonation." Proc., 4th Int. Conf. on Seismic Zonation, EERI, Stanford, California, pp. 149-195.

Hensolt, W.J. and Brabb, E.E. 1990. "Maps showing elevation of bedrock and implications for design of engineered structures to withstand earthquake shaking in San Mateo County, California." USGS Open File Report 90-496.

Housner, G.W., Chairman. 1990. "Competing against time." Report to Governor G. Deukmejian from the Governor's Board of Inquiry on the 1989 Loma Prieta Earthquake.

Idriss, I.M. 1990. Response of soft soil sites during earthquakes. Proceedings, H. Bolton Seed Memorial Symposium, Berkeley, California, Vol. II.

Ishihara, K. 1985. "Stability of Natural Deposits During Earthquakes," Proc., 11th Int. Conf. on Soil Mechanics and Foundation Engr., San Francisco. 1:321-376.

Iwasaki, T., Tokida, K., Tatsuoka, F., Watanabe, S., Yasuda, S. and Sato, H. 1982. "Microzonation for soil liquefaction potential using simplified methods." Proc., 3rd International Conf. on Microzonation, Seattle, WA, Vol. 3, pp. 1319-1330.

Iwasaki, T., Tatsuoka, F., Tokida, K. and Yasuda, S. 1978. "A practical method for assessing soil liquefaction potential based on case studies at various sites in Japan." Proc., 2nd International Conf. on Microzonation, San Francisco, Vol. 2, pp. 885-896.

Jalil, W. 1992. "New French seismic code orientations." Proceedings, 10th World Conference, Earthquake Engineering, Madrid, Balkema, Rotterdam, Vol. 10, pp. 5867-5870.

Japan Road Association. 1980, 1991. "Specifications for highway bridges, Part V - Earthquake resistant design."

Jarpe, S., Hutchings, L., Hauk, T. and Shakal, A. 1989. "Selected strong- and weak-motion data from the Loma Prieta earthquake sequence." Seismol. Res. Letters, 60, pp. 167-176.

Jibson, R. 1987. "Summary of research on the effects of topographic amplification of earthquake shaking on slope stability." U.S. Geological Survey, Open-File Report 87-268, Menlo Park, California, USA.

Kawasumi, H. 1968. (Editor-in-Chief), "General report on the Niigata earthquake of 1964." Tokyo Electric Engineering College Press, March.

Keefer, D.K. 1984. "Landslides caused by earthquakes." Geological Soc. of America Bull., Vol. 95, pp. 406-421.

Keefer, D.K. and Wilson, R.C. 1989. "Predicting earthquake-induced landslides with emphasis on arid and semi-arid environments." Pub. of the Inland Geological Soc., Vol. 2, pp. 118-149.

Keefer, D.K., Wieczorek, G.F., Harp, E.L. and Tuel, D.H. 1978. "Preliminary assessment of seismically induced landslide susceptibility." Proc., 2nd Int. Conf. on Microzonation, San Francisco, Vol. 1, pp. 279-290.

Kuribayashi, E. and Tatsuoka, F. 1975. "Brief review of soil liquefaction during earthquakes in Japan." Soils and Foundations, Vol. 15, No. 4, pp. 81-92.

Kuwano, J., Ozaki, K. and Nakamura, S. 1992. "Simplified method of liquefaction hazard evaluation for subsurface layers." Proceedings, 10th World Conf. on Earthquake Engineering, Madrid, Balkema, Rotterdam, Vol. 3, pp. 1431-1434.

Lee, M.K.W. and Finn, W.D. Liam. 1978. "DESRA-2, dynamic effective stress response analysis of soil deposits with energy transmitting boundary including assessment of liquefaction potential." Soil Mechanics Series No. 38, Dept. of Civil Engineering, University of British Columbia, Vancouver, B.C.

Li, T. 1990. "Landslide management in mountain areas of China." Int. Centre for Integrated Mountain Development, Kathmandu, Nepal, Occasional Paper, 15, 50 p, October.

Liao, S.C. and Whitman, R.V. 1985. "Overburden correction factors for SPT in sand." Journal of Geotechnical Engineering Division, ASCE, Vol. 112, No. 3, pp. 529-572.

National Land Development Technology Center. 1991. "Research report on countermeasures to liquefaction damage of small buildings utilizing liquefaction potential maps." Tokyo, NLDTC (in Japanese).

Nazarian, S. and Stokoe, K.H., II. 1984. "In-situ shear wave velocities from spectral analysis of surface waves." Proc., 8th World Conf. on Earthquake Engineering, San Francisco, California, Vol. 3, pp. 31-38.

Nilson, T.H. and Brabb, E.E. 1979. "Landslides." U.S. Geological Survey Prof. Paper 941-A, pp. A75-A87.

PWRI (Public Works Research Institute). 1986. "Dense instrument array observation of strong earthquake motion." Ministry of Construction, Tsukuba, Japan.

Rice, S.J. 1973. "Geology for planning - Novato area." California Div. of Mines and Geology, 56 p.

Robertson, P.K., Woeller, D.J. and Finn, W.D. Liam. 1992. "Seismic cone penetration tset for evaluating liquefaction potential under cyclic loading." Canadian Geotechnical Journal, Vol. 29, No. 4, August, pp. 686-695.

Schnabel, P.B., Lysmer, J. and Seed, H.B. 1972. "SHAKE: a computer program for earthquake response analysis of horizontally layered sites." Report No. EERC 72-12, University of California, Berkeley, December.

Seed, H.B., Tokimatsu, K., Harder, L.F. and Chung, R.M. 1985. "Influence of SPT procedures in soil liquefaction resistance evaluations." Journal of Geotech. Engineering, Vol. 111, No. 12, pp 1425-1445.

Seed, H.B. and Idriss, I.M. 1983. *Ground Motions and Soil Liquefaction During Earthquakes.* Earthquake Engineering Research Institute, El Cerrito, California.

Seed, R.B. and Harder, L.F. 1990. "SPT-based analysis of cyclic pore pressure generation and undrained residual strength." Proc., H. Bolton Seed Memorial Symp., University of California, Berkeley, California, Vol. 2, pp 351-376.

Seed, R.B., S.E. Dickenson, M.F. Riemer, J.D. Bray, N. Sitar, J.K. Mitchell, I.M. Idriss, R.E. Kayen, A. Kropp, L.F. Harder, Jr. and M.S. Power. 1990. "Preliminary Report on the Principal Geotechnical Aspects of the October 17, 1989 Loma Prieta Earthquake." Earthquake Engineering Research Center, University of CA at Berkeley, UCB/EERC-90/05.

Thenhaus, Paul C., Hanson, S.L., Effendi, I., Kertapati, E.K. and Algermissen, S.T. 1993. Pilot studies of seismic hazard and risk in North Sulawesi Province, Indonesia. Earthquake Spectra, Vol. 9, No. 1, February, pp. 97-120.

Tinsley, J.C., Youd, T.L., Perkins, D.M. and Chen, A.T.F. 1985. "Evaluating liquefaction potential: evaluating earthquake hazards in the Los Angeles region." J.I. Ziony (ed.), US Geological Survey Prof. Paper 1360, pp. 263-316.

Trifunac, M.D. and Brady, A.G. 1975. "On the correlation of seismic intensity scales with peaks of recorded strong ground motion." Bulletin of the Seismological Society of America, Vol. 65, pp. 139-162.

Uniform Building Code. 1989. International Conference of Building Officials, 5360 South Workman Mill Road, Whittier, California, U.S.A.

Varnes, D.J. 1984. "Landslide hazard zonation: a review of principles and practice." Natural Hazards 3. UNESCO and International Association of Engineering Geology, Commission on Landslides and Other Mass Movements on Slopes.

Wakamatsu, K. 1991. "Maps for historic liquefaction sites in Japan." Tokyo, Tokai University Press (in Japanese with English abstract).

Wakamatsu, K. 1992. "Evaluation of liquefaction susceptibility based on detailed geomorphological classification." Proc., Technical Papers of Annual Meeting, Architectural Institute of Japan, Vol. B, pp. 1443-1444 (in Japanese).

Wieczorek, G.F., Wilson, R.C. and Harp, E.L. 1985. "Map showing slope stability during earthquakes in San Mateo County, California." US Geological Survey Map I-1257-E.

Youd, T.L., J.C. Tinsley, D.M. Perkins, E.J. King, and R.F. Preston. 1978. "Liquefaction Potential Map of San Fernando Valley, California," Int. Conf. on Microzonation for Safer Construction, 2nd, San Francisco 1:267-278.

Youd, T.L. 1980. "Ground failure displacement and earthquake damage to buildings." Proc., Conference on Civil Engineering and Nuclear Power, Knoxville, Tenn., ASCE, Vol. 2, pp. 7-6-1 to 7-6-26.

Youd, T.L. 1991. "Mapping of earthquake-induced liquefaction for seismic zonation." Proc., 4th Int. Conf. on Seismic Zonation, Vol. 1, pp. 111-148.

Youd, T.L. and Perkins, D.M. 1978. "Mapping of liquefaction induced ground failure potential." Journal of Geotechnical Engineering, ASCE, Vol. 104, No. 4, pp. 433-446.

Youd, T.L. and Perkins, D.M. 1987. "Mapping of liquefaction severity index." Journal of Geotechnical Engineering, ASCE, Vol. 113, No. 11, pp. 1374-1392, November.

Youd, T.L. and Wieczorek, G.F. 1984. "Liquefaction during the 1984 and previous earthquakes near Westmorland, California." US Geological Survey Open-File Report 84-680.

APPENDIX 1

Simplified Method for Evaluating Liquefaction Potential

The most generally applied method for assessing liquefaction potential is that developed by Seed and his colleagues, and generally referred to as *"the simplified procedure for evaluating liquefaction potential"* (Seed and Idriss, 1983). The geotechnical procedures in the Japan Bridge code are commonly used for liquefaction hazard mapping in Japan (Iwasaki et al., 1982).

The Seed simplified procedure assesses liquefaction potential on the basis of the past performance of soils during earthquakes. The in-situ state of the soil and its resistance to liquefaction are characterized by the standard penetration resistance normalized to a reference overburden pressure of 100 kPa (1 tsf) and a standard energy level of 60% of the free fall energy of the hammer and is designated $(N_1)_{60}$.

The seismic demand on the deposit is characterized by the ratio of the cyclic shear stress to the effective overburden pressure, τ/σ'_v, where τ is an average effective shear stress used to characterize the time history of shear stresses developed by the earthquake. The state of the deposit under investigation $[(N_1)_{60}, (\tau/\sigma'_v)]$ is plotted on the chart shown on Figure A-1. Then the deposit would be judged to liquefy or not depending on which region the state point plots. The updated procedures for applying this method are described by Seed and Harder (1990).

This type of evaluation can also be conducted using the core penetration test, and a chart similar to Figure A-1 is used to determine whether a deposit is likely to liquefy or not. The chart is shown in Figure A-2.

Some soils, such as gravels, are difficult to penetrate, and in addition, the boreholes required for the standard penetration test and the special equipment required to push the cone can make large-scale penetration studies expensive. Measurement of shear wave velocity from the surface using the technique called spectral analysis of surface waves (SASW) gives the distribution of shear wave velocity with depth without having to use boreholes or mobilize expensive equipment (Nazarian and Stokoe, 1984; Addo and Robertson, 1992).

Resistance to liquefaction, is specified by the normalized shear wave velocity, V_{s1}, and τ/σ'_v is determined from the chart in Figure A-3. A very accurate liquefaction potential map can be drawn if a wide distribution of penetration resistances and/or shear wave velocities are available. (Finn, 1993; Tokimatsu et al., 1991; and Robertson et al., 1992).

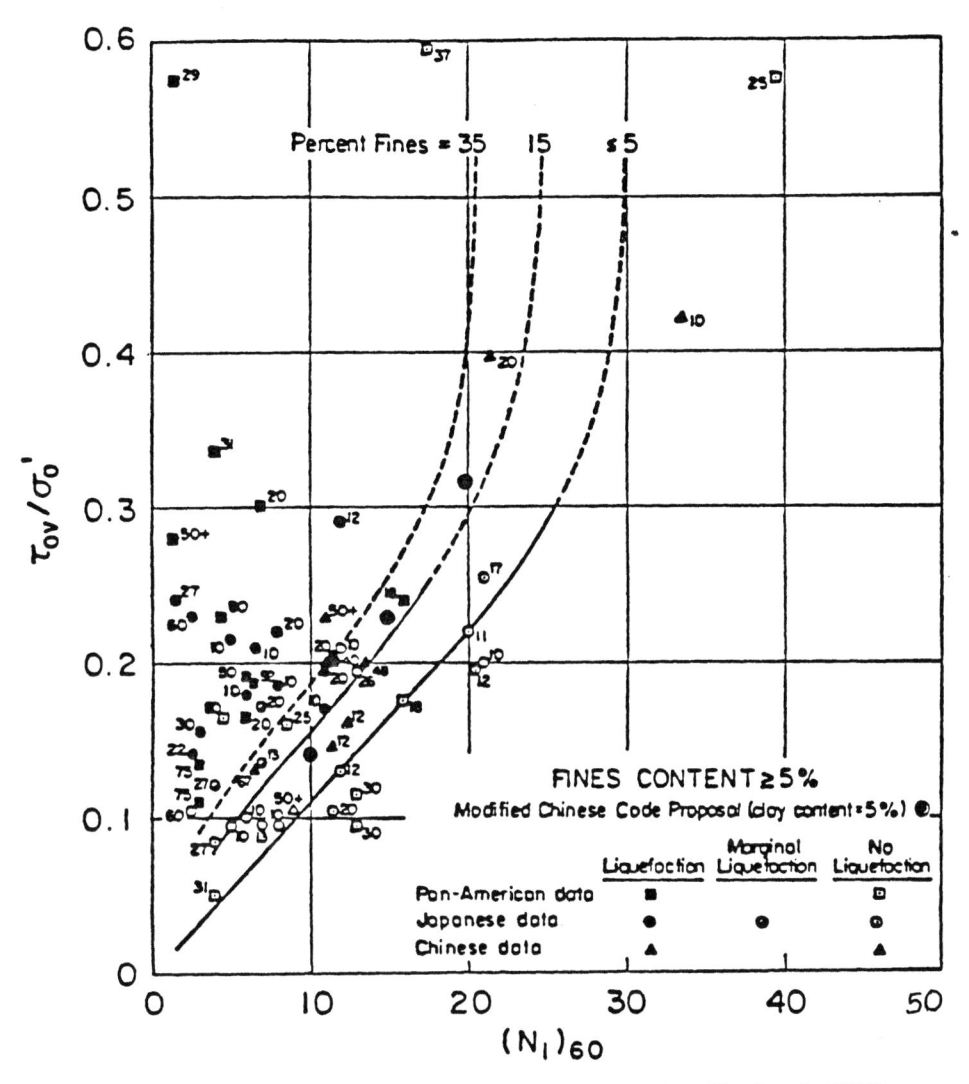

Fig. A-1. Seed's liquefaction resistance chart based on SPT (Seed et al., 1985)

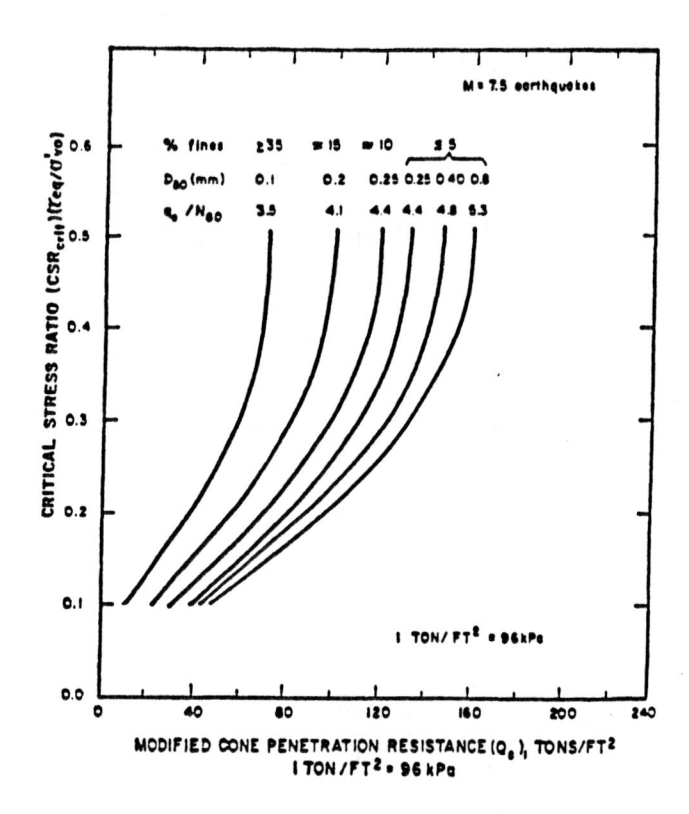

Fig. A-2   Seed's liquefaction resistance chart based on CPT (Seed et al., 1985)

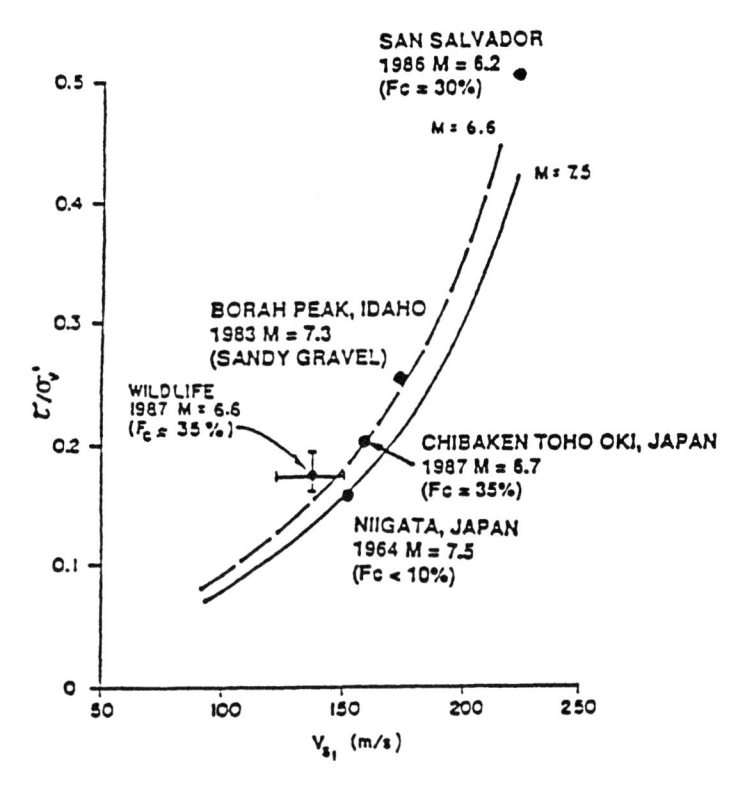

Fig. A-3. Liquefaction resistance chart based on shear wave velocity (Finn et al., 1989)

ESTIMATION OF EARTHQUAKE DAMAGE TO BUILDINGS AND OTHER STRUCTURES IN LARGE URBAN AREAS

CHRISTOPHER ROJAHN
Executive Director
Applied Technology Council
555 Twin Dolphin Drive, Suite 550
Redwood City, California 94065

1. Introduction

Damage to existing structures can be expected to be severe if large earthquakes occur in or near large urban areas. For example, Steinbrugge et al. (1981) estimate that property losses for a magnitude-7.5 earthquake on the Newport-Inglewood fault in the Los Angeles, California, metropolitan area would be $62.2 billion (1980 dollars), excluding losses for communication and transportation systems, dams, military installations, and consequent losses such as unemployment, loss of taxes, shutdown of factories due to loss of supplies, and automobile damage. These large expected dollar losses are of concern to facility owners, insurers, and government officials.

Methods and data for predicting damage to buildings, bridges, dams, utility systems, and other man-made structures have been the subject of extensive research over the past several decades. These research activities have shown that structural damage is dependent upon several key factors and have resulted in the development of several approaches for predicting expected damage.

The intent of the text that follows is to provide an overview of the factors that affect earthquake structural damage potential; an overview of methods to estimate earthquake-induced structural damage, including those provided in the Applied Technology Council (ATC) Report ATC-13, *Earthquake Damage Evaluation Data for California* (ATC, 1985); and an overview of methods to estimate monetary and non-monetary losses resulting from structural damage. The paper concludes with a discussion of the basic steps to be followed in developing estimates of expected structural damage and associated losses in an urban area due to selected scenario earthquake(s). Required resources and available potential supporting information are identified where possible.

2. Factors Affecting Earthquake Damage

The extent to which buildings and other structures are damaged is generally dependent upon the following factors:
- Structural characteristics;
- Ground-shaking severity; and
- Collateral hazards.

These factors, particularly structural characteristics and ground-shaking severity, may interact to enhance or reduce damage potential. A clear understanding of these factors individually as well as their combined impact will improve accuracy in the application of damage-prediction methods. Following are discussions of each of these factors.

2.1 STRUCTURAL CHARACTERISTICS

In general, structural characteristics important for earthquake damage-evaluation purposes include: construction material, soil-foundation material, structural foundation, structural system, configuration, structural continuity, age, proximity to other structures, and exposure to prior earthquakes. Although there are exceptions, good earthquake performers normally have the following characteristics: (1) they are constructed of materials that can undergo long-duration horizontal and vertical shaking without excessive loss of stiffness or strength; (2) they are founded on firm foundation materials (the structure does not tip or settle due to foundation failure); (3) they have regular, symmetrical plan shapes (so that torsional response is not induced in the structure); (4) they are composed of structural elements that are tied together (the structural elements do not separate during earthquake shaking); (5) they are of recent age (designed since the adoption of seismic codes); (6) they are not too close to other structures (so as to avoid pounding of two structures against one another); and (7) they have not previously been damaged by an earthquake.

Damage can generally be expected to be low for structures that have been designed in accordance with earthquake codes or that naturally have earthquake-resistant properties. Examples of the former include recently designed and constructed buildings in urban areas of California that have adopted and enforced seismic design codes; examples of the latter include one-story wood-frame houses that are anchored to their foundations and that have light-weight roof systems.

The existence and implementation of seismic codes, however, does not necessarily assure good seismic performance. Poor performance of buildings designed using codes has been observed in several recent U. S. earthquakes, including the 1964 Alaska earthquake, the 1971 San Fernando, California earthquake, 1979 Imperial Valley, California earthquake (Steinbrugge, 1982), and the 1989 Loma Prieta, California, earthquake near San Francisco (Rojahn, 1990). Conversely, several engineered buildings designed prior to the existence of U. S. seismic building code regulations performed very well during the 1906 San Francisco earthquake (Galloway, 1907). Thus, although age or the existence of seismic design codes may indicate expected performance, they are not all-encompassing criteria for distinguishing earthquake-resistant construction (ATC, 1985).

2.2 GROUND-SHAKING SEVERITY

In general, for a given site and distance from the earthquake source (fracture in the earth's crust along which the earthquake occurs), ground-shaking severity is directly proportional to the magnitude of the earthquake. In other words, the larger the earthquake, the more severe the ground-shaking. In California, where historic earthquakes have been shallow (most are less that 15 km deep), earthquake shaking can be expected to be more severe at shorter distances from the fault rupture zone than it is at larger distances. This point is dramatically illustrated by the strong ground-motion data from the 1979 Imperial Valley, California earthquake (U. S. Geological Survey, 1981) [Figure 1]. In parts of the world where earthquakes occur at greater depth (70 to 700 km), ground-shaking may be very severe at large distances from the earthquake source zone. Such was the case for the September 1985 earthquake along the coast of Mexico that severely damaged hundreds of multi-story buildings in Mexico City and the 1977 earthquake in Romania that collapsed approximately 35 multi-story buildings in Bucharest approximately 100 km south of the earthquake source region.

Earthquake shaking is normally measured in terms of acceleration and, in some cases, in velocity or displacement. The shaking is cyclic in nature, and the most important characteristics of such motions (from an engineering standpoint) are: (1) amplitude; (2) frequency content; and (3) duration. Damaging levels of ground-shaking can be expected to occur for earthquakes of about magnitude 5.5 or greater when the amplitude of ground-shaking in the horizontal direction is approximately $0.20\ g$ (g equals the force of gravity) or greater. In order for such motions to be damaging, however, they need to be of sufficient duration (several cycles of high-amplitude motion) and at a frequency near that of the structure affected. The fundamental frequency of a building is dependent on its height and stiffness (in the horizontal direction). Fundamental frequencies, for example, normally range from 1 to 10 Hz (cycles per second) for 1-5-story buildings, and 0. 5 to 2 Hz for 5-to-10-story buildings. When the fundamental frequency of the building and the dominant frequency of earthquake-induced ground-shaking coincide (or nearly so), motions in the building can be amplified by two or more times (relative to the ground motion). This phenomenon was probably responsible for much of the damage in Mexico City where the fundamental frequencies of damaged multi-story buildings were near that of the dominant frequency of ground motion of lake-bed sites (about 0. 5 Hz).

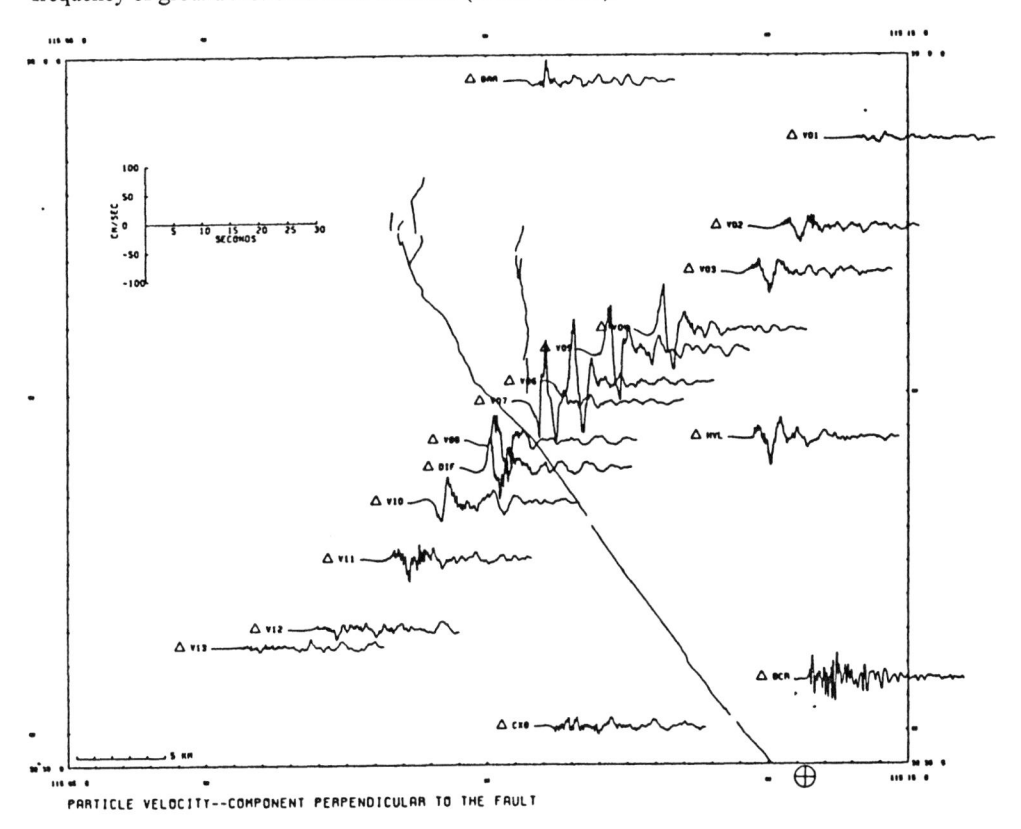

Figure 1.- The horizontal particle velocity recorded at USGS strong-motion sites during the October 15, 1979 magnitude 6.6 Imperial Valley earthquake is plotted in a plan view of the source region.

2.3 COLLATERAL HAZARDS

In addition to strong ground-shaking, several other phenomena, commonly regarded as collateral hazards, have been observed to significantly affect earthquake losses (ATC, 1985). These include:
- Poor ground (loose sands, sensitive clays);
- Landslide;
- Fault rupture;
- Inundation; and
- Fire.

Poor ground includes soil conditions such as loose sands, sensitive clays, and some lightly cemented sands (ATC, 1985). Flow failures and lateral spreading in sensitive clays and settlements of loose dry sand of a few inches to a foot have been observed to be the cause of severe earthquake structural damage. A more common cause of damage, however, is liquefaction, which is a process in which loose saturated sand liquefies during the earthquake and loses its shear strength. Liquefaction is commonly manifested during earthquakes in the form of water spouts (sand and water seething from holes in the ground), the submergence or tilting and differential settlement of structures, and the re-emergence of buried tanks. A structure that did not sustain damage from deformation caused by ground-shaking might sustain substantial damage if it settles differentially or tilts.

A *landslide*, which is a downslope movement of a soil or rock mass due to gravity, can seriously damage a structure that is located on or at the base of the downslope movement (ATC, 1985). The slope of surfaces on which landsliding has been observed to occur during earthquakes varies from quite steep to almost horizontal. In addition, the season of the year is a significant secondary factor in landslide potential because the amount of moisture present influences the load stability of most weak soils. In California, a strong earthquake during the wet winter rainy season can be expected to cause more damage from landslides than an earthquake in the dry summer.

Fault rupture can cause significant damage to structures situated immediately over simple fault breaks and to structures situated in the fault zone of more complex alluvial surficial deposits (ATC, 1985). In the simpler case, the structure is literally sliced in a shearing motion with one portion of the structure moving in one direction and the other moving in the opposite direction. For the more complex case, surficial soil materials are deformed and distorted over a broad area, and structures founded in these zones are subject to potentially serious disruption.

Inundation during earthquakes can result from several causes, including tsunamis, seiches, dam/reservoir failures, and areal subsidence or tilting (ATC, 1985). Tsunamis are transient sequences of long-period sea waves generated impulsively by earthquakes, coastal or submarine landslides, or volcanic phenomena. Seiches are periodic oscillations of enclosed or semi-enclosed bodies of water caused by earthquakes or landslides that disrupt the normal boundaries of lakes, bays, or other such large volumes of water. Inundation resulting from areal subsidence or tilting occurs when there is ground failure adjacent to large bodies of water. Dam/reservoir failures can occur as a result of severe ground-shaking or collateral hazards (ATC, 1985).

Fire is one of the greatest potential dangers immediately following an earthquake and, if unchecked, could lead to a major conflagration under certain circumstances. Important parameters for evaluating possible conflagration include: fire ignitions (the numbers of fires started); fire reporting and response time (the time between fire ignition and reporting); fire spread; and fire suppression (ATC, in preparation). Fire ignitions are clearly related to damage, with the disruptions to the normal functionality of gas and electrical systems being the primary causes. Fire reporting and response time are influenced, respectively, by the time required for citizens to report fires (assuming telephones are inoperable) and the density and extent of damage to buildings, which may impede access to fires by firefighting equipment and personnel. Fire spread is influenced by the availability and layout of fuel supply (construction density and material) and weather conditions (wind and moisture). Finally, fire suppression is influenced by the number and availability of firefighting equipment and personnel.

3. Methods to Estimate Damage to Buildings and Other Structures

Earthquake damage can be expressed in a variety of ways. These include (1) percent financial loss; (2) percent of structures damaged; and (3) other structure-specific parameters, such as percent of building floor area damaged (e.g., Pomonis et al., 1993). In general, the most revealing expression for consistent application to all types of structures and components is percent financial loss. Almost any type of earthquake loss, for any facility or component, can be rationally deduced from this expression. Applied Technology Council defines percent financial loss in terms of "damage factor" as follows (ATC, 1985):

$$\text{Damage Factor (DF)} = \frac{\text{Dollar loss}}{\text{Replacement value}}$$

The mean damage factor for a group of similar structures exposed to the same ground-shaking intensity is defined as:

$$\text{Mean Damage Factor (MDF)} = \frac{1}{n} \sum_{i=1}^{n} \frac{(\text{Dollar loss})_i}{(\text{Replacement value})_i}$$

where n is the number of structures in the sample.

For a given scenario earthquake, damage factor for each individual structure, or mean damage factor for groups of similarly affected structures, can be estimated for the effects of ground shaking and other collateral hazards, as described below.

3.1 ESTIMATION OF DAMAGE FROM GROUND SHAKING

Structural damage resulting from earthquake ground shaking is normally estimated through the use of motion-damage relationships, which are also known as ground shaking vulnerability functions. Historically, for most parts of the world, the motion parameter in available motion-damage (percent loss) relationships has been defined in terms of seismological intensity scales, particularly the Modified Mercalli Intensity (MMI) Scale (Wood and Newmann, 1931), which is summarized in Table 1. An example motion-damage relationship expressing expected damage as a function of MMI for typical Costa Rican building types (Sauter and Shah;, 1978) is provided in Figure 2.

Other seismological intensity scales used worldwide include the Rossi-Forel Intensity Scale (Richter, 1958); Medvedev-Sponheuer-Karnik (MSK) Intensity Scale (Medvedev, Sponheuer, and Karnik, 1965), the Japan Meteorological Agency (JMA) Scale (Barosh, 1969), and the Geofian Scale (Barosh, 1969). A comparison of the MMI, Rossi-Forel, JMA, Geofian, and MSK sales is provided in Figure 3.

Most of the major earthquake loss studies in the United States to date have used motion-damage relationships that are based on the MMI scale. Examples of studies that use this shaking characterization include Algermissen et al. (1978), ATC (1985), ATC (1991), ATC (1992), Benjamin (1974); Blume et al. (1975); Scholl et al. (1982), Steinbrugge et al. (1969), Steinbrugge and Schader (1979), Whitman et al. (1973), and Wiggins (1981). The primary reason for using MMI data is because most of the available data on earthquake effects in the United States are available in this form (as a result of more than fifty years of data collection by the U. S.

Geological Survey and its predecessors). As a result, there is a basic common understanding of the meaning of the MMI scale among earthquake practitioners. In addition, an important positive feature of the MMI scale is that intensity ratings are also based on other phenomena that have a more universal and unvarying basis. These include such items as toppling of grocery-shelved items at low intensity levels, ground failures at intermediate intensities, and the disorientation of persons at high-shaking intensities (Nason, 1980).

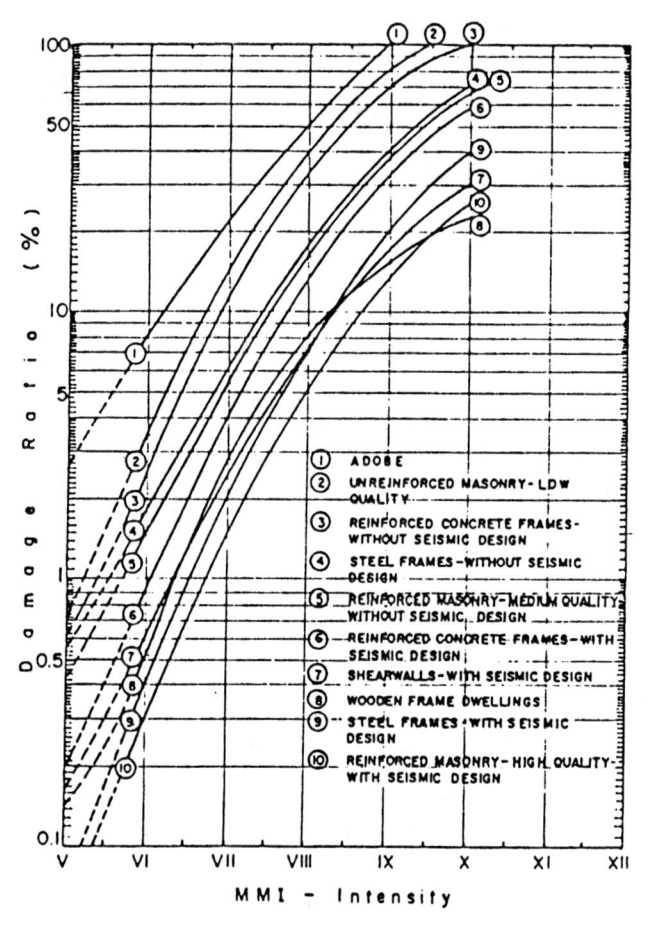

Figure 2.- Estimated motion-damage curves for typical Costa Rican Building Types (Sauter and Shah, 1978).

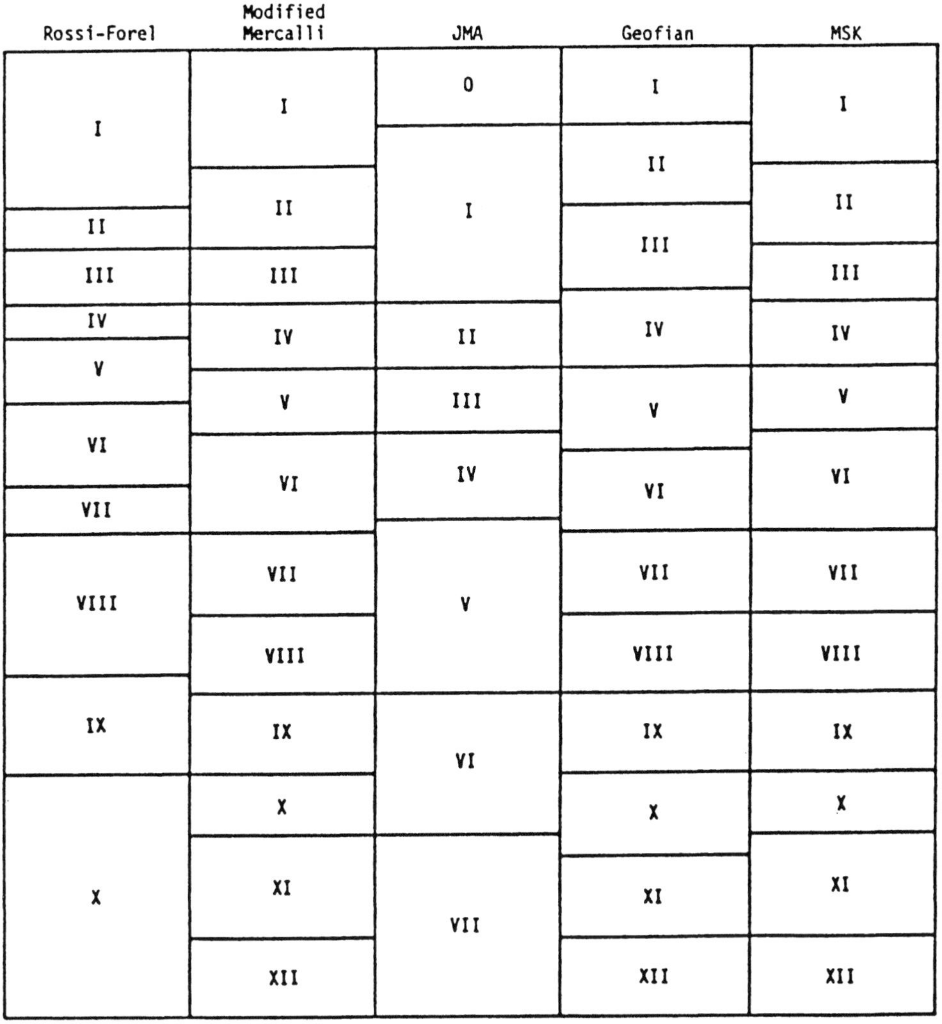

Rossi-Forel	Modified Mercalli	JMA	Geofian	MSK
I	I	0	I	I
			II	
II	II	I		II
			III	
III	III			III
IV	IV	II	IV	IV
V				
	V	III	V	V
VI				
	VI	IV	VI	VI
VII				
VIII	VII	V	VII	VII
	VIII		VIII	VIII
IX	IX	VI	IX	IX
	X		X	X
X	XI	VII	XI	XI
	XII		XII	XII

Figure 3.- Graphic comparison of seismological intensity scales (ATC, 1985).

It is recognized, however, that the MMI scale and other seismological intensity scales have several shortcomings that render them less than ideal for earthquake loss studies for large urban areas. In the case of MMI, for example, the shortcomings include: (1) the scale is subjective in nature and may be interpreted differently by different users; (2) the scale is not ideally suited to new types of construction as it is based largely on the performance of unreinforced masonry buildings and chimneys and other types of older construction, and (3) the scale combines long- and short-period structural damage at given intensity levels and is therefore biased by earthquake magnitude and distance. For these reasons, current research is focusing on the development of motion-damage relationships that use ground motion parameters that are instrumentally based.

Instrumentally based ground motion parameters that could be used for motion-damage relationships include: (1) peak ground acceleration (largest absolute value of recorded acceleration); (2) Response Spectrum Intensity (Housner, 1952); (3) effective peak acceleration (ATC, 1978); (4) root-mean-square acceleration (McCann and Shah, 1979); (5) Arias Intensity (Arias, 1970); (6) Engineering Intensity (Blume, 1970); and (7) various linear-elastic spectral response parameters, particularly acceleration and velocity response. To date, none of these instrumentally based parameters has been used to consistently develop motion-damage relationships for all of the major classes of structure types (e.g., buildings, bridges, dams, power plants) existing in most urban areas throughout the world. Some motion-damage relationships for specific building types based on spectral response parameters (acceleration, velocity and displacement), however, have been developed and are available. These include the spectral acceleration based motion-damage relationships developed by Ordaz et al. for Mexico City structures (written commun., M. Ordaz, 1993), which are based on extensive field observational data resulting from the 1985 Mexico earthquake, and the spectral acceleration-MMI based acceleration response spectra developed by Pomonis et al. (1993), which are based on a database of 2,300 buildings worldwide from 15 surveys. Other available response spectra motion-damage relationships for buildings include the acceleration and velocity response relationships of Wong (1975) for high-rise buildings in the Los Angeles, California, region, the acceleration, velocity, and displacement response relationships of Scawthorn et al. (1981) for mid-rise buildings in Sendai, Japan, and the acceleration response relationships of URS/Blume (1981) for selected buildings in several US locations.

3.2 ESTIMATION OF DAMAGE FROM COLLATERAL HAZARDS

In addition to damage caused by strong ground-shaking, collateral hazards such as ground failure, fault rupture, landslide, inundation, and fire can also cause serious damage to buildings and other structures. Other than the work done by the Applied Technology Council (ATC, 1985), there is little information in the literature regarding estimation of structural damage from these collateral hazards. Following is a description of the ATC-13 approach (ATC, 1985) for estimating damage caused by poor ground, landslide, fault rupture, and inundation. The estimated damage from each of these four collateral causes is defined in terms of Mean Damage Factor, which is the same form used to describe damage due to ground shaking (see previous section). A method for estimating the damaging effects of fire following earthquake is being developed in the currently ongoing ATC-36 project (ATC, in preparation).

The procedure for estimating damage caused by *liquefaction/poor ground* is based on (1) observations of ground failure from liquefaction during the 1906 San Francisco earthquake, in which case damage on poor ground was 5 to 10 times greater that on firm ground (Wood, 1908), and (2) regional probabilistic estimates of liquefaction potential developed by Legg et al. (1982). The procedure is as follows:

$$MDF(PG) = MDF(S) \times P(GFI) \times 5$$

for surface facilities and

$$MDF(PG) = MDF(S) \times P(GFI) \times 10$$

for buried facilities, where MDF(S) = mean damage factor caused by ground shaking; MDF(PG) = mean damage factor caused by poor ground; and P(GFI) = probability of a given ground failure intensity, taken directly, noncumulatively, from Table 2, for a given shaking intensity.

The procedure for estimating damage caused by *landslide*, which uses the slope failure concept proposed by Legg et al. (1982), is as follows:

$$MSF(LS) = PSFI \times CSF_{SFS}$$

where MDF(LS) = mean damage factor caused by landslide; PSFI = probability of a given slope failure intensity; and CDF$_{SFS}$ = central damage factor for a given slope failure state (Table 3).

The method for estimating damage caused by *fault rupture* assumes that structures astride the fault or in the drag zone can be significantly damaged as a result of fault rupture. Various mean damage factors for fault rupture, MDF(FR), are given in Table 4. The fault zone (Table 4) is assumed to be 100 meters each side of the fault (to be reduced to 5 meters if the fault trace is known) and the drag zone is assumed to be 200 meters each side of the fault (to be reduced to 100 meters if the fault trace is known). The surface fault slip displacements of 0.2, 0.6, 1, 3.5, and 10 meters (Table 4) are approximately representative of the slip for earthquakes of magnitudes 6, 6.5, 6.75, 7.5, and 8, respectively (ATC, 1985).

The procedure for estimating damage caused by *inundation* assumes that depth and velocity of water are the primary factors affecting damage caused by inundation. The damage factor for facilities exposed to high-velocity water is given in Table 5. The mean damage factors for inundation, MDF(I), specified in this table apply only for ground-surface structures less than 10 meters high.

The total mean damage factor for a facility is conservatively the sum of the mean damage factors for ground-shaking, poor ground/liquefaction, landslide, fault rupture, and inundation.

These methods for estimating structural damage resulting from collateral hazards are currently under review in the ongoing ATC-36 project, "Development of Earthquake Loss Evaluation Methodology and Data for Utah." Revisions in methodology resulting from this review will be documented in the final ATC-36 Report (ATC, in preparation), which is scheduled for publication in 1994.

4. Methods to Estimate Losses Resulting from Structural Damage

Normally, earthquake losses can be classified into three main categories:
- Losses due to direct physical damage;
- Indirect economic losses; and
- Deaths and injuries (casualties)

Losses due to direct physical damage (L$_{Direct}$) to a given structure simply equal the financial cost to repair or replace that structure. If damage is defined in terms of percent financial loss (damage factor), losses due to direct physical damage to a given structure, i, are calculated as:

$$L_{Direct} = (Replacement\ Value_i) * (DF_i)$$

Replacement value is a function of the cost of materials and labor, and structure size, location and use. When one is calculating replacement value for a given structure, the cost to replace structural components, nonstructural components, and contents, and their respective percent damage, must be considered.

Indirect economic losses are consequent losses resulting from damaged facilities being out of service for some finite time after an earthquake. Indirect economic losses include:

- Lost production or sales by economic establishments housed in damaged facilities;

- Production or sales lost by facilities unable to obtain critical supplies from damaged facilities;

- Production or sales lost as a result of damaged lifelines (electric, water, transportation, gas and liquid fuel supply, and communication systems) serving economic establishments;

- Costs to temporarily or permanently replace injured or deceased personnel;

- Relocation costs for economic establishments housed in damaged structures that must be vacated;

- Increased unemployment compensation costs to a community resulting from temporary or permanent layoffs by economic establishments housed in badly damaged structures; and

- Loss of tax revenue from economic establishments housed in badly damaged structures (e.g., major hotels).

As suggested by this list, indirect economic losses are dependent primarily upon the extent of structural damage and the extent of time that damaged structures are out of service. There is little information in the literature that discusses indirect economic losses and the means to estimate them. Partial treatment of the issue is provided by the ATC-25 Report, *Seismic Vulnerability and Impact of Disruption of Lifelines in the Conterminous United States* (ATC, 1991), which recommends a first-order-approximation method for estimating indirect economic losses resulting from damage to lifeline systems (i.e., buildings and industrial structures are not considered). The ATC-25 methodology consists of a two-step process:
1. Development of estimates of interruption of lifelines as a result of direct damage.
2. Development of estimates of economic loss as a result of lifeline interruption.

ATC-25 estimates of lifeline interruption are derived from similar expert-opinion data in the ATC-13 Report (ATC, 1985), which contains interruption, or loss of function, estimates for 65 facility types, classified by social function. Example ATC-25 time-to-restoration (interruption) curves for airport terminals in California designed to resist earthquakes are provided in Figure 4 as a function of MMI.

ATC-25 indirect economic loss estimates are derived from expert-opinion based tables defining the reduction in value added (value of shipments/products less the cost of materials, supplies, contract work and fuels used in the manufacturer or cultivation of a product) for each major sector of the economy as a function of lifeline interruption. Other indirect effects are not considered, nor are structure types other than lifelines.

Deaths and injuries are caused primarily by damage to structures and are normally a function of the degree of damage, the weight of collapsed structural materials, and collapse mechanisms. Collapsed reinforced concrete structures, for example, normally injure and kill a higher percentage of occupants than do collapsed wood structures. ATC-13 derived expected injury and death rates, expressed as a function of "Damage Factor" are provided in Table 6. These estimates are currently being evaluated and updated in the ATC-36 project (ATC, in preparation).

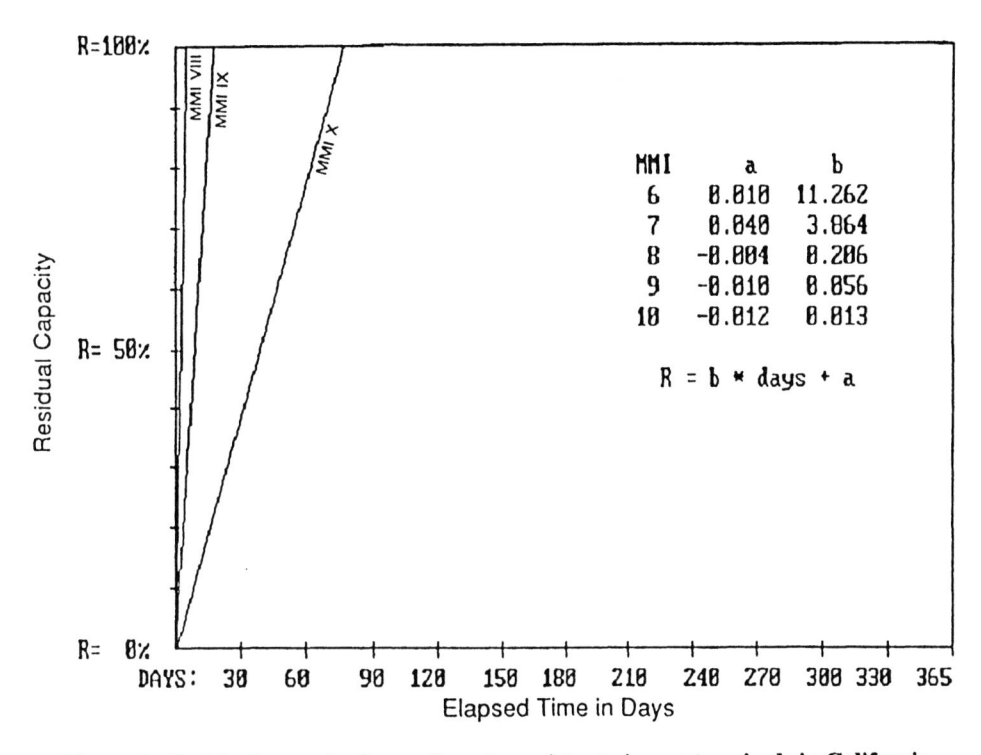

Figure 4.- Residual capacity for earthquake-resistant airport terminals in California (ATC, 1991).

5. Development of Structural Damage Estimates for Scenario Earthquakes

Prior to undertaking a comprehensive investigation of the expected structural damage and associated losses in an urban area due to selected scenario earthquake(s), it is useful to review the steps involved in conducting such a study and the resources required. Following is a preliminary, but not necessarily exhaustive, list of the basic steps that are followed in developing earthquake damage and loss estimates for large urban areas. Required resources and available potential supporting information (known to the author) are identified where possible.

STEP 1. DEFINE THE MAGNITUDE, MECHANISM, AND LOCATION OF SCENARIO EARTHQUAKE(S) TO BE CONSIDERED

Resources Needed.

• Prior seismicity studies that define earthquake source zones and the size, location, and occurrence intervals of earthquakes that could cause damaging earthquakes for the urban area under consideration.

• Panel of seismological experts knowledgeable in the seismicity of the region who can collectively define the scenario earthquake(s) to be considered.

STEP 2. ESTABLISH THE BASIS FOR ESTIMATING THE SEVERITY OF EARTHQUAKE GROUND SHAKING AT SPECIFIC SITES WITHIN THE URBAN AREA

Resources Needed.
• A ground shaking characterization (e.g., Modified Mercalli Intensity scale; acceleration response spectra) that (1) is compatible with the vulnerable models selected in Step 6, (2) can be easily incorporated in a model that spatially distributes shaking severity, and (3) incorporates or is capable of incorporating site amplification effects. Traditionally, seismological intensity scales, such as MMI, have been useful in areas where few strong ground motion data have been collected.

• Panel of engineering seismology and structural engineering experts knowledgeable in (1) the results of prior earthquake damage investigations in the region and (2) the advantages and limitations of the various available alternative ground shaking characterizations.

STEP 3. ESTABLISH THE BASIS FOR ESTIMATING THE OCCURRENCE OF COLLATERAL HAZARDS AT SPECIFIC SITES WITHIN THE URBAN AREA (FAULT RUPTURE, GROUND FAILURE/ LIQUEFACTION, LANDSLIDE, INUNDATION, FIRE FOLLOWING EARTHQUAKE)

Resources Needed.
• Maps of scenario-earthquake active faults, if within the urban area; maps of liquefaction potential for the urban area; maps of landslide potential for the urban area; maps of inundation potential for the urban area that include potential inundation from failed dams, from regional tilt, and from tsunami; models for estimating fire ignition, spread, and suppression.

• Panels of consultants and advisors to develop the needed collateral hazard data/methodology if the information identified above is not available and there are substantiated technical reasons to believe such hazards are possible and potentially damaging.

Potential Available Supporting Information. ATC-36 fire following earthquake model (ATC, in preparation)

STEP 4. DEFINE THE STUDY AREA (GEOGRAPHICAL AREA IN WHICH DAMAGE TO STRUCTURES IS TO BE ESTIMATED; NORMALLY THE AREA WITHIN THE LEGAL URBAN BOUNDARIES AND POSSIBLY THE IMMEDIATE SURROUNDING REGION)

STEP 5. CONDUCT A PRELIMINARY INVENTORY OF STRUCTURES IN THE URBAN AREA TO DETERMINE THE EARTHQUAKE RESISTANT FEATURES OF TYPICAL STRUCTURES, AND DEVELOP A STRUCTURAL CLASSIFICATION SCHEME.

Resources Needed.
• Earthquake engineering researchers and practitioners who can identify earthquake resistant features of buildings and lifelines within the urban area; engineering students who can assist in the inventory process; suggested structural classification schemes that could be adapted for use in the urban area under consideration (that are compatible with available seismic vulnerability functions (see Step 6)); panels of consultants and advisors to finalize the recommended structural classification scheme.

Potential Available Supporting Information. ATC-13 earthquake engineering classification scheme (Table 7) for structures in California (ATC, 1985), ATC-36 earthquake engineering classification scheme for structures in Utah (ATC, in preparation); structural classification scheme used in the Quito, Ecuador, earthquake scenario study conducted under the auspices of GeoHazards International (in preparation).

STEP 6. ESTABLISH/ SELECT GROUND SHAKING VULNERABILITY FUNCTIONS FOR EACH STRUCTURE TYPE IN THE SELECTED STRUCTURAL CLASSIFICATION SCHEME

Resources Needed.
- Earthquake engineering researchers and practitioners who (1) understand the extent to which the identified structure types will be damaged as a function ground shaking; (2) can adapt available ground shaking vulnerability functions from other regions (models that define the relationships between ground shaking and damage for given structure types) for use in the study area; and/or (3) can develop any needed ground shaking vulnerability functions. It is important to emphasize that the vulnerability functions used must be based on ground shaking characterizations that are compatible with the characterizations used in Steps 2 and 5 above.

- Panel of earthquake engineering experts to review and approve all selected/developed vulnerability functions. It is important that the functions used reflect a broad spectrum of engineering opinion so that they are readily acceptable by the community at large.

Potential Available Supporting Information. ATC-13 ground shaking vulnerability functions for structures in California (ATC, 1985), ATC-36 ground shaking vulnerability functions for structures in Utah (ATC, in preparation); vulnerability functions used in the Quito, Ecuador Earthquake Scenario study conducted under the auspices of GeoHazards International (in preparation); other vulnerability functions (see above discussion on estimation of damage from ground shaking, including references cited).

STEP 7. ESTABLISH/ SELECT METHODS FOR ESTIMATING DAMAGE RESULTING FROM COLLATERAL HAZARDS (FAULT RUPTURE, GROUND FAILURE / LIQUEFACTION, LANDSLIDE, INUNDATION, FIRE FOLLOWING EARTHQUAKE)

Resources Needed.
- Available methods for estimating damage resulting from collateral hazards (fault rupture, ground failure/liquefaction, landslide, inundation, fire following earthquake) that can be adapted for use in the study area, considering the quality and type of information developed in Step 3 above.

- Earthquake engineering researchers and practitioners who understand and can apply available methods, considering any constraints imposed by the results from Step 3 above; panel of earthquake engineering experts to review results from application of the methods used.

Potential Available Supporting Information. ATC-13 methods for estimating damage resulting from fault rupture, ground failure/liquefaction, landslide and inundation (ATC, 1985), ATC-36 collateral hazard damage estimation methods, which are updated versions of those provided in ATC-13 and which include a method for estimating the effects of fire following earthquake (ATC, in preparation).

STEP 8. CONDUCT AND DOCUMENT A DETAILED INVENTORY OF STRUCTURES TO IDENTIFY ALL STRUCTURES IN THE STUDY AREA BY STRUCTURE TYPE (IN ACCORDANCE WITH THE SELECTED STRUCTURAL CLASSIFICATION SCHEME)

Resources Needed.
- Computer equipment and relational database software that can be used to store and manipulate detailed inventory information; methods for inferring structure type by design date, use and other attributes on the basis of available, local, expert opinion; earthquake engineering specialists who can identify the types of inventory information needed, direct inventory data acquisition, and oversee input of inventory information in the selected computerized relational database(s); engineering students capable of conducting field surveys and obtaining electronic inventory information from available sources; graduate students/consultants capable of developing integrated electronic inventories from various sources (if local technology permits).

Potential Available Supporting Information. ATC-36 discussion of methodology used to develop a detailed electronic inventory of structures in the Salt Lake City, Utah, region (ATC, in preparation); *Regional Seismic Hazard And Risk Analysis Through Geographic Information Systems* (King, in preparation).

STEP 9. ESTIMATE STRUCTURAL DAMAGE TO ALL STRUCTURES IN THE STUDY AREA

Resources Needed.
- Computer equipment, and relational database and geographic information system software that can be used to combine inventory information, ground shaking vulnerability functions, and collateral hazard damage estimation methodology to estimate damage to all structures in the study area; earthquake engineering specialist who can direct the development of the structural damage estimates; graduate students/consultants capable of designing and implementing the needed software and output reports; panel of earthquake engineering specialists to review and approve the results.

Potential Available Supporting Information. *Regional Seismic Hazard And Risk Analysis Through Geographic Information Systems* (King, in preparation).

STEP 10. ESTIMATE MONETARY LOSSES (DIRECT AND INDIRECT) AND CASUALTIES (DEATHS AND INJURED) FOR THE STUDY AREA

Resources Needed.
- Computer equipment, and relational database and geographic information system software that can be used to combine the damage estimates developed in Step 9 with (1) structure size, use and replacement value information to develop estimates of financial loss resulting from direct damage and (2) death and injury models to estimate casualties; earthquake engineering specialist who can direct the development of loss information; graduate students/consultants capable of designing and implementing the needed software and output reports; panel of earthquake engineering specialists to review and approve the results.

- It is important to note that the art and science of developing indirect monetary losses is not well developed and will probably be beyond the scope of most earthquake damage and loss studies.

Potential Available Supporting Information. Regional Seismic Hazard And Risk Analysis Through Geographic Information Systems (King, in preparation).

References

Algermissen, S. T., Steinbrugge, K. V., and Lagorio, H. J., 1978, "Estimation of Earthquake Losses to Buildings (Except Single Family Dwellings)," US Geological Survey Open-File Report 78-441, 161 pp.

Anagnostopoulos, S. A., and Whitman, R. V., 1977, "On human loss prediction in buildings during earthquakes," *Proceedings of the Sixth World Conference on Earthquake Engineering*, New Delhi, India.

Applied Technology Council, 1978, *Tentative Provisions for the Development of Seismic Regulations for Buildings*, ATC-3-06 Report. Palo Alto, California.

Applied Technology Council, 1985, *Earthquake Damage Evaluation Data for California*, ATC-13 Report. Redwood City, California.

Applied Technology Council, 1991, *Seismic Vulnerability and Impact of Disruption of Lifelines in the Conterminous United States*, ATC-25 Report. Redwood City, California.

Applied Technology Council, 1992, *A Model Methodology for Assessment of Seismic Vulnerability and Impact of Disruption of Water Supply Systems*, ATC-25-1 Report. Redwood City, California.

Applied Technology Council, in preparation, *Earthquake Loss Evaluation Methodology and Databases for Utah*, ATC-36 Report, Redwood City, California.

Arias, A., 1970, "Measures of Earthquake Intensity," *Seismic Design for Nuclear Power Plants*, MIT Press, Cambridge, Massachusetts.

Barosh, P. J., 1969, *Use of Seismic Intensity Data to Predict the Effects of Earthquakes and Underground Nuclear Explosions in Various Geologic Settings*, U. S. Geological Survey Bulletin 1279.

Benjamin, J. R., 1974, *Probabilistic Decision Analysis Applied to Earthquake Damage Surveys*, Earthquake Engineering Research Institute, Berkeley, California.

Blume, J. A., 1970, "An engineering intensity scale for earthquakes and other ground motion," *Seismol. Soc. Am. Bull.* 60.

Blume, J. A., Wang, E. C. W., Scholl, R. E., and Shah, H. C., 1975, *Earthquake Damage Prediction: A Technological Assessment*, Stanford University Report No. 17, Stanford, California.

Galloway, J. D., Chairman, 1907, R*eport of Committee on Fire and Earthquake Damage Buildings*, Transactions ASCE LIX: 223.

Housner, G. W., 1952, *Intensity of Ground Motion During Strong Earthquakes*, Earthquake Research Laboratory Report, California Institute of Technology, Pasadena, California, August.

King, S. A., in preparation, *Regional Seismic Hazard And Risk Analysis Through Geographic Information Systems*, PhD Dissertation, Department of Civil Engineering, Stanford University, Stanford, CA.

Legg, M., J. Slosson & R. Eguchi, 1982, "Seismic hazard for lifeline vulnerability analyses," In *Proceedings of the Third Int'l. Conference on Microzonation*, Seattle, WA.

McCann, M. W., Jr., and Shah, H. C., 1979, "RMS Acceleration for Seismic Risk Analysis: An Overview," *Proceedings of the Second U. S. National Conference on Earthquake Engineering*, Stanford University, Stanford, California, August 22-24.

Medvedev, S. V., W. Sponheuer, & V. Karnik, 1965, "Seismische Skala," *Proceedings of the Third World Conference on Earthquake Engineering*, New Zealand.

Nason, R., 1980, *Damage in San Mateo County, California in the Earthquake of 18 April 1906*, U. S. Geological Survey Open-File Report 80-176

NOAA, 1972, *A Study of Earthquake Losses in the San Francisco Bay Region*, prepared for the Office of Emergency Preparedness by the U. S. Department of Commerce, National Oceanic & Atmospheric Administration, 220 pp.

Pomonis, A., Coburn, A., & R. Spence, 1993, "Seismic Vulnerability, Mitigation of Human Casualties and Guidelines for Low-Cost Earthquake Resistant Housing", *Stop Disasters: Newsletter of the United Nations International Decade for Natural Disaster Reduction.* No. 12.

Richter, C. F., 1958, *Elementary Seismology*, W. H. Freeman and Company, San Francisco, California.

Sauter, F. and H. C. Shah, 1978, "Studies on Earthquake Insurance," *Proceedings of the Central American Conference on Earthquake Engineering*, Vol. II. San Salvador, El Salvador.

Scholl, R. E., Kustu, O. Perry, C. L. and Zanetti, J. M., 1982, *Seismic Damage Assessment for High-Rise Buildings*, URS/Blume Engineers Report URS/JAB 8020, URS/John A. Blume & Associates, San Francisco, California, 300 pp.

SEAOC, 1991, *Reflections on the Loma Prieta Earthquake October 17, 1989*, Ad Hoc Earthquake Reconnaisance Committee, Structural Engineers Association of California, Sacramento

Steinbrugge, K.V., 1982, *Earthquakes, Volcanoes, and Tsunamis: An Anatomy of Hazards*, Scandia American Group, New York.

Steinbrugge, K. V., and Shader, E. E., 1979, *Mobile Home Damage and Losses - Santa Barbara Earthquake August 13, 1978*, Report to the California Seismic Safety Commission.

Steinbrugge, K. V., F. E. McClure & A. J. Snow, 1969, *Studies in Seismicity and Earthquake Damage Statistics: Appendix A*, U.S. Dept. of Commerce, Coast and Geodetic Survey. Washington, DC.

Steinbrugge, K. V., S. T. Algermissen, H. J. Lagorio, L. S. Cluff and H. J. Degenkolb, 1981, *Metropolitan San Francisco and Los Angeles earthquake loss studies: 1980 assessment*, USGS Open File Rep. 81-113.

U.S. Geological Survey, 1981, *Earthquake hazards reduction program 1979-80*, U.S. Geol. Surv. Open File Rep. 81-41, Menlo Park, California.

Whitman, R. V., Cornell, C. A., and Taleb-Agha, G., 1975, "Analysis of earthquake risk for lifeline systems," *Proceedings of the U. S. National Earthquake Engineering Conference*, Ann Arbor, Michigan.

Whitman, R.V., J. W. Reed & S. T. Hong, 1973, "Earthquake damage probability matrices," In *Proceedings of the Fifth World Conference on Earthquake Engineering*, Int'l. Assn. for Earthquake Engineering, Rome, Italy.

Wiggins, J. H., 1981, "Seismic Performance of Low-Rise Buildings - Risk Assessment," *Seismic Performance of Low-Rise Buildings - State-of-the-Art and Research Needs*, A. Gupta, ed., Proceedings of ASCE Workshop held at the Illinois Institute of Technology, Chicago, Illinois.

Wood, H., 1908, "Isoseismals: Distribution of Apparent Intensity. The California Earthquake of April 18, 1906," *Report of the State Investigation Commission*: 220-254. Carnegie Institution of Washington.

Wood, H. O. and F. Newmann, 1931, "Modified Mercalli Intensity scale of 1931," *Seismol. Soc. Am. Bull.* 21: 277-283.

TABLE 1

Modified Mercalli Intensity Scale
(Richter, 1958)

I. Not felt. Marginal and long-period effects of large earthquakes.

II. Felt by persons at rest, on upper floors, or favorably placed.

III. Felt indoors. Hanging objects swing. Vibration like passing of light trucks. Duration estimated. May not be recognized as an earthquake.

IV. Hanging objects swing. Vibration like passing of heavy trucks; or sensation of a jolt like a ball striking the walls. Standing motor cars rock. Windows, dishes, doors rattle. Glasses clink. Crockery clashes. In the upper range of IV wooden walls and frames creak.

V. Felt outdoors; direction estimated. Sleepers wakened. Liquids disturbed, some spilled. Small unstable objects displaced or upset. Doors swing, close, open. Shutters, pictures move. Pendulum clocks stop, start, change rate.

VI. Felt by all. Many frightened and run outdoors. Persons walk unsteadily. Windows, dishes, glassware broken, knickknacks, books, etc., off shelves. Pictures off walls. Furniture moved or overturned. Weak plaster and masonry D cracked. Small bells ring (church, school). Trees, bushes shaken (visible, or heard to rustle).

VII. Difficult to stand. Noticed by drivers of motor cars. Hanging objects quiver. Furniture broken. Damage to masonry D, including cracks. Weak chimneys broken at roof line. Fall of plaster, loose bricks, stones, tiles, cornices (also unbraced parapets and architectural ornaments). Some cracks in masonry C. Waves on ponds; water turbid with mud. Small slides and caving in along sand or gravel banks. Large bells ring. Concrete irrigation ditches damaged.

VIII. Steering of motor cars affected. Damage to masonry C; partial collapse. Some damage to masonry B; none to masonry A. Fall of stucco and some masonry walls. Twisting, fall of chimneys, factory stacks, monuments, towers, elevated tanks. Frame houses moved on foundations if not bolted down; loose panel walls thrown out. Decayed pilings broken off. Branches broken from trees. Changes in flow or temperature of springs and wells. Cracks in wet ground and on steep slopes.

IX. General panic. Masonry D destroyed; masonry B seriously damaged. (General damage to foundations.) Frame structures, if not bolted, shifted off foundations. Frames racked. Serious damage to reservoirs. Underground pipes broken. Conspicuous cracks in ground. In alluviated areas sand and mud ejected, earthquake fountains, sand craters.

X. Most masonry and frame structures destroyed with their foundations. Some well-built wooden structures and bridges destroyed. Serious damage to dams, dikes, embankments. Large landslides. Water thrown on banks to canals, rivers, lakes, etc. Sand and mud shifted horizontally on beaches and flat land. Rails bent slightly.

XI. Rails bent greatly. Underground pipelines completely out of service.

XII. Damage nearly total. Large rock masses displaced. Lines of sight and level distorted. Objects thrown into the air.

(TABLE 1, continued)

Masonry A. Good workmanship, mortar, and design; reinforced, especially laterally, and bound together by using steel, concrete, etc.; designed to resist lateral forces.

Masonry B. Good workmanship and mortar; reinforced, but not designed in detail to resist lateral forces.

Masonry C. Ordinary workmanship and mortar, no extreme weaknesses like failing to tie in at corners, but neither reinforced nor designed against horizontal forces.

Masonry D. Weak materials, such as adobe; poor mortar; low standard of workmanship; weak horizontally.

TABLE 2
Ground Failure Probability Matrix for Poor Ground
(ATC, 1985)*

Probability of Ground Failure Percent by MMI and Soil Type

Zone	Type of Deposit	VI	VII	VIII	IX	X	XI	XII
1a	Stream channel, tidal channel	5	20	40	60	80	100	100
1b	San Francisco Bay mud and fill over bay mud	3	15	30	40	60	80	90
2a	Holocene alluvium, water table shallower than 3m (10 ft.)	2	10	20	30	40	60	80
2b	Holocene alluvium, water table deeper than 3m (10 ft)	.5	2	5	7	12	25	40
3	Late Pleistocene alluvium	.1	.5	1	2	4	7	10

* Estimates are based on consensus of ATC-13 advisory Project Engineering Panel (PEP).

TABLE 3
Relation between Landslide Severity and Facility Damage Factor
(ATC, 1985)*

Central Slope Failure State	Damage Factor (Percent)
Light	0
Moderate	15
Heavy	50
Severe	80
Catastrophic	100

* Estimates are based on consensus of ATC-13 PEP.

TABLE 4
Damage Factors for Fault Rupture
(ATC, 1985)*

Mean damage factor (Fault Rupture) in Percentage for Various Fault Displacements

Facility type and Location	0.2m	0.6m	1.0m	3.5m	10m
Subsurface structures					
In fault zone	50	80	100	100	100
In drag zone	20	40	60	80	100
Surface structures					
In fault zone	10	30	70	100	100
In drag zone	0	0	2	10	20

* Estimates are based on consensus of ATC-13 PEP.

TABLE 5.
Damage Factor Caused by Inundation (High Velocity Water)
(ATC, 1985)[a]

Depth of Water (meters)	Mean Damage Factor (Inundation)[b] (Percent)
1	10
2	20
3	50
4	80
5	100

[a]Estimates are based on consensus of ATC-13 PEP.
[b]Applies only to ground surface structures less than 10 meters high. For higher buildings, use 50% of the values in the table.

TABLE 6.
Injury and Death Rates
(ATC, 1985)*

Damage State	Central Damage Factor (%)	Fraction Injured Minor	Fraction Injured Serious	Fraction Dead
1	0	0	0	0
2	.5	3/100,000	1/250,000	1/1,000,000
3	5	3/10,000	1/25,000	1/100,000
4	20	3/1,000	1/2,500	1/10,000
5	45	3/100	1/250	1/1,000
6	80	3/10	1/25	1/100
7	100	2/5	2/5	1/5

*Estimates are based on consensus of the ATC-13 PEP and data adapted from NOAA (1972), Anagnostopoulos and Whitman (1977), and Whitman et al. (1975) and are for all types of construction except light steel construction and wood-frame construction. For light steel construction and wood-frame construction, multiply all numerators by 0.1..

TABLE 7.
ATC-13 Earthquake Engineering Facility Classification

A. *Building Types*
- Wood Frame--low rise
- Light metal--low rise
- Unreinforced masonry (bearing wall)--low, mid rise
- Unreinforced masonry (with load-bearing frame)--low, mid, high rise
- Reinforced concrete shear wall (with moment-resisting frame)--low, mid, high rise
- Reinforced concrete shear wall (without moment-resisting frame)--low, mid, high rise
- Reinforced masonry shear wall (without moment-resisting frame)--low, mid, high rise
- Reinforced masonry shear wall (with moment-resisting frame)--low, mid, high rise
- Braced steel frame--low, mid, high rise
- Moment-resisting steel frame (perimeter frame)--low, mid, high rise
- Moment-resisting steel frame (distributed frame)--low, mid, high rise
- Moment-resisting ductile concrete frame (distributed frame)--low, mid, high rise
- Moment-resisting nonductile concrete frame (distributed frame)--low, mid, high rise
- Precast concrete (other than tilt-up)--low, mid, high rise
- Long-span--low rise
- Tilt-up, low rise
- Mobile homes

B. *Bridges*
- Conventional (less than 500-foot spans)--multiple span, continuous
- Major (greater than 500-foot spans)

C. *Pipelines*
- *Underground*
- *At grade*

D. *Dams*
- Concrete
- Earthfill and Rockfill

E. *Tunnels*
- Alluvium
- Rock
- Cut and Cover

F. *Storage Tanks*
- Underground--liquid, solid
- On ground--liquid, solid
- Elevated--liquid, solid

G. *Roadways and Pavements*
- Railroad
- Highways
- Runways

TABLE 7.   ATC-13 Earthquake Engineering Facility Classification (Continued)

H. *Chimneys (High Industrial)*
- Masonry
- Concrete
- Steel

I. *Cranes*

J. *Conveyor Systems*

K. *Towers*
- Electrical Transmission Line--conventional, major
- Broadcast
- Observation
- Offshore

L. *Other Structures*
- Canals
- Earth Retaining Structures (over 20 feet high)
- Waterfront Structures

M. *Equipment*
- Residential
- Office (furniture, computers, etc.)
- Electrical
- Mechanical
- High Technology and Laboratory
- Trains, Trucks, Airplanes and Other Vehicles

CREATING THE SCENARIO AND DRAFTING EARTHQUAKE HAZARD REDUCTION INITIATIVES

GEORGE G. MADER
AICP
Spangle Associates, Urban Planning and Research
Portola Valley, California, USA

1. Introduction

An understanding of the potential damage probable earthquakes can cause is fundamental to determining the most appropriate types of measures or initiatives for reducing earthquake damage. It has been common practice for years for experts in emergency response, such as fire fighters, police, doctors, and others to participate in exercises whereby damage caused by earthquakes or other hazards is described in a compressed time period and the participants are required to make emergency response decisions. Such exercises, in effect, test the emergency response system and allow improvements to be made. These tests have proven to be very worthwhile.

More recently, there have been significant attempts in at least two countries to describe the potential effects of earthquakes and present these descriptions or scenarios to those individuals who are concerned not with emergency response but rather with how cities can be reconstructed after a major earthquake or how changes can be made in cities before an earthquake to make a city more resistant to earthquake damage. This is a rapidly-developing field and one which holds promise for reducing the earthquake vulnerability of cities. The major result of such activities is the development of recommendations which will lead to fewer injuries and deaths and less property damage. I would prefer to describe this field of activity with the emphasis on recommendations. We might describe these activities as: Exercises to Develop Earthquake Hazard Reduction Initiatives Based on Earthquake Damage Scenarios. For simplicity, I will use the term exercise to refer to this concept.

My comments are meant to help bring closure to the presentations of the previous speakers. I will describe the purpose of an exercise, what is included in an exercise, the different levels of scenario information and then briefly describe two examples. Finally, I will describe the ingredients which contribute to a successful exercise.

2. Purpose of an Exercise

Those cities which have experienced serious, damaging earthquakes in recent years should not need to conduct an exercise in order to get public attention, locate the vulnerable parts of the city or decide what corrective actions need to be taken to reduce damage in the future. Unfortunately, however, they may still need assistance in formulating initiatives. It is in those cities which have not had recent damaging earthquakes, however, that the need is great to consider an exercise. Such an exercise can educate the city as to the realistic damages which might occur in a future earthquake. This educational function is the first and foremost purpose of an exercise. The information developed in the exercise will most likely be news to the majority of residents, businessmen and public officials. Members of the seismological and geological fields, if involved, will undoubtedly have significant knowledge, but that knowledge will probably not have been disseminated in an effective way. A properly-conducted exercise is one way in which to begin to communicate this information. One purpose of an exercise, therefore, is to communicate the potential damages from a possible earthquake to those who have major responsibilities for the safety of persons and structures and eventually to the general public.

A second purpose is to provide a context within which major recommendations can be made for improvements in the city which will reduce deaths, injuries and damage. If these recommendations—called initiatives in this paper—are carefully prepared and involve responsible persons, it is likely that government and business will listen and take mitigation actions. The initiatives can range from the very general to the very specific. A general initiative might be to develop a program for public education about measures to take to reduce hazards to families in their homes. A more specific initiative might be to develop zoning to guide development away from unstable ground. An even more specific initiative might be to undertake a program to retrofit fire stations so that they are usable in the event of an earthquake.

A third purpose is to identify those scientific and engineering subject areas where ·more information is needed in order to more fully understand the earthquake phenomenon and the engineering measures that can be taken to reduce vulnerability. For example, an initiative might be to undertake mapping to determine the areas in a city subject to liquefaction. Another initiative might be to conduct a study of certain classes of buildings, such as concrete frame buildings with brick infill walls, to determine their vulnerability to earthquakes, and to develop improvements in the building code to make future buildings stronger.

3. Components of an Exercise

The first part of an exercise is to develop a scenario of expected earthquake damage. The scenarios can vary greatly in terms of detail and scientific accuracy. The more detailed the scenario, the more realistic it should be. The most detailed and accurate scenario, for example, would be that conducted for a single building. This is often done for very expensive buildings or those whose failures would cause major damage to surrounding properties. These studies require the combined talents of seismologists, geologists, soils engineers, and structural engineers. Such expensive studies cannot be made for entire cities due to the high cost involved. When dealing with an entire city, it is necessary to first select an anticipated earthquake or earthquakes as to magnitude, location, and characteristics. Next, is necessary to know the geology from the focus of the earthquake to the location of expected damage. Finally it is necessary to know the soil characteristics at the site, and the structural design and construction of the structures under consideration. Detailed scenarios of this type are desirable since they can lead to specific recommendations; however, in preparing a scenario for a city, it is necessary to generalize each of these variables.

For the purpose of developing recommendations for those activities which should be undertaken to reduce vulnerability, it is also possible to develop less scientifically-driven scenarios which I will call general scenarios. These are most feasible in those areas where there is already some general understanding of earthquakes and the types of damage they can cause. It has been possible in these situations for a group of city officials such as building officials, engineers, planners, firemen, and policemen, to describe within a short period of time, their impression of what damage an earthquake might cause. In these situations, participants are given an overview of the types of earthquake hazards they might expect in their community and the types of structural damage they might expect. The participants then describe on a map their best judgment of the types and distribution of expected damage. While this map will not be as accurate as the more scientifically generated map, it will likely portray the general range and extent of damage. This can be sufficient to convince the participants that they have an earthquake threat and that some actions will need to be taken. One advantage of creating the scenario in this manner is that the participants will have to respond to the damage they predicted rather than damage someone else has presented to them.

4. Example of an Exercise Based on a General Scenario

A general scenario approach has been tested in three cities in the US: Pleasanton, California; Los Angeles, California; and Evansville, Indiana. Each of these exercises took place in a single eight-hour day. Top city officials, including planners, administrators, public works directors, attorneys, building officials, policemen, firemen, participated in the exercises. The persons conducting the exercise, using a strict schedule, led the group through an exercise in which they prepared a damage scenario, and then simulated the several post-earthquake tasks involved in recovery and reconstruction. The officials were forced to deal with the types of decisions they would likely have to face in the aftermath of a real earthquake. Finally, they were then asked to recommend those pre-earthquake measures which they thought would have reduced the damage they predicted the earthquake caused.

The exercises consisted of twelve tasks starting with the creation of the scenario. The tasks are as follows:
- Damage Scenario
- Emergency Shelter
- Closure and Relocation
- Damage Assessment
- Geologic Evaluation
- Demolition
- Temporary Business Sites
- Temporary Housing
- Permit Processing
- Reconstruction Planning
- Timing
- Mitigation

A major concept of the exercise is to immerse the players in the damage and the difficulties of dealing with the damage and then lead them to mitigation—what they might do now to avert damage or to make the recovery tasks more manageable.

The person conducting the exercise introduces each task with a slide presentation, giving the players background and examples of recovery from recent earthquakes and examples of mitigation actions to help them carry out the tasks. For the damage scenario task, the introduction illustrates the kinds of lands and buildings prone to earthquake failure. It describes, with examples, damage from ground shaking, fault rupture, landslides, liquefaction, and tsunami. The introduction also shows different types of buildings which are prone to earthquake damage, including structures made of unreinforced masonry and reinforced concrete, tilt-ups, wood-frame buildings, mobile homes, other buildings such as high-rises, irregularly-shaped structures and buildings with "soft-stories."

With this general information about earthquake hazards, the players are told that a damaging earthquake has struck their city and their job is to create a damage scenario. The players sit around a table with a base map of the city in front of them. They have handouts summarizing the slide presentation on earthquake hazards and potentially hazardous buildings.

Based on the information provided, and knowledge of the community, they are asked to identify, on the base map and within 45 minutes, the following:

a) at least one area where they expect landslides, liquefaction failures, and/or fault ruptures. These areas should be outlined in brown and numbered.

b) at least two groups of blocks (one commercial and one residential) where they expect concentrated building damage. These areas should be outlined in orange and numbered.

c) major facilities such as hospitals, schools, government buildings, and high-rise buildings that they think are now at least temporarily unusable. The facilities should be identified in blue.

d) highway overpasses, roads, and other transportation facilities which might have collapsed or been left impassable by the earthquake. These facilities are marked in black.

In the three exercises conducted so far, players successfully completed the full set of tasks. Consensus is that participants benefited significantly from the process. It forced them to think for the first time in a systematic way about what would happen if a damaging earthquake struck their community. They saw forcefully how hard recovery would be and how they could act now to reduce the burdens of recovery.

5. Example of an Exercise Based on a Detailed Scenario

Detailed scenarios have been prepared for a number of areas, such as for the city of Los Angeles and the San Francisco Bay Area. Scenarios of varying types also have been prepared in other countries such as Japan, and efforts have been underway in the Seismed Project of UNDRO in the Mediterranean area as well. I would like to dwell, however, on the Quito, Ecuador project, since it represents a scenario which included the development of initiatives in addition to identifying potential damage.

In this project, OYO Pacific (now Geohazards International) over a year and one-half period, worked with Ecuadorians to organize a major project. The objective was to develop initiatives which would be adopted for Quito and lead toward the reduction of potential earthquake hazards.

The project can be described as having four major components: 1) initial contacts, 2) establishment of an organizational framework, 3) development of information and 4) development of initiatives through a workshop. The final component, yet to be completed, will include the adoption and implementation of the initiatives. It would appear that these basic steps would be applicable in other countries.

5.1 INITIAL CONTACTS

A first step was to identify the key persons who need to be involved in the project. This required getting to know key elected officials, key government staff persons, and scientists and engineers who could develop needed information. In the case of Quito, the backing of the mayor, a top aide to the mayor, and scientists and engineers at the university were crucial to the success of the project.

5.2 ORGANIZATION

Next it was necessary to establish formal organizations to carry out the project. The overseeing body was the Conseil d'Affaires Sociales et Economiques (CASE) which was made up of prominent international and local representatives of government, business and industry. This group had the responsibility to review the policy aspects of the entire project and make the final recommendations on initiatives to the city government. The second body was the Grupo de Trabajo (working group) which was charged with developing the seismological, geological and structural information basic to the scenario. This group included Ecuadorian scientists and engineers and scientists from Ecuador, France, Japan, and the United States. The third and final group was the International Technical Advisory Committee (TAC). This group was made up of international experts in the scientific, engineering and public policy areas. The group was to provide technical guidance to the overall project.

5.3 DEVELOPMENT OF INFORMATION

The major effort of the project was the preparation of the technical basis for an earthquake damage scenario. It involved close cooperation between the Ecuadorian experts and OYO experts along with the review and advice of the TAC. This part of the project occupied the better part of the 18-month project. The scenario which was finally developed provided information in far greater detail than in the general scenario previously described.

5.4 DEVELOPMENT OF INITIATIVES THROUGH THE WORKSHOP

In order to understand the exercise, the following aspects need to be discussed: exercise participants, pre-exercise education, exercise outline and output initiatives.

5.4.1 *Exercise Participants.* First, many of the individuals who had participated on the CASE, GT or TAC were invited to participate. They already had considerable background and expertise that would be needed in the workshop. It became obvious, however, that these individuals did not completely represent the five working groups which were to be involved in the workshop, which were:
- Banking and Insurance
- Emergency Services
- Lifelines

- Buildings and Planning
- Industry

To solve this problem other responsible persons from government and the private sector were asked to participate. Also, foreign resource persons were invited to participate with each group. They were to provide background and make suggestions, but not to make decisions for the Ecuadorian members. Finally, others were invited to observe the workshop.

5.4.2 Pre-Exercise Education. It was realized that many of the persons in the agencies identified to attend the exercise had not participated in the project but would be asked to respond in the workshop to the scenario which had been prepared. It was decided, therefore, that a team should visit these agencies in advance of the workshop to explain the types of hazards which the scenario would portray. This would allow the participants to consider the information and begin to formulate their particular concerns. This step proved to be invaluable in that the participants came to the workshop with a rather good understanding of the earthquake threat. Also, the initial concerns evidenced by these individuals were noted and reported at the workshop to help provide context. This step was not followed in the previously described exercises based on a general scenario because of time constraints and in general a somewhat greater prior exposure of participants to basic earthquake problems. Also, the level of detail in the scenarios was far less in those exercises.

5.4.3 Workshop Outline. The workshop covered two full days. The schedule for the workshop is set forth below.

QUITO EARTHQUAKE SCENARIO AND MITIGATION WORKSHOP SCHEDULE

Day One

9:00 – 9:30	**Welcome and Introductions**
9:30 – 10:00	**Technical Summary** Part One: Quito's Earthquake Hazards *frequency, magnitude, locations of faults, faulting, intensified ground shaking, landslides, liquefaction failures (maps and slides)*
10:00 – 10:30	Part Two: Effects on Structures *vulnerable buildings and infrastructure* *(maps and slides)*
10:30 – 11:00	**Break**
11:00 – 11:30	Part Three: Effects on Government and Business *results of interviews conducted with governmental and business leaders on potential impacts*
11:30 – 12:00	Part Four: Examples of Types of Earthquake Damage

12:00 – 1:00   **Lunch**

1:00 – 1:30   **Introduction to Group Sessions**

1:30 – 3:30   **Group Sessions: Assessing Impacts**
Groups meet to review and refine scenario information presented and to explore the implications of the damage on Quito's banking and insurance, emergency services, industry, building and planning, and lifelines.

3:30 – 4:00   **Break**
Group leaders and recorders prepare for general assembly presentations.

4:00 – 6:00   **General Assembly: Assessing Impacts**
Group leaders describe the outcomes of their sessions and add information to the scenario map.

Day Two

9:00 – 9:30   **Recap of Scenario**

9:30 – 10:00   **Organization and Program for Reducing Earthquake Risk – An Example from California**
The formation and structure of the California Seismic Safety Commission and its earthquake risk reduction program, "California at Risk".

10:00 – 10:30   **Example of a City Program to Reduce Earthquake Risk**
The earthquake risk reduction program of the City of Los Angeles.

10:30 – 11:00   **Break**

11:00 – 11:30   **Introduction to Group Sessions**
Mitigation examples and instructions for group sessions.

11:30 – 1:00   **Group Sessions: Reducing Impacts**
Groups meet to recommend actions for reducing the impacts identified in Day One sessions. The groups also assign priorities to the actions and indicate the schedule, resources needed, and agencies responsible for carrying them out.

1:00 – 2:00   **Lunch**

2:00 – 3:30   **Group Sessions: Reducing Impacts (continued)**

3:30 – 4:00	**Break**	
	Group leaders and recorders prepare for general assembly presentations.	
4:00 – 6:00	**General Assembly: Reducing Impacts**	
	Group leaders present the list of actions developed in the group sessions.	
6:00 – 6:30	**Summary and Wrap-up**	

The first half of the first day was taken up by technical presentations to all participants on seismology, geology, structural damage and examples of the types of earthquake damage. Also, the results of the interviews with participants were presented. By noon, the participants had a rather good introduction to the damage scenario. They had an appreciation of shaking, ground failure, building damage, effects on government and business and had seen convincing slides showing the types of earthquake damage that might be expected in Quito.

In the afternoon of the first day, they met in groups to explore the types of damage they might anticipate for their particular areas of concern. They were in effect further defining the damage scenario by identifying structural damage they could expect and impacts of such damage on providing goods and services. This detailed self-examination was then presented to the entire body of participants at the end of the day.

After the exhausting first day, the participants returned the second day, hopefully refreshed, to determine what programs or initiatives they should propose in order to reduce the damage produced by the theoretical earthquake. The first part of the day was spent with presentations from foreign experts on the types of initiatives that had been developed in US projects. This part could also have drawn on initiatives from other countries. In the afternoon, the participants met again in groups to develop initiatives which they thought would reduce potential damage. Finally, the most important initiatives from each group were presented to the assembled complete group.

Each group was to present their top five recommendations to the group. A few examples will serve to illustrate the recommendations:

Banking and Insurance

• Establish contingency plans for operation in the event of an earthquake including back-up information, cash supply, emergency equipment.

• Control policy capacity of insurance companies so as to not exceed resources.

Emergency Services

• Diagnose earthquake vulnerability of buildings and resources needed after an earthquake.

• Undertake programs to increase public awareness of the realities of the earthquake threat.

Lifelines

• Evaluate infrastructure systems as to their risk of damage in an earthquake.

• Train personnel and coordinate with utility companies, civil defense and armed forces to help ensure availability of lifelines after an earthquake.

Buildings and Planning

• Develop a program to evaluate and reinforce existing development and historic buildings.

• Revise city zoning based on seismic intensity maps.

6. Comparison of Approaches

The two examples described represent the approximate ends of a spectrum of possible scenario exercises. One is very general and takes a short amount of time. The other is very detailed and takes a comparatively long period of time and more resources. There are many other possible intermediate designs for such exercises. It would appear that each exercise should be tailored to the particular situation. The major point is that something worthwhile can be developed with minimal effort while a much larger effort can bring greater rewards.

7. Ingredients for Success

When considering an area as a candidate for a scenario workshop, it is important to determine to what extent the conditions that exist will enhance the chances for a successful project. The following eight items appear to be important to the success of such a project.

7.1 A CHAMPION

There needs to be at least one person in a responsible position who is interested in the project and will provide support and leadership. In most instances this should be a major elected official. It might also be a key staff person or a leader outside of government, but these are less likely to be successful.

7.2 INTEREST OF HIGH GOVERNMENT OFFICIALS

Those persons in charge of major departments of government need to participate, support and be interested in the process. If they are not so involved, the recommendations that are developed will not be sufficiently realistic and may not be carried out.

7.3 INTEREST OF BUSINESS OFFICIALS

Since much of the success of a project will depend on the business community taking actions to reduce their vulnerabilities, they need to be convinced that the project will be of benefit to them. Also, they can be important in supporting the government's involvement in the exercise.

7.4 TECHNICAL INFORMATION

As has been pointed out, the amount of technical information can vary and a successful project can still be carried out. A project cannot, however, be carried out without some basic understanding of the earthquake threat and potential damages.

7.5 GOVERNMENTAL ORGANIZATION

To the extent that local government is well-organized, the project and implementation will be better assured. An organization where different departments cooperate with each other on a regular basis is highly beneficial. A disorganized government will not be able to reach decisions and take necessary actions.

7.6 EMERGENCY RESPONSE ORGANIZATION

Emergency response is a major concern. The emergency response organization needs to be involved in the scenario workshop. Quite simply, if they are well organized, properly-staffed and trained, and have the necessary equipment, they can carry out many of the mandates that will result from a scenario workshop.

7.7 PLANNING ORGANIZATION

Planning departments normally control the types and distribution of development. They can do much to make certain that hazardous areas are kept free of development or that development is of a low intensity in such areas. Without such controls, hazardous areas will be built upon.

7.8 BUILDING CONSTRUCTION CONTROLS

Of course, the single set of controls that will most reduce the loss of life and injury is building codes. This is a difficult area due to economic limitations and cultural variations. Nonetheless, a city needs adequate controls and personnel to enforce the controls.

8. Benefits

The foregoing descriptions indicate that an exercise which is properly designed and carried out can provide a number of benefits. Those benefits include the following:

- Provides a comprehensive assessment of vulnerability.
- Identifies specific initiatives which provide direction to government and the private sector.
- Identifies specific research and engineering needs.

- Serves an alert or educational function.
- Develops an informed commitment on the part of participants.
- Involves a wide variety of interest groups.
- Can be sized to fit the resources and needs of an area.
- Can be developed to accommodate a small or rather large group of participants.
- Is a process which is attractive to participants.

THE QUITO PROJECT: TECHNICAL ASPECTS AND LESSONS

JEANETTE FERNANDEZ[1]
JORGE VALVERDE[1]
HUGO YEPES[1]
GONZALO BUSTAMENTE[2]
JEAN-LUC CHATELAIN[3]
[1]*Escuela Politécnica Nacional*
[2]*Illustrious Municipality of Quito*
[3]*ORSTOM*

1. Introduction

The Quito Project was a pilot project launched to determine if it were possible to conduct an earthquake damage scenario study in a developing country. The technical work involved institutions from Ecuador, Canada, France, Japan, and the United States.

The Project was justified in that the city has experienced many earthquakes in the past: 23 earthquakes have produced intensities of VI and above during the 460 years of Quito's written history. The most damaging events occurred in 1859 and 1868 (Intensity IX[1]). In 1587 and 1755, intensities VIII were reported, while in 1797, 1923, and 1987, the city experienced intensities around VII. History shows that for intensities VIII and above, the average recurrence time is 115 years; for intensities VII and above, 65 years; and for intensities VI and above, 20 years. These numbers are averages, and it should be noted that damaging earthquakes can occur in a short interval of time (nine years between the 1859 and 1868 events), or they can be separated by as much as 168 years (between the 1587 and 1755 events). During the last century, the city of Quito has not experienced earthquakes with intensities of VIII and above, but this does not mean that such an earthquake cannot be experienced in the future. For instance, no damaging earthquakes occurred during the 18th century, and then three damaging earthquakes occurred in the following 113 years. Thus the relative quiescence that the city is experiencing now does not mean that the possibility of a future damaging earthquake can be definitively ruled out. Moreover, Quito's extremely rapid growth makes it more vulnerable to earthquakes now than in the past, and the Quito of tomorrow will respond to repetitions of the large historical earthquakes in very different ways than the Quito of the past. It is highly expected that because of Quito's growth and earthquake history, the city's vulnerability will increase in the future unless concerted action is taken.

Adequate existing data in the fields of seismology and soils engineering, as well as a geographic information system (GIS) and a computerized city database at the Municipality made the project feasible. Although the municipality database is quite comprehensive, its building data were inadequate for structural engineering analysis, and further work was needed in this field. It is

---

[1]For a chart comparing the various scales of intensity (i.e., Modified Mercalli, MSK, JMA, etc.), please see Figure 3 in C. Rojahn's "Estimation of Earthquake Damage to Buildings and Other Structures in Large Urban Areas."

important to note that this work would not have been possible without the financial and technical participation of OYO Corporation, and therefore, while the Quito project proved that such work could be undertaken in a developing country, its special circumstances cannot necessarily be extrapolated to other cities.

Aspects to consider for developing earthquake damage scenarios in urban areas have been discussed in this meeting by experts from various fields, who are part of the Quito Project. We will first present how this methodology has been used in the Quito Project, and then share the lessons that we have learned from this experiment.

2. Technical Aspects

2.1 ESTABLISHING SEISMIC INTENSITY DISTRIBUTION MAPS

The historical seismicity of Quito and the distribution of seismogenic structures around the city were used to determine hypothetical (but plausible) earthquakes and intensity-distance relationships.

Out of ten possible destructive earthquakes, three were chosen because of time limitations for performing this experiment: (1) A subduction earthquake (M=8.4), with an epicentral distance of 200 km to the city, because such an event has been given a 60% or higher probability of occurring before the year 2000; (2) An upper-plate continental event (M=7.3) at an epicentral distance of 80 km, as the 1868 earthquake; and (3) A local earthquake (M=6.5), 25 km from downtown Quito, modeling the 1587 earthquake that may have occurred on an active north-south trending fault.

Attenuation relations were established for Ecuador, using 23 earthquakes with intensities of VI and above, recorded in Quito's 460 years of written history (which includes 1104 seismic intensity observations). These relations were then corrected for application to the city of Quito.

Soil characteristics were obtained from over 2000 drillings from various sources (e.g. private consultants, municipality files, EPN studies). It was decided, because of time constraints, to restrict the study area to the current city limits. Based on topography, soil characteristics and surface geology, the city was divided into 20 zones. For each of these zones, a representative soil column was established down to a depth of 20 meters, usually not reaching the base rock, the location of which is still unknown in many parts of the city.

Intensity in the 20 zones was then computed, using soil models and seismic responses, peak accelerations, and soil amplifications, for the three hypothetical earthquakes, yielding the following results:

	Intensity range (MSK scale)	Average intensity (MSK scale)
Subduction earthquake	5.60 - 6.13	6.12
Inland earthquake	6.10 - 6.90	6.63
Local earthquake	6.33 - 7.99	7.20

The results were checked by comparing computed intensity with observed ones for the 1987 event.

2.2 DISTRIBUTION OF ESTIMATED STRUCTURAL DAMAGE

The three most relevant aspects considered in the criteria for assessing structural damage are: (1) the classification of structures by their main lateral force-resisting system; (2) the location and distribution of different structural types throughout the city; and (3) the expected behavior of each structural type under seismic conditions.

Because of the size of the city and the time limitations of the project, only the most populated portion of the city was studied in detail, using urban blocks as basic inventory elements in order to enter the data in the Municipality's GIS. The study revealed that, according to their lateral load resistant system, there were fifteen main types of structures in Quito. Among those, the nine most common were subdivided in three categories, according to building heights: (1) low-rise buildings up to three stories; (2) medium-rise buildings from four to seven stories; and (3) high-rise buildings, over eight stories high. Then every block was assigned to the predominant structural system, i.e., the one with the largest covered area per square meter of the block. We made three findings: (1) the most common types of structure in the city are RC frames with flat slabs; (2) there is a very important belt around the city of non-engineered structures and unreinforced masonry structures; and (3) there is a dense concentration of adobe buildings in the old part of the city (Centro Historico).

In order to estimate the vulnerability of the structures, some buildings that we considered as representative of each type of structure were evaluated according to the Quick Check recommendations of ATC-21 and ATC-22. Also, four adobe structures located in the Centro Historico were revised mainly under shear conditions (two of them have two stories, the others one and four stories, respectively). Special structures such as hospitals, schools, and industrial facilities, as well as the sewage system, water reservoir tanks, transmission towers, gas and oil stations near the city, and the airport, were inspected individually and in greater detail.

Estimates of percent of physical damage caused by ground shaking were estimated using the damage probability matrices method provided by ATC-13, i.e., using a relationship between the damage factor versus MMI scale. The expected loss in US dollars caused by ground shaking for each facility has been estimated using damage probability matrices. In the method used, the damage factor is defined as the ratio of the estimated cost due to earthquake damages, divided by the facility replacement value. We considered seven states of damages: none, slight, light, moderate, heavy, major, destroyed. After the damage estimation, the recovery times for lifelines were estimated.

The estimation of damages to the structures and lifelines has been computed for each of the three hypothetical earthquakes. As for the establishment of the seismic intensity distribution maps, the method was tested by comparing computed, estimated damages to observed damages for the 1987 earthquake.

2.3 DESCRIPTION OF THE EFFECTS OF AN EARTHQUAKE ON THE CITY

In order to have a better estimate of how the city is functioning, officials from 17 different organizations which deliver public services in Quito were interviewed. In a multi-stage process, the interviews were written up and returned to the interviewees for review. The revisions then underwent further scrutiny at a meeting of all the participants in the interviews.

From this process, a great many details about public services and the functioning of the city were discovered. The facts found in the interviews were used to estimate the consequences of a destructive earthquake in Quito from a different angle than the estimations obtained using the technical method described above.

The final stage of the study consisted in the establishment of the scenario, portraying the consequences of a destructive earthquake on the city of Quito. The seismic hazard and damage assessments as well as the interviews were combined subjectively in order to develop a vivid description of the possible consequences of the local earthquake (1) at the time of the earthquake, (2) one hour later, (3) two days later, (4) one week later, and (5) one month later.

2.4 EARTHQUAKE MITIGATION PROJECTS

After the technical analyses were reviewed by the international technical advisory committee, international and Ecuadorian specialists from the fields of business and industry, city government, city planning, emergency services, and lifelines met for a two-day workshop in Quito. After estimating the effects on such factors as production capacity, employment, sales, and services, the participants developed lists of specific recommendations within their field of expertise for reducing earthquake risk in Quito. For each recommendation, the participants described the steps necessary, possible funding sources for expertise and equipment, the responsible agencies, and expected start and completion dates. The primary recommendation was for the establishment of a Seismic Safety Advisory Board, whose responsibility would be to review, revise and then administer the Earthquake Hazard Reduction Project. Other high-priority recommended actions concern existing facilities, new facilities, earthquake planning, earthquake recovery, and further research needed to improve the findings of these projects.

3. Lessons Learned

In addition to describing the technical aspects of developing an earthquake damage scenario, we believe it is our responsibility to tell future scenario committees about some of the complexities we faced in working on a pilot project, as well as in collaborating with international, multidisciplinary teams based in different cities around the world, all on a very tight schedule and without a known budget.

Thanks to this project, different scientific groups worked together toward a common goal. This would not have happened otherwise. This was an opportunity to assess the state of earthquake-related data in the country and to determine the strong and weak points of these data. It was probably the first time in city history that multidisciplinary teams were put together, whose backgrounds ranged from specific sciences (e.g., seismology, geology) and applied engineering (e.g., civil engineering, soil mechanics) to policy-makers and politicians, in order to pursue a specific goal and gather measurable results in a very short time. We are not saying that no other multidisciplinary workshops or conferences were previously held in Quito to discuss other city problems, but that something similar to this Project was never attempted or successfully completed in the past. The Project made it possible to (1) integrate existing research data and results that were scattered among different institutions, and (2) to present them to leading experts in each field, seeking their knowledge and advice. Without this Project, a scenario for Quito would have taken a much longer time or would not have been put together at all. This does not imply that it is impossible to achieve such a project locally, but that it is always hard to get the necessary coordination to perform a complex task such as this one in such a short time. For a society with very limited economic resources like ours, aspects such as training, possibilities to attend seminars such as this one or the World Conference of Earthquake Engineers in Madrid, and opportunities to be in contact with experts from other countries were also very valuable.

Another benefit of this Project was that we were able to convince the Mayor that the city faced a seismic threat, a fact that no previous local government was interested in.

The scenario produced by this Project is by no means the last one possible—rather, it is only the first. The Project results represent the existing data available in Ecuador for evaluating Quito's seismic risk, but more importantly, we believe they provide a diagnosis of the research needed for developing a more reliable earthquake damage assessment and improving the awareness of seismic hazard in the city, from the local government and utility managers to the private sector to the residents.

The results of the Project are credible because they represent years or decades of work in the fields of seismology, soils engineering and structural engineering. They also represent at least five years of tremendous effort made by the Municipality to put the city data into a computer with ORSTOM's GIS "Savane." Without all this previously-existing information, it might have been impossible to accomplish our job or to achieve reliable results in only one-and-a-half years.

When a job has been successfully finished, you may feel satisfied, as we do, but you should never feel that you have done enough, especially when you work with scenarios. The scenarios should always be improved, since you are dealing with the people's security. We will now try to describe the difficulties we encountered in working on this Project.

One of the main obstacles we found throughout the experiment was communication, and not only because Quito's communication system was deplorable even before an earthquake. It turned out to be very hard to coordinate tasks being performed simultaneously in Quito, Tokyo, San Francisco, and Vancouver. Some of the tasks were interdependent, so lack of communication resulted in delays or in truncated results because part of the available information was missing. Lack of a common language also contributed to the communication problem; Spanish, English, Japanese, and French were used during this Project. Because of time constraints, people carrying out some crucial aspects of the Project had to read or write in a foreign language, and this impacted the ideas groups attempted to express with each other. This problem also affected Project meetings, though it was attenuated, but not eliminated, by simultaneous translation. So far, the Project reports are written in or translated into English, but they will mostly be read by Spanish-speaking people. We recommend that the whole process be conducted in the native language of the city where it takes place.

At the same time, it is also very important that the technical aspects of the seismic hazard and damage assessment and scenario itself be performed locally. This is the only way to ensure an effective transfer of technology to the local people and the only way for the local variations and cultural attitudes to be taken into consideration. This is a very sensitive issue and could result in the success or failure of the whole effort. Some of the methodologies used, especially for formulating the scenario, should be reviewed by local people and adapted to the local conditions. For the same reason, the project requires a full-time coordinator who knows the technical aspects for performing a scenario and possesses and understanding of the local people and the way the city operates.

In our case, being a pilot project, no funds were available for hiring people, and the financial costs in expert and equipment time and information have been high for Escuela Politecnica Nacional, the Municipality, and ORSTOM. Furthermore, there was no time to establish what each party had to contribute and expect from the other parties. Everything was based on the good will of the others, and while it worked well, this may not always be the case. It will be very important for other cities to establish in a transparent way what each party will contribute and receive.

Finally, one of the main goals of the Project was to raise awareness about the seismic threat in Quito. This goal will be reached only if condensed versions of the scenario are broadly distributed to the most diverse audiences. Once people have been made conscious of the earthquake threat, their reactions will provide valuable input for improving the next earthquake damage scenario in Quito or other cities in Ecuador or anywhere else.

GOVERNMENTAL ASPECTS OF THE EARTHQUAKE DAMAGE SCENARIO PROJECT OF QUITO, ECUADOR

TEODORO ABDO S.
Municipality of Quito
Quito, Ecuador

1. Introduction: Quito's Vulnerability to Earthquakes

The city of Quito, capital of the Republic of Ecuador, has a population of more than one millon people and is growing at an annual rate of 3.3% [I. Municipio de Quito, 1992]. Its geographical location in a valley has made the city grow in a longitudinal way. The current city limits are more than 40 kilometers in length and from 3 to 5 kilometers in width.

Explosive population growth and the form that this growth has taken have made the city very vulnerable to earthquakes. The new neighborhoods which have appeared in recent years are mainly inhabited by poor migrants from the smaller towns of the country. Their homes are built in dangerous areas, for example the steep and deforested slopes of the neighboring volcano Pichincha, and are constructed with non-engineered technologies.

In addition, the lifeline systems have a topology with very little redundancy. Large areas of the city could be left without access to drinking water, the road network, the airport, electricity, or communications if only one of the connections were to fail. To complicate matters, emergency services such as police, fire protection and medical services are concentrated in only a few older areas of the city. If roads or communications were unusable, these emergency service agencies would have trouble doing their tasks at a time when they would be most needed—during emergencies.

The history of seismic activity in Quito and its surrounding areas shows that the probability of a significant seismic event with damage potential for the city is high. However, public opinion polls indicate that the citizens are not aware of the danger, despite the damage caused by the 1987 earthquake (MSK VII) or the 1990 earthquake (MSK V).

The central government has the responsibility, by law, to manage disasters and to protect the population. However, as in other areas of central government responsibilities, Quito's city government feels that there is a need for direct work by both local officials and the community in order to protect the population and public facilities investments from earthquake damage. The municipal government believes that it has this responsibility, despite the fact that the public would likely not even notice the lack of a defined policy to mitigate disaster effects.

Before embarking on the Quito Earthquake Risk Management Project, the Municipality of Quito possessed some limited information, including maps of risk, regarding dangers to the city. It was clear, however, that more research was needed. The new administration, which began its term in August 1992, found this project to be in its beginning stages and decided to promote it.

2. The Central Government and The Municipality

In order to analyze the implications of the involvement of the local government in the project, it is necessary to clarify the division of responsibilities in the Ecuadorian government's political structure.

In the unitary system of Ecuador, many of the public services are centralized and depend on the national budget. As an example related to the project, the national agency of Civil Defense is by law the institution that must manage disasters. Police and Fire services are also managed at the national level, though they are not administratively connected to Civil Defense. Only during responses to disasters can Civil Defense command these institutions.

The public health and educational systems are also run by ministries at the national level. The telephone system and the supply of natural gas and gasoline are controlled by publicly-owned enterprises. Electrical power generation is also a public monopoly, although the municipality of Quito owns a minority share of the company and therefore has some influence in decision-making for this sector. Only water and sewage are purely local responsibilities in Quito.

Despite these laws which restrict action by Ecuadorian cities, the demands of the community for better public services frequently come directly to the mayor of Quito. Because of the the city's limited political and financial resources available to respond to these demands, Quito's city government has at times found itself in the difficult position of not being able to respond adequately to the demands of its citizens. This situation is exacerbated by Ecuador's tax structure, which is also very centralized. The only significant local revenues come from property taxes. Nonetheless, the municipality has invested some of its own resources in public health and education in order to meet some of the unsatisfied demand.

With respect to disaster mitigation, the only clear responsibility of the local government (other than water and sewage) lies in its urban planning department, which attempts to steer the growth of the city away from areas of high risk. Unfortunately, growth has been so rapid that it has not always been possible to control where new settlements have been or will be built. Furthermore, already-settled and organized communities do not easily accept arguments regarding future risk as a reason to destroy their communities, especially if they do not clearly perceive the danger.

The local administration has nonetheless decided to work in this area by taking more responsibilities for disaster mitigation because this vulnerability must be faced in the short term, and because not much was being done to confront this vulnerability.

3. The Scenario Project: The Need for Leadership

The project had just begun when the new administration took office, and the new administration brought about two main positive changes for the development of the project. The first resulted from the positive attitude of the new administration to the project. The new mayor's approach was that, even if the plan could not be completed by the end of his term, both the research program and the implementation of some important initiatives could be carried out in the four-year period. Of course, it is much easier to start this type of project at the beginning than at the end of a term. This is especially true in Ecuador, because its political system prohibits re-election. (As a result, it is not unusual for important initiatives to suffer delays due to both transition periods between administrations and changes in priorities by the new government.) The mayor's strong leadership was necessary to bring together all of the different sectors involved in the project because the spread of responsibilities is so wide and crosses administrative jurisdictions.

The second circumstance that helped the project was the positive image of the new mayor and the Municipality itself. The mayor was able to bring together many of the high officials from different national and local agencies, including two national ministers. (In this regard, it was helpful that the national government is located in Quito.) Many people from local private organizations (e.g., banks, insurance companies and production chambers) saw the mayor as a natural leader to conduct this project.

During the execution of the project, it was necessary for the mayor to indicate his continued interest by insisting on the cooperation of some of the people involved, who in this stage of the project were not high officials but the technical staff of the agencies and organizations.

4. The Risks and Opportunities of Raising Awareness

When the mayor decided to lead the project, he knew the implications of assuming the role of promotor of such a study. The response that the Municipality must bring in the short and long term, after raising the awareness of the population to the earthquake threat, was taken into account in this decision.

Though the research project is now coming to an end, it is still difficult to evaluate the implications of the results, in terms of future work and responsibilities for the local government. There is, however, political will to implement many of the recommendations of the last workshop, where some initiatives to reduce the impact of earthquakes were proposed.

In fact, the Municipality is now working for the first time to develop an emergency preparedness plan. In the near future, after the final report of the research work is completed, the development of a general plan for the city is foreseen. Very little technical knowledge currently exists in the country to help in preparing such plans; therefore the cooperation of USAID in Quito will be of paramount importance in filling this gap.

Various institutions from the private sector, especially banks and insurance companies, have started working together to form an organization to deal with their particular vulnerabilities. These institutions recognize the need for the Municipality's active participation, both at this stage and in the future. In order to set joint priorities and long-term objectives for private sector and local government, they also feel that political leadership is indispensable.

5. The Future

As explained above, the complexity of Ecuador's political system and its division of responsibilities will impose serious restrictions on the work that the local government can do.

This administration has two possible strategies to face this problem. The more feasible and potentially less-confrontational approach would be for the Municipality to work only inside its current, legislated boundaries. With this strategy, the Municipality would at least be able to reduce the impact of a damaging earthquake on the water and sewage systems and, possibly, the transmission of power. The small capabilities of the Municipality in public health and security could also be prepared to respond to disasters. In addition, the project results would be used to implement long-term policies in such areas as construction codes and urban planning, which can be changed at the local level. Under this strategy, the planning and operation of many vital programs such as communications, rescue, fire suppression, public health, and shelter, would be left to their legally-authorized units within the country.

The second strategy would incorporate all the measures of the first strategy but would also involve applying significant political pressure to national agencies, in order to force them either to complete their work or to transfer responsibilities to the local government. This strategy should be understood in the context of global and continuous negotiations between the Municipality and the central government, not only on this issue but also on many other issues. The viability of this approach therefore depends strongly on the political situation at the time action from the different agencies is needed.

The formation of a multi-sectorial commission in the Municipality is an alternative that would help to put more pressure on the different agencies, and it would also lend legitimacy to the work. Multiple alternatives for the composition of this commission are now being studied at the administrative level.

6. Final Remarks

It is quite clear that finding the best technical approach to the mitigation of earthquake damage in Quito will only be part of the preparation process. As with all public (and private) sector activity, technical solutions can never be more effective than the ability to implement them.

In Quito, the implementation (as well as the development of the technical plan) will involve bringing together various parties not normally accustomed to working together, and not legally required to do so. In such a setting, political leadership will undoubtedly be required; it seems that Quito at present has that leadership.

The only real power that the mayor has, with respect to agencies over which he has no legal authority, is that of being the elected representative of the citizens. His power, then, ultimately resides in the citizens themselves. In order to use that power, the municipal government has felt the need to increase public awareness of the danger of earthquakes. Of course, once the government has raised this awareness, the government must be prepared to answer the heightened concern with effective policy, or else it risks provoking the anger of its constituents.

A lesson for Quito is that once it has started down the road of action on this problem, it must be ready to follow it to its conclusion. We have started down that road.

I. Municipio de Quito, 'Quito en Cifras', Dirección de Planificación, Plan Distrito Metropolitano #7, 1992.

DEVELOPING A COMPREHENSIVE EARTHQUAKE DISASTER MASTERPLAN FOR ISTANBUL[1]

MUSTAFA ERDIK
Department of Earthquake Engineering
Kandilli Observatory and Earthquake Research Institute
Bogaziçi University
Istanbul, Turkey

1. Introduction

1.1 BACKGROUND AND GENERAL CONSIDERATIONS

The City of Istanbul, situated astride the Bosporus in both Europe and Asia, and on both sides of the Golden Horn, is a natural inlet of the Bosporus and a deep-water harbor. Istanbul is the largest city in Turkey, with a history that extends over twenty-six centuries, and it is believed to be the oldest continuously-occupied metropolis in the world. Istanbul has been the only city continuously among the largest ten in the world for 1000 years, except for a short period in the 15th century (Jones, 1992). Its geographical location, at the point where the only land route between Europe and Asia crosses the only sea route between the Black Sea and the Mediterranean, has always lent it great strategic and commercial importance and ensured continued historical status. In recent decades the city has experienced unprecedented growth. Between 1950 and 1990 the population has increased at least eight-fold, from about one million to eight million. Today, Istanbul houses about one-eighth of the total population and one-half of the industrial potential of Turkey. Istanbul has experienced numerous earthquakes. In recent decades the earthquake disaster risks have increased due to overcrowding, faulty land-use planning and construction, inadequate infrastructure and services, and environmental degradation.

Because of recent destructive events in Mexico City (1985, M8.1); Spitak, Armenia (1988, M7.0); Loma Prieta, USA (1989, M7.1); Manjil, Iran (1990, M7.7), the Philippines (1990, M7.8); and Erzincan, Turkey (1992, M6.9), the earthquake risk in urban areas has attracted increasing attention. This is especially relevant for Istanbul, where earthquake records spanning two millenia indicate that at least one major earthquake has occurred in every century, the most recent of which took place in 1894.

In addition, scientists predict a credible seismic gap along the extension of the North Anatolian Fault that passes just 25 km south of the city. Many people voice serious concerns about the earthquake performance of the building stock and the infrastructure of Istanbul.

Urbanization in earthquake-prone countries creates a corresponding increase in the earthquake vulnerabilities and risk. Urbanization is a complex socioeconomic process which controls the spatial distribution of industry, trade, settlements, lifelines, etc., not necessarily with adequate

---

[1] This paper is a focused version of the report prepared for the World Bank as a commissioned paper under the "Environmental Degradation and Urban Vulnerability" study. I have also freely quoted from the ongoing study for developing a "Disaster Master Plan for Istanbul."

consideration of the earthquake hazard to which they are being exposed. Ultimately the society accepts a compromise between the exposure to earthquake hazard and the economic necessities. In Turkey the rapid urban growth (about 4.4% annually in general, 5.3% in Istanbul) and spread of population in earthquake-prone areas are rapidly contributing to the mounting costs of disasters in terms of loss of life and property damage. The rate of expansion of slums and squatter settlements around major urban centers is about twice the average urban rate. This high rate of population growth in urban areas places severe strains on infrastructures, public services and housing, adversely affects the environment, and increases the risk due to natural hazards. The destruction of property in disasters creates severe hardships in the economy of developing countries where capital goods are scarce and the potential for investment is limited. The loss of production during post-disaster recovery imposes additional hardships. The urban centers in Turkey—and also in countries with similar levels of development—are comparatively more vulnerable to earthquakes than those in the more developed countries, since the scarce capital resources of the former do not generally encourage construction practices to the same degree of earthquake safety.

The inevitability of the occurrence of earthquakes in earthquake-prone urban centers makes it imperative that certain preparedness and emergency procedures be contrived in the event of and prior to an earthquake disaster. A disaster, as defined by the U.N. Ad Hoc Group of Experts, is a disruption of the human ecology which exceeds the capacity of the community to function normally, unless disaster preparedness and mitigation measures are in place. In urban centers the impact of disastrous earthquakes is best portrayed and quantified through the preparation of "earthquake damage scenarios." The first ingredient of such scenarios is the assessment of the hazard.

Earthquake hazard in urban areas is usually quantified and portrayed in terms of microzonation maps. The vulnerabilities and the damage statistics of lives, structures, systems, and the socioeconomic structure constitute the second ingredient. Earthquake damage scenarios are based on the intelligent consideration and combination of the hazard, secondary hazards and the vulnerabilities. The preparation of such scenarios can also be based on extrapolations from past disasters experienced in the urban area. For disasters with long return periods, this procedure may not produce reliable assessments since it entails the introduction of the current physical conditions, development, socioeconomic structure, and the population distribution into the extrapolation.

The only available earthquake damage scenario for Istanbul was prepared by the Earthquake Research Division of the General Directorate of Ministry of Public Works and Settlement. As a first study, it provides insight into the nature of such documents. However, it encompasses the whole province (i.e. it is not specifically concerned with the city of Istanbul), the hazard is extrapolated from the 1894 earthquake, and it is only concerned with global damage to building type structures and the associated casualties.

Preparation of an "earthquake disaster master plan" in urban areas is the most essential step towards the mitigation of the risk. For the proper preparation of disaster response and management plans, realistic earthquake hazard scenarios constitute the prerequisite elements. The earthquake disaster master plan for an urban area must reflect the general consequences of the anticipated earthquake. It should be a compendium of concerted counter-disaster mitigation strategies formulated for the specific city and the disaster. In the preparedness phase, the mitigation strategies should concentrate on awareness-building and training. The weak areas in the urban infrastructure may be specified to be retrofitted. The performance of the critical structures and systems may also need strengthening.

There does not exist any earthquake disaster master plan for Istanbul, or for that matter, any other Turkish city. In recognition of its importance and expedience, the governor of Istanbul has commissioned Kandilli Observatory and the Earthquake Research Institute with the preparation of an earthquake disaster master plan.

This preliminary report will review the relevant subjects and the critical issues being encompassed in the preparation of the earthquake disaster master plan, in light of the appropriate experience from disasters in Turkey, and provide applications and examples from Istanbul.

1.2 EARTHQUAKE DISASTER MASTER PLAN FOR ISTANBUL

The preparation of the earthquake disaster master plan for Istanbul has been undertaken by the Kandilli Observatory and Earthquake Research Institute of Bogaziçi University in association with other relevant government, academic and private organizations. The scope of the master plan will be enlarged in the future to cover all disasters, including those related to the environmental degradation. The development of the master plan will encompass the following steps:

1.2.1 *Assessment of the Earthquake Hazard in Istanbul*

A. Compilation of data on all historical earthquakes that affected Istanbul. Detailed analysis of the damage in relation to the urban settlements, building types and population densities at the time of the earthquake. Evaluation of intensity distributions and isoseismal maps.

B. Compilation of all available topographical, geological, geotectonic, geotechnical data, and bore-hole logs.

C. Carrying out geophysical surveys if and where needed.

D. Preparation of soil maps with typical cross-sections at critical locations.

E. Compilation and evaluation of recent microearthquake and earthquake data.

F. Probabilistic and deterministic assessment of the earthquake hazard in terms of expected accelerations and response spectral levels at the bedrock.

E. Assessment of the modification of ground motion due to site response and preparation of microzonation information in the form of GIS database and maps. The information will include surface ground motion and liquefaction, subsidence, and landslide susceptibility.

1.2.2 *Vulnerability Investigations*

A. Compilation of demographic information, and preparation of population density maps for different times of the day.

B. Compilation of lifeline and infrastructure (major roads, railroads, bridges, overpasses, public transportation, power distribution, water, sewage, telephone, and natural gas distribution systems) information including their nodal points (stations, pumps, switchyards, storage systems, transmission towers, treatment plants, airports, marine ports etc.) in the form of maps (1:10,000 scale) and a GIS database.

C. Study of the current building stock, its typification, and preparation of building type distribution maps.

D. Investigation of dams, power plants, major chemical and fuel storage tanks in terms of their primary safety and secondary hazards.

E. Compilation and analytical-empirical assessment of vulnerabilities for lifelines and infrastructure in consideration of their physical layout and redundancies.

F. Assessment of the vulnerabilities of the typified building stock in terms of physical damage, loss of small businesses, and loss of life.

G. Specific investigations of the earthquake safety of local government buildings and the buildings of major or critical police, firefighting, medical, communication, and civil defense centers.

H. Specific investigation of the earthquake safety of cultural and historical monuments and museums, including displayed treasures and artwork.

1.2.3 *Earthquake Disaster Scenarios*

A. Preparation of earthquake disaster scenarios for two earthquakes of medium and high intensity corresponding to return periods of approximately 100 and 500 years.

B. Sensitivity analysis of these scenarios in the light of uncertainties in physical and social parameters of hazard, vulnerability and the date of earthquake. Extrapolation of results to future.

1.2.4. *Assessment Measures for Mitigation*

A. Determination of and the sequencing of rational, realistic, and cost-effective pre- and co-earthquake disaster measures at the following sectoral levels: family and household; schools of all grades; medical services; firefighting; police; civil defense; transportation; water and power supply; resource management and stockpiling; mobile command and communication operations; heavy construction machinery.

B. The planning of awareness-raising, training and education programs for each sector. Coordination with governmental, municipal, private, and voluntary organizations.

C. Determination of legislative and regulatory aspects of mitigation in conformity with the development plans and the local authority. (Development of procedures for effective control of earthquake-resistant design and construction; identification and development of methods for retrofitting; hazardous buildings and facilities; regulations for hazardous material management.)

D. Determination of measures regarding emergency management. (Institutional base; capacity development and training for rescue, evacuation, demarcation, debris removal, and emergency health care; coordination with government, municipal, private, and voluntary organizations.)

2. Earthquake Hazards in Urban Areas

Earthquake hazard means the occurrence, within a specific period of time in a given area, of a potentially damaging earthquake. In this chapter a brief review of the earthquake hazards that can affect an urban area will be provided, with emphasis on assessment and with examples from Istanbul.

Apart from the vibratory ground motion, earthquakes manifest themselves in several hazards to be considered in urban areas. Kockelman (1985) classifies these hazards as follows:
- Critical: Ground rupture due to faulting; Tsunamis and seiches; Liquefaction
- Sometimes Critical: Ground shaking; Subsidence; Slope stability problems; Soil creep
- Less Critical: Expansive soils; Compressible and/or collapsible soils.

In urban areas with no danger of tsunami inundation, modification of the ground motion, soil failures and terrain movements, and the surface fault ruptures merit specific consideration.

2.1 MODIFICATION OF THE GROUND MOTION BY SITE CONDITIONS

Weak formations overlying rigid substrata, such as sedimentary basins and alluvial valleys, can trap the earthquake vibrations, modifying them both in amplitude and frequency content. Earthquake engineers have documented and recognized that the damage distribution is generally site-dependent, and the structures founded on unconsolidated material frequently sustain greater damage in earthquakes. Due to urbanization, reclaimed land near the coast has spread rapidly. The earthquake response of the reclaimed land and the soft alluvium can be much more amplified than that of the consolidated deposits. As evidenced by observations in several earthquakes, the immediate vicinity of lateral discontinuities and contact zones between highly-contrasting formations are usually the zones of amplification. Amplification due to topography has been identified in theoretical as well as empirical studies. The tops of isolated hills, elongated crests, edges of plateaus, and cliffs are usually zones of amplification due to diffraction and focusing.

For the study and assessment of the ground motion modification due to site conditions, simplified methods, strong-motion records and determination of the transfer functions (numerical, experimental—i.e, microtremor recording) are currently being used. Several researchers have shown that, for layers of a given thickness, the relative shaking response will be greatest where the surface geologic units have the lowest impedance values and where the impedance contrast between the surface layer and the underlying one is the greatest. Thus the maps that broadly characterize the near-surface shear-wave velocities of geological materials provide a first approximation of relative shaking response. Shear-wave velocities can be correlated with soil texture (mean grain size), standard penetration resistance, rock hardness and lithology. Joyner and Fumal (1985) have incorporated the local shear-wave velocity of near-surface geologic materials in the assessment of site effects for the attenuation of peak ground acceleration (PGA) and velocity (PGV). They have further shown that, unlike PGV, for PGA the correlation with the shear-wave velocity is not statistically significant at the 90% level, or, in other terms, the site dependence of the PGA is not statistically significant. However this finding is not shared by others. For example, Fukushima et al. (1985) have computed, on the basis of a worldwide strong-motion data set, the residuals between the observed and the predicted PGAs and the mean values for each ground classification (rock, hard-, medium-, and soft-soil ground). The observed PGAs are about 40% lower at the rock site and about 40% higher at the soft-soil site than those predicted, but the differences are insignificant for the hard- and medium-soil grounds.

In a recent study (Boore at al.,1993) of the estimation of response spectra and peak accelerations, the sites were characterized by dividing the site geology into four classes, depending on the average shear-wave velocity in the upper 30 m. The site class A includes sites where the average shear-wave velocity is greater than 750 m/s; site class B sites where the velocity is between 360 and 750 m/s; site class C sites where the velocity is between 180 and 360 m/s; and site class D where the velocity is less than 180 m/s. Analysis of 271 two-component data from 20 western American earthquakes indicates significant differences between site classes at all periods.

2.2 EARTHQUAKE-INDUCED SOIL FAILURES AND TERRAIN MOVEMENTS

The most important earthquake-induced soil failures are liquefaction, loss of strength, and densification. Liquefaction refers to the following (EERI, 1986): (1) flow failures (massive displacement of completely liquefied soil); (2) lateral spreads (lateral displacement of surface soil layers over a liquefied layer); (3) slumps (in steep banks underlain by liquefied sediments); and

(4) loss of shear strength (tipping or bearing failure of aboveground structures, buoyant rise of underground structures). Techniques to evaluate the liquefaction potential are well-established and generally involve the preparation of two types of maps, one showing the liquefaction susceptibility and the other expressing the opportunity for critical levels of shaking. These two maps are merged to depict the real liquefaction potential (Youd et al., 1979). Areas with cohesionless, granular, or water-saturated materials imply high to moderate susceptibility.

The lithology, geologic age, depth to groundwater, and standard penetration test data are important estimators for the liquefaction susceptibility. Holocene and Pleistocene cohesionless deposits have respectively high and low susceptibilities. Where the water table is less than 3 m and the soil consists of less-than-1000-year-old fluvial silt and sand, the susceptibility is very high. Susceptibility depends critically on the depth to groundwater, a parameter that may be affected through land use, precipitation, groundwater extraction, and/or recharging in the urban areas. Tokida (1990) lists the following criteria for liquefaction susceptibility (a conglomerate of Japanese criteria for bridges, water supply, sewage, and buildings): (1) Saturated alluvial sandy layers within 20 m from ground surface; (2) Ground water level within 10 m from ground surface; (3) D50 values between 0.02 and 2 mm in grain size acccumulation curve; and (4) Standard penetration test blow count N≥30.

Earthquake-induced terrain movements include landslides, rockfalls, and subsidence. The areal limits of landslides that can be triggered by earthquakes should be indicated on microzonation maps. Materials that are most susceptible to earthquake-induced landslides are: (1) weakly-cemented, weathered and/or fractured rocks; (2) loose, unsaturated sands; (3) saturated sand and gravel with layers; and (4) sensitive clay. Experience indicates that most earthquake-induced landslides involve materials that have not previously failed, and existing landslides are seldom reactivated (EERI, 1986). The probability that a landslide will occur on a particular slope during a particular earthquake is a function of both the pre-earthquake stability of the slope and the severity of the earthquake ground motion. The stability of the slope is controlled by the strength of the material and the morphology of the slope. Methods to estimate the potential for earthquake-induced slope failures are not sufficiently developed to permit a reliable regional appraisal of the problem. According to Newmark (1965), one of the measures of slope stability under seismic shaking is the acceleration required to initiate an irreversible displacement of the soil mass. Strength loss in sensitive clay during strong earthquake shaking may involve failures similar to liquefaction and specifically can initiate large landslides, as was the case in the 1964 Alaska earthquake (EERI, 1986). In urban areas, landslides (rock falls, soil slides, lateral spreads and slumps) can cause massive property damage. Transportation lines are blocked. The lifelines are damaged, causing the disruption of community services. Prudent siting, involving adequate setbacks from steep slopes, flattening cut slopes, and avoidance of unstable areas can mitigate the hazard. For massive landslide problems, the risk can be accepted with appropriate emergency response preparedness.

2.3 TECTONIC RUPTURES (SURFACE FAULT RUPTURES)

The information about the movements and the surface expressions of possible active faults should be included in the microzonation maps to avoid (or to accommodate) their effects on structures and systems. It is generally considered appropriate to designate faults that show evidence of Quaternary motion as active with a possibility of rupture.

2.4 EARTHQUAKE HAZARD ASSESSMENT

A rational earthquake hazard assessment methodology for urban areas should provide for uncertainties associated with the input parameters and be based on appropriate stochastic models. The earthquake hazard is usually depicted as annual probabilities of exceedance for given ground motion levels or earthquake intensities. The earthquake hazard assessment for Istanbul will be used as an example.

Figures 1 and 2 are provided to describe the relative position of Istanbul in the seismic setting of Turkey. They illustrate the epicenters of earthquakes respectively with magnitudes (Ms) greater than or equal to 4.0 for the last 30 years and greater than or equal to 5.0 for the last two millenia.

Figure 3 indicates the active tectonic elements in the Marmara Graben to the south of Istanbul. Superimposed on the same figure are the epicenters of earthquakes in the last 30 years with magnitudes greater than 3. The North Anatolian fault zone which extends from the Karliova junction (about 40 N - 41 E) in eastern Turkey to as far as mainland Greece, splits into three strands (southern, middle and northern) to the west of about 30.5 E longitude. The northern strand passes through the northern half of the Marmara Sea, forming a series of discontinuous pull-apart basins and ridges, collectively referred to as the "Marmara Graben." The Marmara Graben is recognized as the source of major earthquakes that could seriously affect Istanbul.

In Istanbul, earthquake records spanning two millenia indicate that, on average, at least one medium intensity (I_o=VII-VIII) earthquake has affected the city every 50 years (Ambraseys, 1991). The average return period for high intensity (I_o=VIII-IX) events has been 300 years. The temporal distribution of the earthquakes has not been uniform. The fact that there has not been a major earthquake in the 20th century represents an atypical development. It should also be noted that the region to the immediate south (about 20 km) of Istanbul has been identified as a seismic gap (Barka and Toksöz, 1989).

The epicentral distribution of the major historical earthquakes is indicated in Figure 4. In the most recent earthquake (July 10, 1894), the damage in Istanbul was widespread and in places very serious. The average intensity was about VIII. Many public buildings, mosques, and houses were shattered and left on the verge of collapse, while most of the older constructions fell down. Water distribution was disrupted. Slumps occurred in Sirkeçi and Eminönü. There was general panic. People left the city or stayed outside their houses for a long period. Deaths are estimated at a few thousand (about 0.3-0.4% of a total population of 900,000).

The probabilistic hazard assessment methodology that has been employed for Istanbul is elaborated in Erdik et al. (1985). Subsequent to the acquisition of pertinent geotectonic and seismologic data and seismic source modeling, the methodology can be described on the basis of the following key steps and parameters:

1. The source seismicity information step involves the construction of recurrence relationships after an artificial homogenization process to account for the deficient data.

2. Development of intensity-based local attenuation relationships.

3. Careful evaluation, identification and adoption of a strong-motion data-based ground motion attenuation relationship.

4. Use of a proper stochastic model for recurrence forecasting.

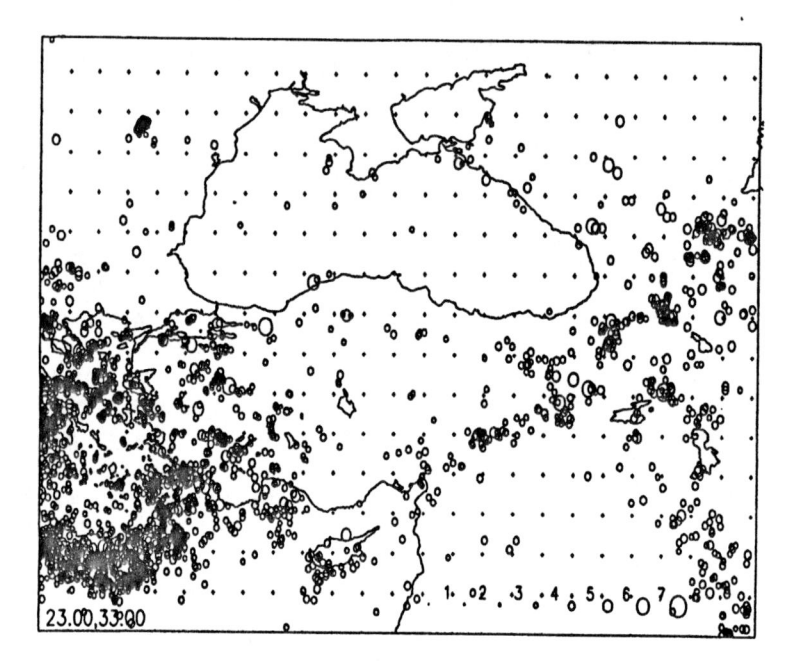

Figure 1. Seismicity of Anatolia and vicinity over the time period 1960-1989 for earthquakes of magnitude (Ms) greater than or equal to 4. The legend for magnitudes is in the lower right corner.

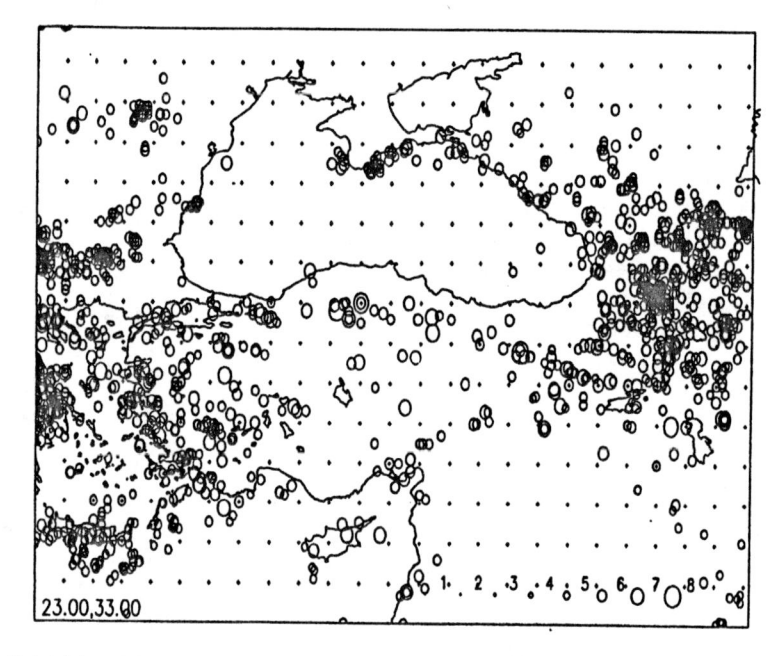

Figure 2. Seismicity of Anatolia and vicinity over the last two millenia for earthquakes of magnitude (Ms) greater than or equal to 5. The legend for magnitudes is in the lower right corner.

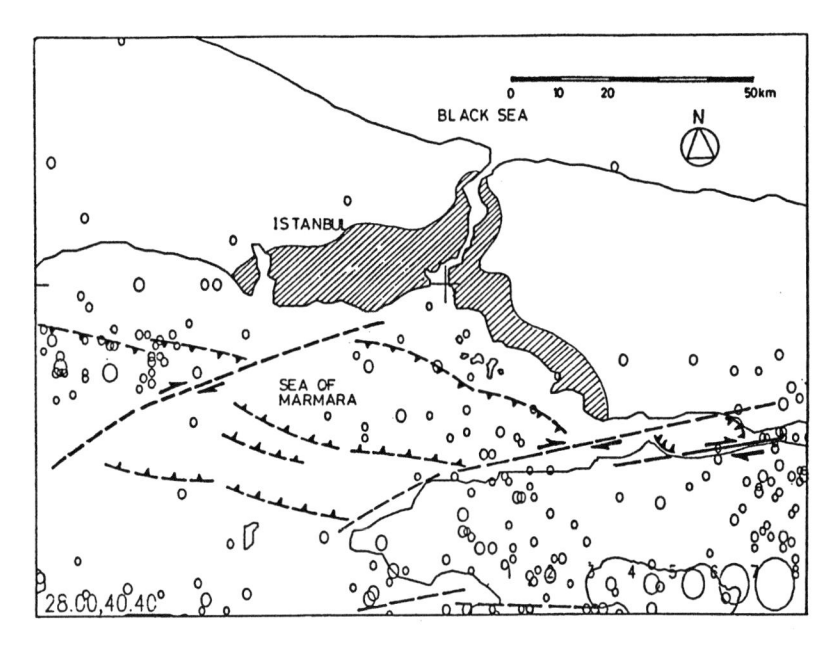

Figure 3. Active tectonic elements around İstanbul. Figure also illustrates the earthquake activity of last 30 years. The legend for magnitudes is in the lower right corner. (After Erdik and Barka, 1992).

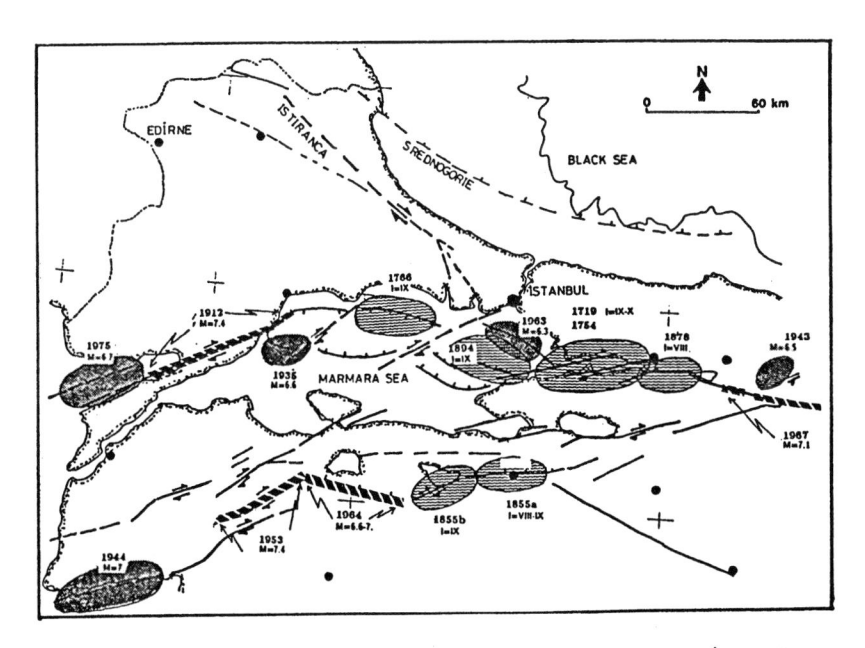

Figure 4. Epicentral distribution of the historical earthquakes around İstanbul (After Erdik and Barka, 1992).

For the last step, a homogenous Poisson process has been employed. The pros and cons of this process and its comparison with others are documented in Erdik et al. (1983 and 1985). For the attenuation of intensities, the regional intensity-based attenuation relationships developed by Erdik et al.(1983 and 1985) were utilized. For the attenuation of peak ground acceleration (PGA), the relationships developed by Campbell (1981) and Joyner and Boore (1981) were found to be appropriate on the basis of existing comparisons with limited Turkish data (Erdik et al., 1985). Figures 5 and 6 illustrate the variation of MSK Intensity and the PGA (at prominent rock outcrops) in Istanbul for different annual probabilities of exceedence. Figure 7 indicates the synthetic isoseismal maps to be expected in case of magnitude (Ms) 6 or 7 earthquakes occurring along the Marmara Graben System.

2.5 MICROZONATION MAPS

Seismic microzonation maps can be defined as maps providing estimates of parameters needed for the siting and the earthquake-resistant design of civil engineering structures and systems. A seismic zonation map uses empirical and observational data in connection with a physical model to provide these estimates. Microzonation for earthquake hazards has the potential to serve as a guide for safer land use and construction. The microzonation procedures should combine a multitude of information to achieve an optimum solution to the problem of mitigating earthquake hazards. Microzonation studies, as such, involve the determination of pertinent site characteristics and their incorporation in the design of structures and systems. The necessary information that should be conveyed by an earthquake-hazard-based microzonation map are: (1) Modification of the strong ground motion by site conditions; (2) Earthquake-induced soil failures and terrain movement; and (3) Tectonic surface ruptures. These subjects have been covered in the previous section.

The quantitative incorporation of the "Modification of the Ground Motion by Ground Conditions" in the microzonation maps is a controversial issue. In this regard there have been analytical, empirical and experimental (observational) approaches. Analytical procedures have been employed for the determination of the site response. These methods range from simple one-dimensional calculations to three-dimensional, linear/non-linear, time/frequency domain and finite difference/element computations. Especially in Japan, several microzonation maps have been prepared using one-dimensional non-linear analytical procedures. These maps yield the input motion, amplitude-dependent predominant periods and the peak surface accelerations (e.g., Sugimura et al, 1982). The success of these analytical procedures awaits the collection of an adequate sample of strong-motion data.

By use of microtremor data, the relative response of different points in the region has been claimed to be adequately assessed. From analyses of microtremor records obtained at over 1000 locations in a wide variety of soil conditions, Kanai et al. (1961, 1966) discovered that the time and frequency domain wave shapes of microtremors are distinctly different in different soil conditions ranging from hard rock (Zone I) to loose fill (Zone IV). The Fourier spectra of microtremors for Zone I indicates a sharp peak, whereas almost no distinct peaks can be seen for Zone IV. On the basis of these findings Kanai proposed two methods for the purpose of microzoning. One method attempts to delineate the four soil zones on the basis of the largest period and the mean period of the microtremors. The other one does the same using the largest amplitude and the predominant period of the microtremor measurements. The critics of this

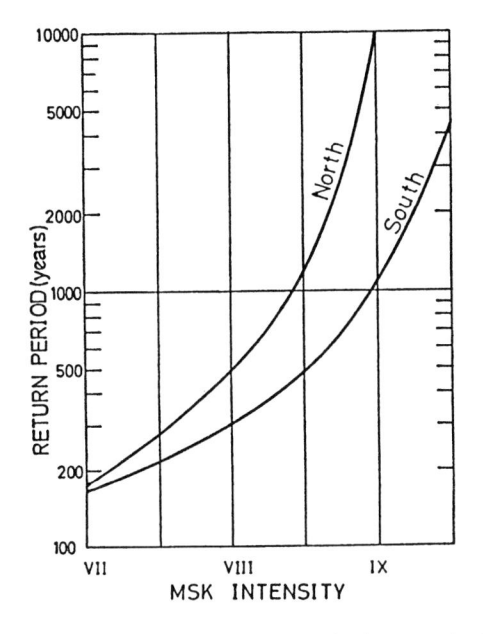

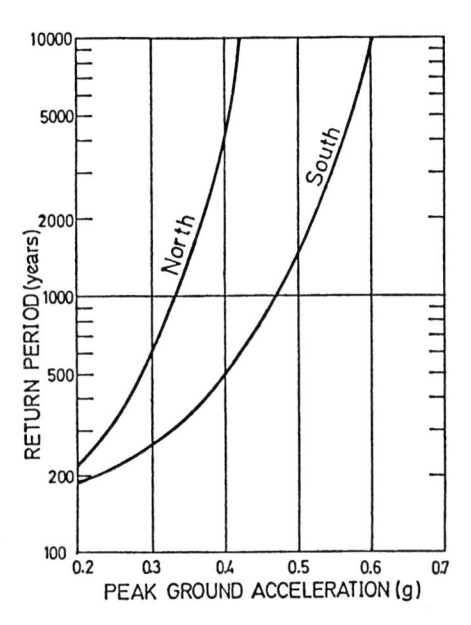

Figure 5. Variation of MSK Intensity (at competent rock outcrop) in İstanbul for different annual probabilities of exceedence.
(After Erdik and Barka, 1992).

Figure 6. Variation of PGA Intensity (at competent rock outcrop) in İstanbul for different annual probabilities of exceedence.
(After Erdik and Barka, 1992).

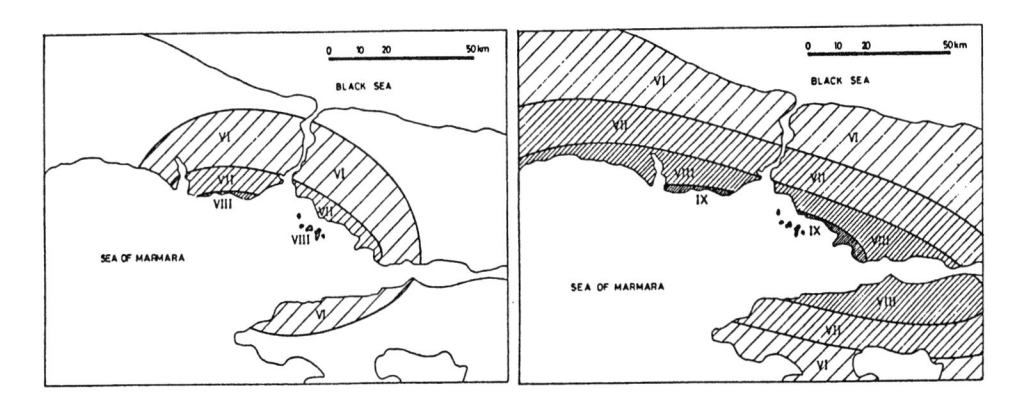

Figure 7. Synthetic iso-seismal maps corresponding to Ms=7 and Ms=7.5 magnitude earthquakes occurring along the Marmara Graben.

method claim that the microtremors originate at shallower depths and follow different propagation paths compared to earthquakes. In certain cases the microtremor measurements may be nonstationary over different periods of the day and yield more information about the excitation than the transfer characteristics of the media. Furthermore, the microtremor data can provide information only on the low-strain behavior of the medium and require careful interpretation in assessing the strong-motion response characteristics of the site.

In several microzonation exercises, the modification of the ground motion is expressed in terms of intensity changes, empirically correlated with the ground conditions. Medvedev (1965) attempted to relate the increments of seismic intensity to seismic site rigidity (product of the longitudinal seismic wave velocity by the density of the geologic layer) and to the elevation of the water table. The intensity increments for different sites with respect to granite are found to vary from 1 (moderately firm ground) to 3 (loose fills). An additional intensity increment of 1 unit is considered for cases where the water table lies directly below the structural foundation level.

In the western United States, the tables providing changes of intensity have been used for microzonation purposes (Evernden and Thomson, 1985). For California the relative intensity values for different ground characteristics vary between -3 (Granite, metamorphic rocks) and +0.5 (saturated or near-saturated alluvium). For Quaternary sedimentary deposits with a depth to water table less than 10 m, no change in the intensity is prescribed. Figure 8 indicates a provincial-scale microzonation map of Istanbul indicating expected intensity changes due to soil conditions.

2.6 MICROZONATION FOR ISTANBUL

As an example of microzonation, we can focus on the central part of Istanbul, the area within the ancient city walls (Erdik et al., 1991).

The regional intensity map (analogous to an isoseismal map) of the 1894 earthquake (estimated M=6.7, I=VIII) and the preliminary damage zoning map for Istanbul are presented respectively in Figures 9 and 10. In the damage zoning map [Figure 10], the black dots represent major structures with heavy to total damage and the encircled areas indicate the general damage zones. Most of the damage took place on the Fatih-Beyazit ridge and slumping was observed in Eminönü. The Grand Bazaar had heavy damage due to its "loose fill" type ground condition.

The preliminary microzonation for Istanbul, provided in Figure 11, is based on the morphology, geology, the distribution of artificial fills, damage distribution in significant historical earthquakes, and other geophysical and geotechnical data (Erdik et al, 1991). The map identifies four earthquake hazard zones. The stable rock zone delineates certain parts of the Carboniferous rock (where the artificial cover is little or none) and the late Miocene Mactra Limestone. The semistable zone represents mostly late-Miocene sand and gravel, and clay and marl. In this zone ground shaking hazard is somewhat increased and slopes are prone to landsliding. The zone encompassing the thick artificial cover will be subjected to an increase in ground shaking. The zone of mud and fill delineates potential ground failures such as liquefaction, fissuring and slumping. In Figure 11 the locations of potential earthquake-induced landslides are also illustrated.

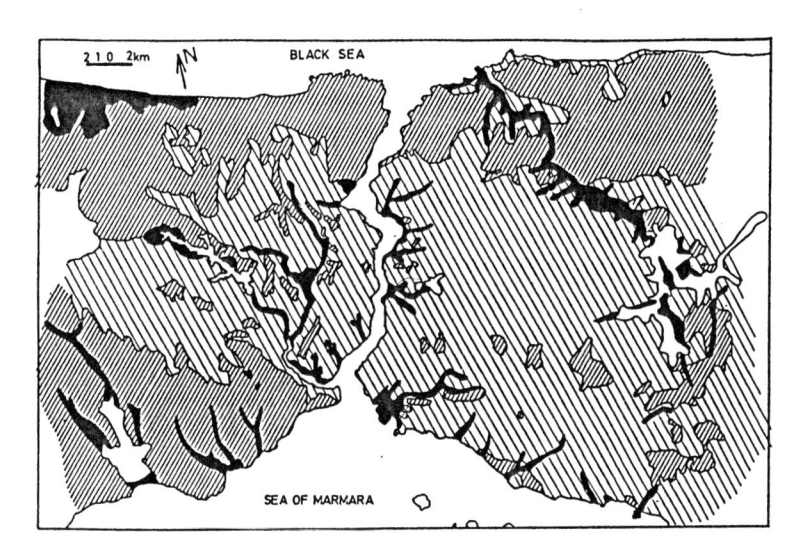

Figure 8. A provincial scale microzonation map of İstanbul indicating expected intensity changes due to geological conditions. Black regions can account for intensity increases between 1.5 and 2 units. Dark shaded areas can account for intensity increases between 0.5 and 1 units. The light shaded areas correspond to stable rock regions where no intensity increase is accounted for.

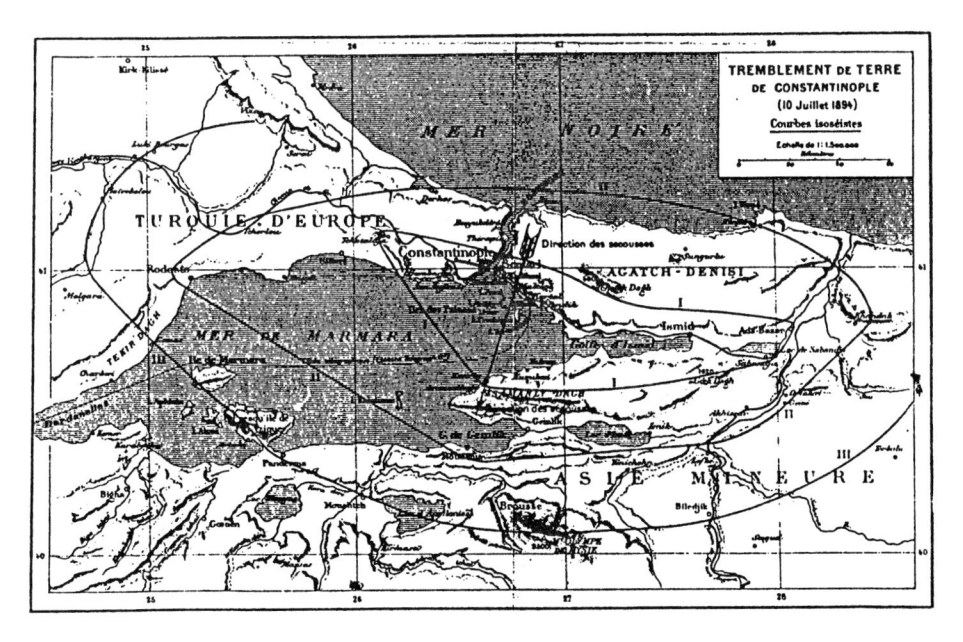

Figure 9. The regional intensity map (analogous to an iso-seismal map) of the 1894 earthquake (estimated Ms=6.7, I=VIII)

138

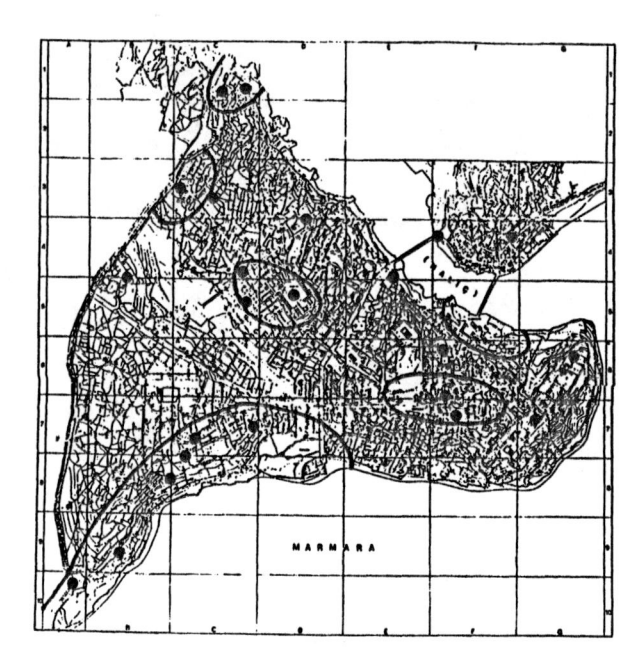

Figure 10. A preliminary damage zoning map for İstanbul. (The black dots represent major structures with heavy to total damage and the encircled areas indicate the general damage zones)

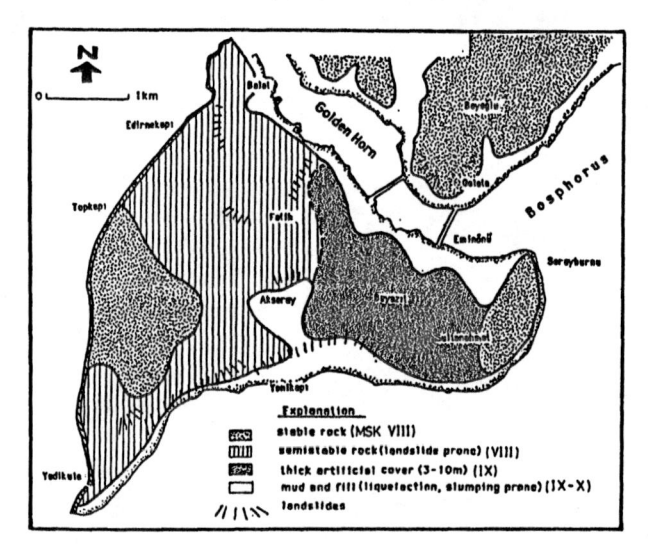

Figure 11. An earthquake microzonation map for İstanbul Peninsula (After Erdik and Barka, 1992).

3. Vulnerability Analysis

In this chapter the vulnerability analysis for earthquakes will be covered with examples from Turkish experience.

The vulnerability analysis and hazard assessment together constitute the key elements of the risk assessment. A probabilistic risk analysis for a given natural hazard takes into account the uncertainties inherent in the earth sciences (hazard) and the engineering information (vulnerability) that are essential ingredients of the analysis. Vulnerability is defined as the degree of loss to a given element at risk, or a set of such elements, resulting from the occurrence of a hazard. Vulnerability analysis involves the elements at risk (physical, social and economic) and the type of associated risk (such as damage to structures and systems and human casualties). Physical vulnerabilities are associated with urban infrastructure, lifelines and buildings. These vulnerabilities are agent- and site-specific. Furthermore they also depend on design, construction and maintenance particularities. The vulnerability analysis of cultural and historical monuments and of urban facilities vital to post-disaster activities deserves special attention. Socioeconomic vulnerabilities measure risks to socioeconomic assets and systems, such as damages to social infrastructure, impacts on production and employment, changes of wealth distribution, and inflation.

The vulnerabilities and the damage statistics of lives, structures, systems, and the socioeconomic structure are the main factors influencing the earthquake risk in urban areas. The process of rapid urbanization, the attendant socioeconomic development, large-scale construction, and the provision of infrastructural services will expose larger populations and valuable elements to earthquake hazards and risks. Seismic risk analysis attempts to calculate the probability of the adverse physical, economic and social effects of an earthquake or series of earthquakes in a given urban center. Vulnerability assessments are usually based on past earthquake damages (observed vulnerability), on laboratory testing and, to a lesser degree, on analytical investigations (predicted vulnerability).

In urban areas, the vulnerability assessments for engineered structures (i.e. residential, governmental and commercial buildings, bridges, dams, port and harbor structures, lifelines, and utilities) and non-engineered structures (mainly squatter settlements) need to be differentiated. The vulnerability of the engineered structures depends on the siting, design and construction essentials and defies generalizations. On the other hand, generalizations can be made about the earthquake vulnerability of different building classes of non-engineered construction. Figure 12 provides the deterministic vulnerability functions for different types of structures (After Akkas and Erdik, 1984). This figure is based on worldwide data and indicates the average damage ratios to be expected for adobe (non-engineered structure) and brick unreinforced masonry structures, and properly designed reinforced concrete framed and shear wall structures for different exposures to earthquake hazard, quantified in Modified Mercalli intensities (MMI). Damage ratio is defined as the cost of repair divided by the cost of rebuilding the same structure. As can be seen, the unreinforced adobe and brick masonry structures are at least 4 to 5 times more vulnerable to receive damage than properly designed reinforced concrete structures under the same earthquake exposure. Algermissen et al. (1978) provide a set of vulnerability functions which gives the average loss, as percent of total value of the structure, against different intensities (MMI) for different building classes. For specific structures and systems, the loss functions are not appropriate due to their general coverage, and structure-specific damage probability matrices may need to be developed.

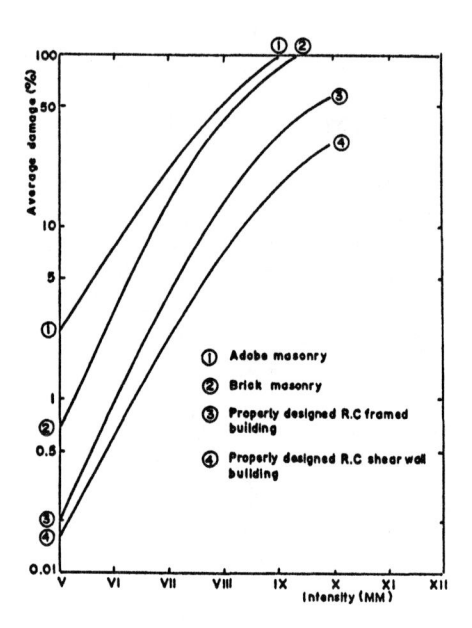

Figure 12. Deterministic vulnerability functions for different types of structures (After Akkas and Erdik, 1984).

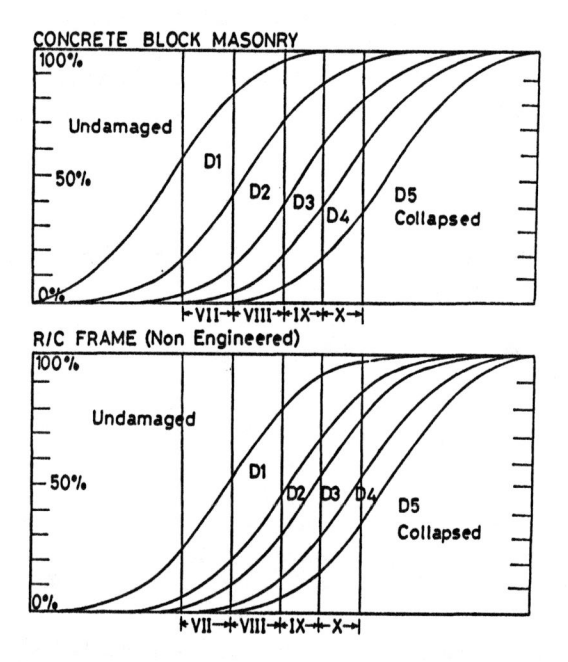

Figure 13. Vulnerability functions for "Concrete Block Masonry" and "R/C Frame - Non Engineered" type buildings (After Coburn and Spence, 1992)

The MSK-81 (Medvedev-Sponheuer-Karnik Intensity Scale, 1981 Revision) classifies the structures into three types: A, B and C. In a Turkish context, these types correspond respectively to rural, masonry and multi-story R/C structures. The same scale classifies the damage to buildings under five grades: Slight Damage; Moderate Damage; Heavy Damage; Destruction; and Total Damage. Coburn and Spence (1992) associate these damage grades with the following definitions:

Table 1
MSK-81 DAMAGE GRADES

Damage Grade	Masonry Building	R/C Framed Building
D1-Slight	Hairline cracks	Infill panels damaged
D2-Moderate	Cracks 0.5-2cm	Structural cracks <1cm
D3-Heavy	Cracks >2cm, material dislodged	Damage to structural members
D4-Destruction	Collapse of individual wall or roof	Failure of structural or major deflection
D5-Collapse	Most wall or roof collapsed	Collapse of structural members and floor

Figure 13 indicates the vulnerability functions for "Concrete Block Masonry" and "R/C/ Frame" type non-engineered (non-seismic) buildings as adopted from Coburn and Spence (1992) and are based on extensive dataset compilations. The horizontal axis indicates the MSK intensities with a range and the vertical scale indicates the percentage loss. Both of these types are common buildings in Turkish urban areas.

The 1992 European Macroseismic Scale (ESG, 1993), an updated version of the MSK scale, differentiates the structural vulnerabilities into six classes (A to F) as indicated in Figure 14. Although this new version has not been fully endorsed or subjected to a worldwide critical review, there is merit in its elaboration in this report. In EMS 1992, the vulnerabilities of reinforced concrete structures are distinguished with respect to the level of anti-seismic design (ASDi), level of earthquake resistance (Quality), and level of regularity. Figure 15 provides the expected damage grades for reinforced concrete structures with different levels of anti-seismic design, low quality, and medium regularity for different intensity levels. Properly designed and constructed with respect to the earthquake resistance code, a reinforced concrete structure in Istanbul should qualify for ASD8. However due to deficiencies in design, concrete quality and workmanship, the bulk of the reinforced concrete building stock deserves to be considered only with an ASD7 level anti-seismic design. The damage grades are illustrated in Figure 16.

In this century only a limited number of earthquakes in Turkey have affected urban areas. The available data will be provided in the following tables. However, much more empirical data will be needed to arrive at statistically significant vulnerability matrices for urban building structures. The following vulnerability matrix for low rise (<=5 story) reinforced concrete frame buildings, the most common building type in urban centers in Turkey, has been compiled on the basis of damage data obtained from the 1976 Denizli and 1971 Bingöl earthquakes (Bayülke, 1982).

142

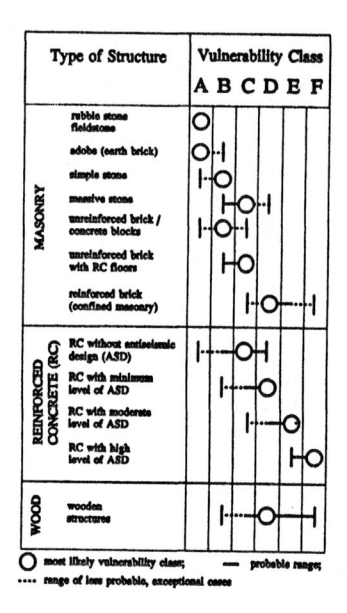

Type of Structure	Vulnerability Class A B C D E F
MASONRY — rubble stone, fieldstone	
adobe (earth brick)	
simple stone	
massive stone	
unreinforced brick / concrete blocks	
unreinforced brick with RC floors	
reinforced brick (confined masonry)	
REINFORCED CONCRETE (RC) — RC without antiseismic design (ASD)	
RC with minimum level of ASD	
RC with moderate level of ASD	
RC with high level of ASD	
WOOD — wooden structures	

○ most likely vulnerability class; — probable range;
···· range of less probable, exceptional cases

Figure 14. The EMS 1992 differentiates buildings into six classes of vulnerabilities, depending on the design, materials and the construction type.

INTENSITY	TYPE	LEVEL Q Damage Grade				
		1	2	3	4	5
VI	ASD₁	◐	◓			
	ASD₂	◐				
	ASD₃	◓				
VII	ASD₁	◖	◐			
	ASD₂	◖	◓			
	ASD₃	◐				
VIII	ASD₁	◐	◖	◐		
	ASD₂	◖	◖	◓		
	ASD₃	◖	◐			
IX	ASD₁				◐	
	ASD₂		◖	◖	◓	
	ASD₃		◖	◐		
X	ASD₁				◖	◐
	ASD₂		◐	◖	◖	◓
	ASD₃			◖	◐	
XI	ASD₁					◖
	ASD₂				●	◖
	ASD₃				◖	◐
XII						

◓ very few   ◐ few   ◖ many   ● most

Figure 15. Relation of damage grades to intensity degrees for ASDi type buildings of low level quality Q1 and medium level of regularity Rm.

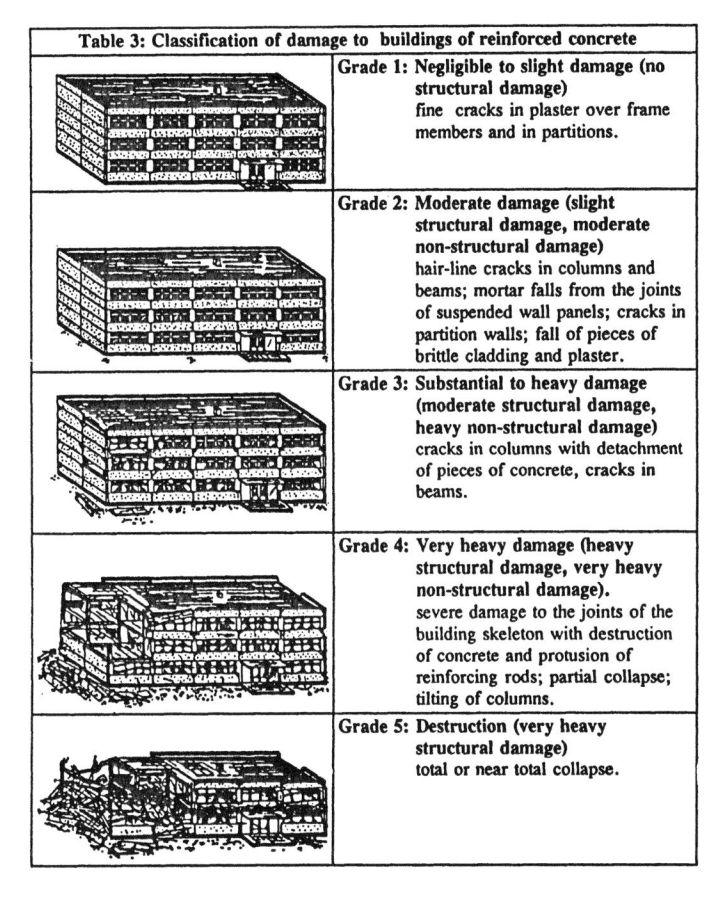

Table 3: Classification of damage to buildings of reinforced concrete	
	Grade 1: Negligible to slight damage (no structural damage) fine cracks in plaster over frame members and in partitions.
	Grade 2: Moderate damage (slight structural damage, moderate non-structural damage) hair-line cracks in columns and beams; mortar falls from the joints of suspended wall panels; cracks in partition walls; fall of pieces of brittle cladding and plaster.
	Grade 3: Substantial to heavy damage (moderate structural damage, heavy non-structural damage) cracks in columns with detachment of pieces of concrete, cracks in beams.
	Grade 4: Very heavy damage (heavy structural damage, very heavy non-structural damage). severe damage to the joints of the building skeleton with destruction of concrete and protusion of reinforcing rods; partial collapse; tilting of columns.
	Grade 5: Destruction (very heavy structural damage) total or near total collapse.

Figure 16. Illustration of the classification of damage grades depicted in Figure 15.

Table 2
VULNERABILITY OF R/C FRAME BUILDINGS

Earthquake	Degree of Damage				
	No	Slight	Medium	Heavy	Collapse
1976 Denizli (MSK VII)	40%	38%	17%	5%	0%
1971 Bingöl (MSK VIII)	16%	27%	36%	15%	6%

The 1976 Adapazari earthquake (MSK VIII) totally damaged (collapsed) seven reinforced concrete buildings (three of which were under construction) out of about 60 buildings (Ambraseys et al., 1968), producing a collapse rate of about 6%, similar to the 1971 Bingöl earthquake.

Data obtained from the 1992 Erzincan earthquake (Sengezer, 1993; Kandilli, 1992) for urban buildings at MSK intensity VIII yield the following vulnerability matrix:

Table 3
VULNERABILITY OF BUILDINGS IN 1992 ERZINCAN EARTHQUAKE

Damage Level	Repair Cost Ratio	Reinforced Concrete	Brick	All
No Damage	0.0	25%	78%	58%
Light Damage	0.2	25%	20%	24%
Moderate Damage	0.4	20%	1%	7%
Heavy Damage	0.6	17%	1%	7%
Partial Collapse	0.8	5%	0%	3%
Total Collapse	1.0	8%	0%	1%

In this matrix the repair-cost ratio refers to the money to be spent for repair as the ratio of the money for rebuilding the same structure. The 1992 Erzincan earthquake has shown that roughly 50% of all 4-6 story R/C structures need to be demolished and rebuilt or to be totally retrofitted after an intensity VIII earthquake.

The ratio of the number of people killed to the number of buildings collapsed is called the "Lethality Ratio." This ratio depends on the average number of people per building, time of earthquake, number of occupants trapped, and capability of rescue first-aid services. Ambraseys and Jackson (1981), using data from Turkey and Greece, provide the following statistics regarding the number of people killed per 100 houses destroyed by earthquakes, M=>5: Rubble masonry houses= 17; Adobe houses= 11; Masonry and reinforced adobe houses= 2; Timber and Brick houses= 1; Reinforced Concrete Frame houses= < 1, but data are lacking. Erzincan data further indicate that there has been 1 death and 3 hospitalized injuries per heavily-damaged or collapsed R/C building.

In addition to buildings, many other engineered urban structures, infrastructures, lifelines and services are vulnerable to the effects of earthquakes. Landslides, rock falls and fault ruptures can block highways and railways or damage pipelines. Strong shaking can cause transmission lines and bridges to fail. Liquefaction can cause failure of port and harbor structures. The earthquake vulnerabilities of these structures and systems are not generally known in explicit formats. In any case these vulnerabilities are highly case- and site- specific and defy generalization. However, the following observations acquired from past urban earthquakes (EERI, 1986), supplemented by the Turkish experience from the 1992 Erzincan earthquake can be used as a guide to assess their physical vulnerabilities.

Bridges. Slope instability, liquefaction and settlement can move the abutments of bridges and tilt the piers, causing extensive damage. The bridge girders can fall off of their supports. Experience from the 1976 Tangshan and 1971 San Fernando earthquakes indicate that about 10-30% of the single span bridges and 5-10% of the continuous girder bridges will receive heavy damage in an intensity VIII earthquake. In the 1992 Erzincan earthquake, one simply-supported single-span highway bridge was heavily damaged and two others lightly damaged.

Underground Pipelines. The greatest damage to pipelines occurs in zones of faulting, poor ground, liquefaction and landslide. Ruptured gas lines lead to leaks and fire hazard. Worldwide data indicate about 0.5-1 pipe breaks per one kilometer pipe in intensity VIII earthquakes, depending on the soil and pipe conditions. In the 1992 Erzincan earthquake, there were at least 4 failures in the main cast-iron pipes and 25 failures in the main branch pipes. Altogether there were 270 failures in total pipe length of 272 km. A large part of the city did not receive any water for about a week and, two weeks after the earthquake, the water supply in the city was only restored to a level of about 70% of the normal value.

Electrical Transmission and Distribution Systems. High-voltage porcelain insulators, bushings and supports are most vulnerable to earthquakes. Damage generally occurs in improperly-anchored electrical equipment. The most vulnerable components of the system are the generators and the transformers. In the 1992 Erzincan earthquake, although the main switchyard was intact, many of the secondary transformers, located on steel poles throughout the city, failed. About forty transformers fell to the ground. 1.8 km of 32 km of underground cables and 4.0 km of 50 km aboveground cables needed repair. The electricity was restored within a few days.

Telecommunications. The anchorage of switching and battery racks against lateral displacement and toppling is an essential measure to avoid earthquake damage. In the 1992 Erzincan earthquake, the interruption of the telephone services was due to the falling-over of the unanchored switching racks. Most telephones in Erzincan were operational within a few days by using emergency power. The underground cables, passing through concrete pipes, sustained little damage, but the connecting lines between the poles and buildings were destroyed with the damage to buildings.

Building Contents. Modern buildings can suffer major functional and economic loss by damage to the equipment and furniture they house, even though the structures experience little damage. Especially in research laboratories, hospitals and offices, unanchored equipment is highly vulnerable to earthquake damage. The same also applies to exhibited pieces in museums and in art galleries. These losses would constitute a substantial portion of the physical losses even in a small earthquake in Istanbul.

In addition to these physical vulnerabilities, the social vulnerability of the urban population may also need to be assessed for a comprehensive earthquake risk assessment. The past earthquake disaster experiences indicate that single-parent families, women, handicapped people, children, and the elderly constitute the most vulnerable social groups.

4. Risk Assessment

4.1 GENERAL CONSIDERATIONS, ELEMENTS AT RISK AND SPECIFIC RISK

Risk, in the context of disaster management, can be defined as the losses that can result from the occurrence of a hazard. In urban areas the population, structures, utilities, systems, and socioeconomic activities constitute the "Elements at Risk."

To indicate various elements at risk in Istanbul, Figures 17 to 22 are provided to illustrate land use, population density, transportation, freeways and bridges, the water supply system, and the sewer system.

The term "Specific Risk" will be adopted to indicate the expected degree of loss due to a particular hazard. The concept of risk, in the parlance of UNDRO, means expected levels of loss due to a specific hazard for a given geographical area over a specific time interval, given the average return period of the hazard. For example, to compute the expected number of casualties due to earthquakes in a given region within a given period of time, the following factors need to be combined: (1) The earthquake intensity return period for the given region; (2) Distribution of building types in the given region; (3) The vulnerability function of each building type; and (4) The number of casualties expected for each damage degree for each building type. The risk is assessed as the combination of hazard, vulnerability and the elements at risk. Any change in one of these ingredients affects the risk directly.

The total number of houses damaged by various hazards in Turkey in the last 70 years is approximately 600,000. About 66% of this total is due to earthquakes, 15% to flooding, 10% to landslides, 7% to rock falls and 2% to meteorological events and snow avalanches. According to the official earthquake hazard zoning map of Turkey, about 43% of the total land, 51% of the population, 75% of the industry and 31% of the dams are located in the first two most hazardous zones, with expected earthquake intensities equal to or greater than MSK VIII. Among all disasters, earthquakes claimed the major share for casualties and physical damage. The statistics of the last 70 years indicate that the average annual earthquake-related losses constitute about 0.8% of the total gross national product, whereas all other natural disasters' share is only 0.2%. Noting that since the 1960s, the vulnerabilities due to flooding and landslides have been highly reduced through river regulations and land-use planning, earthquakes will be the most significant hazard affecting the nation in the future.

With few exceptions, the earthquake disasters in Turkey have taken place in rural areas. Thus the database exists for the estimates of earthquake risk in the rural context. Figure 23 provides the annual expected number of heavily damaged and collapsed rural houses and the percentage of the national budget to be spent for rural reconstruction. Assuming a homogeneous Poisson process, there exists about a 63% probability of spending 0.07% every year or 1.4% of the national budget for every 20 years. Figure 24 provides the earthquake risk in terms of the expected annual number of lives lost for all Turkey. One can expect a 63% annual probability of losing about 10 lives every year, about 800 lives in every 5 years, or about 8,000 lives in every 10 years in rural areas with adobe and stone masonry buildings. As it can be assessed, the losses for wooden frame house stock are much lower.

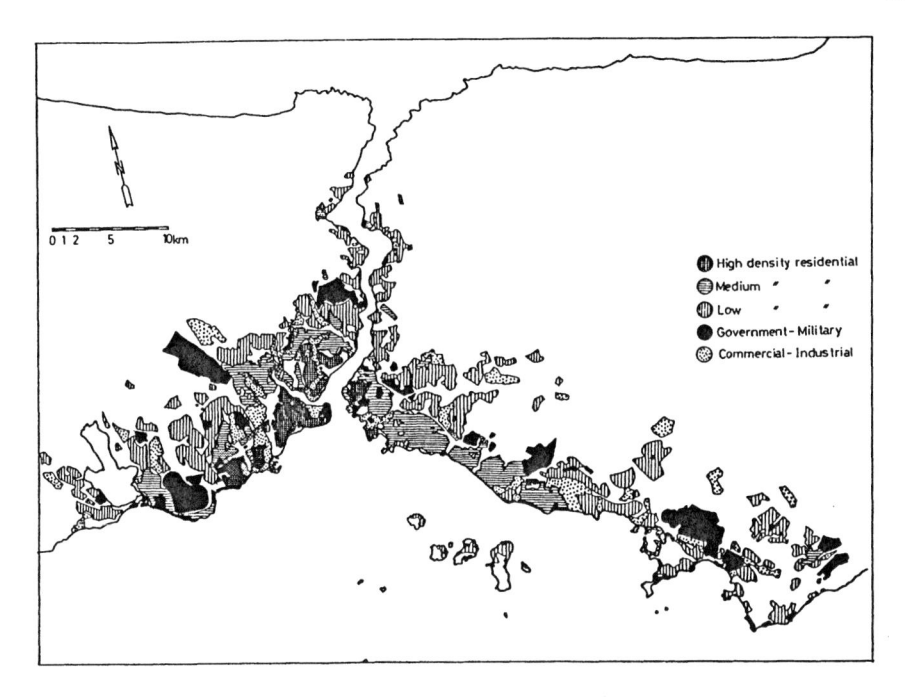

Figure 17. Schematic land use map for İstanbul.

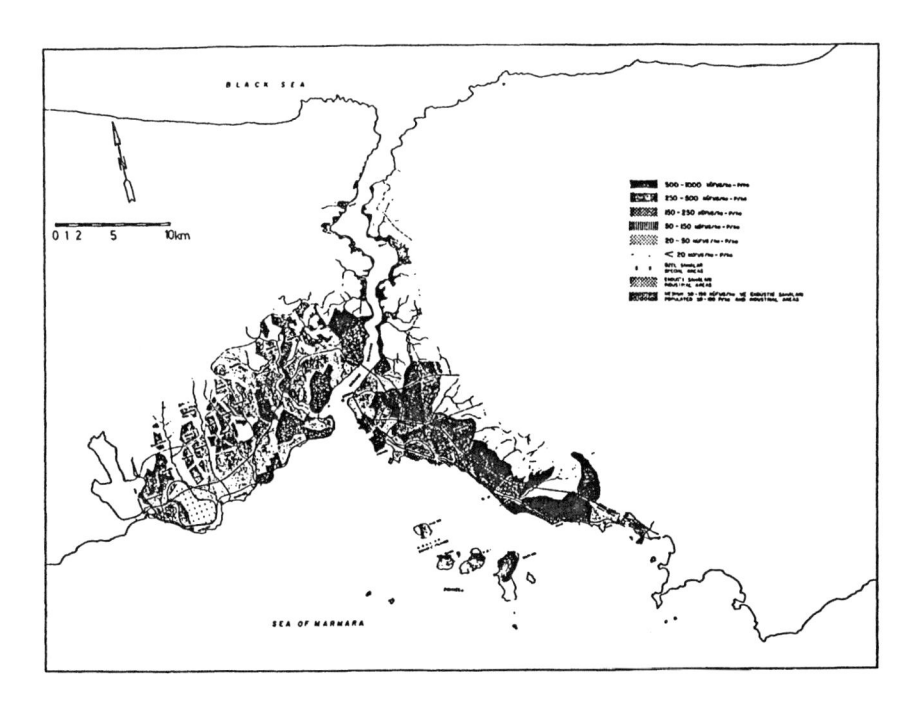

Figure 18. Population density map for İstanbul (projected for year 2000).

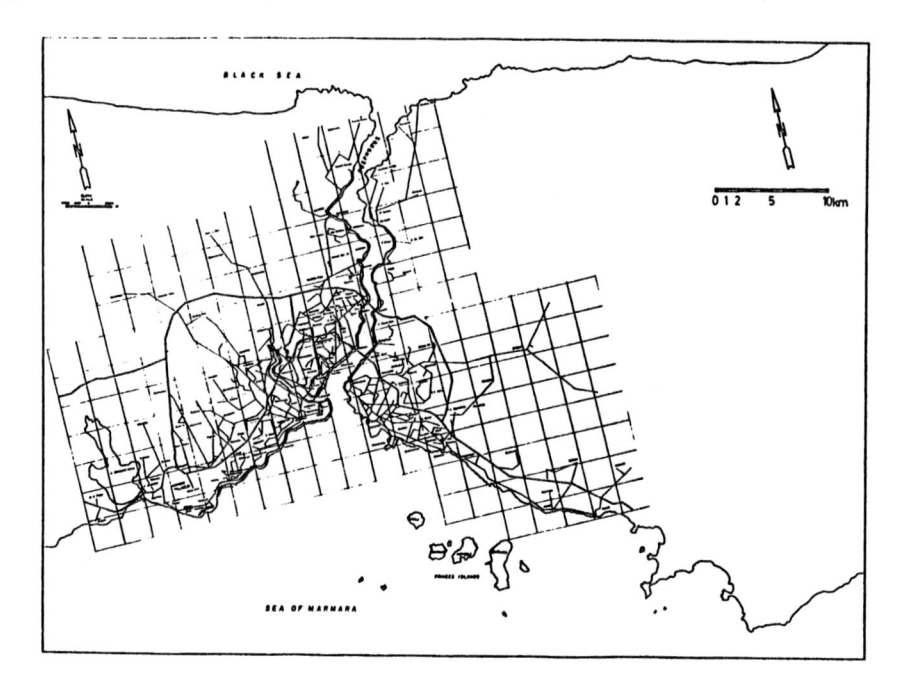

Figure 19. İstanbul municipal transportation network.

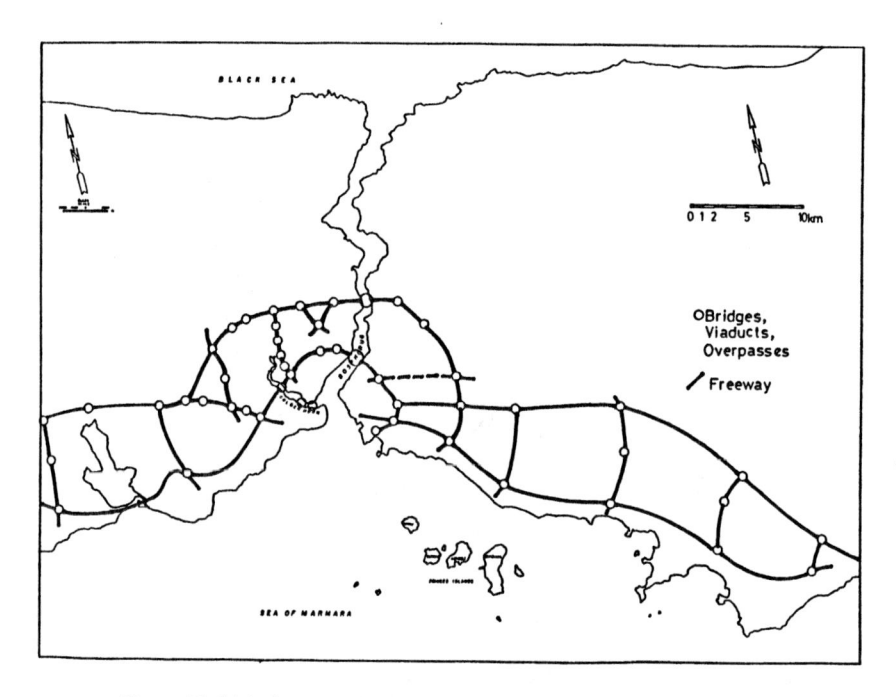

Figure 20. Main freeways, bridges, viaducts and overpasses in İstanbul

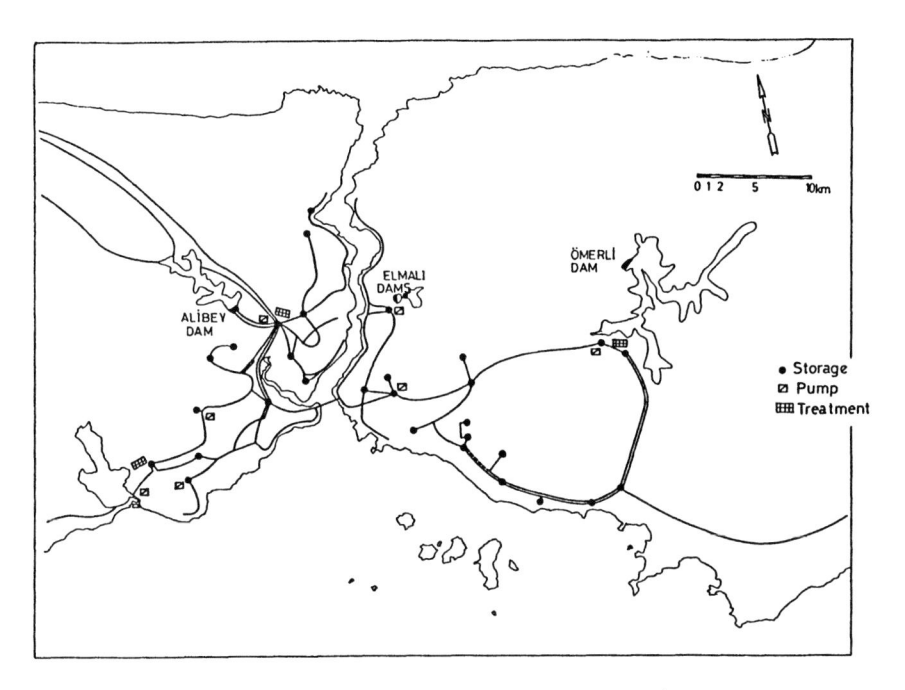

Figure 21. Main elements of the water supply system in İstanbul.

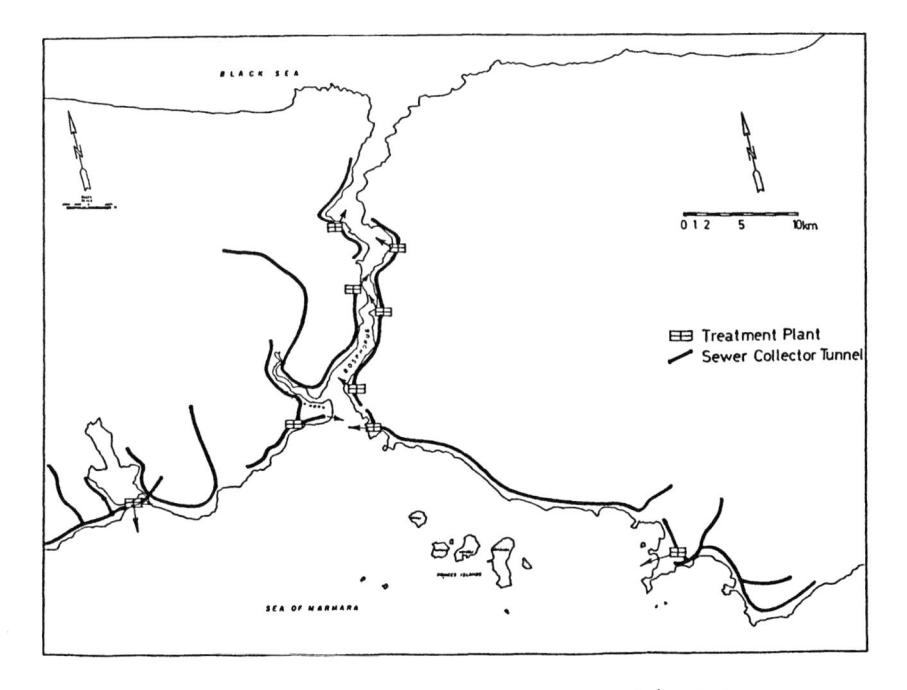

Figure 22. Main elements of the sewage system in İstanbul.

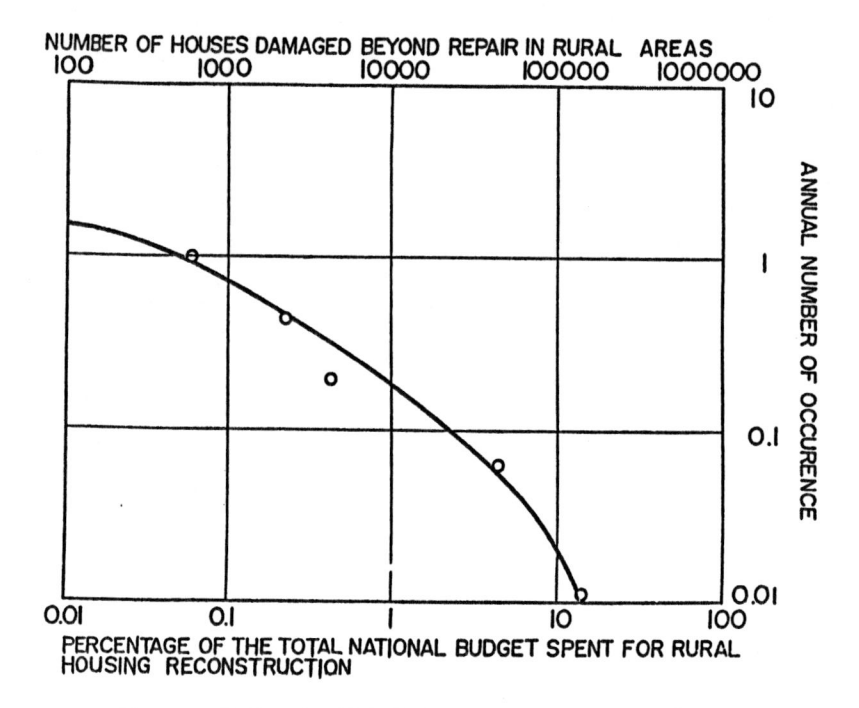

Figure 23. Earthquake risk in Turkey in terms of the annual number of heavily damaged and collapsed rural houses. (After Erdik, 1984).

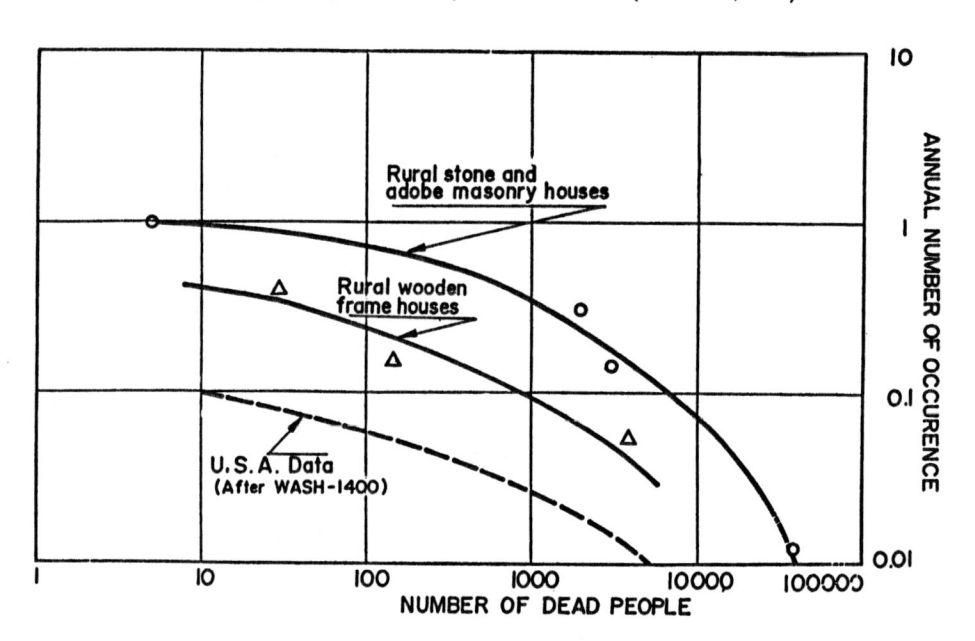

Figure 24 Earthquake risk in Turkey in terms of the annual number of lives lost (after Erdik, 1984).

As an exercise, we can estimate the losses in Istanbul (metropolitan area) using the current population and building distribution and the vulnerability matrices provided.

In Istanbul, assuming a total population of 9,000,000, a total number of reinforced concrete (R/C) buildings of 70,000, and a magnitude 7-7.5 earthquake creating an average intensity of VIII, the estimated direct losses in the metropolitan area to building damage only will be as follows:

- Number of deaths: 21,000
- Number of hospitalized injuries: 63,000
- Number of R/C buildings with substantial to heavy damage: 12,000
- Number of R/C buildings with very heavy damage to partial collapse: 4,000
- Number of R/C buildings with total collapse: 5,000

These physical losses would amount approximately to 30 Billion US$. For proper evaluation of the risk in Istanbul these losses should be compared to those experienced in the 1992 Erzincan earthquake with the same intensity of VIII in the municipal area.

- Number of deaths (in city): 353
- Number of hospitalized injuries: 850
- Number of R/C Buildings with heavy damage to collapse: 315
- Expenditure for repair and reconstruction of buildings: 0.35 Bn US$

4.2 SOCIETAL IMPACT AND SOCIOECONOMIC ASPECTS: 1992 ERZINCAN EARTHQUAKE

The Turkish earthquake experience and data regarding urban events are very limited. In urban areas risks associated with community services, infrastructure, housing, and economic activities require special consideration. The acquisition of the database for vulnerability and risk for these elements will need careful observation. In this regard the 1992 Erzincan earthquake provided valuable data.

The 1992 Erzincan earthquake provided an important test of the ability of urban communities in Turkey to withstand the effects of earthquakes. Prior to the 1992 Erzincan earthquake, almost all of the Turkish experience with earthquake disasters was limited to rural areas. This earthquake has shown that human response to disasters in large cities is affected by factors of size and complexity. In urban areas the economic impacts and the physical damage can be large and protracted. The following include a brief treatment of major socioeconomic problems that surfaced after the Erzincan earthquake (Report of H.E. Mr. Recep Yazicioglu, Governor of Erzincan, May 26, 1992; Erdik at al., 1992) and an assessment their implications for similar situations.

The 1992 Erzincan earthquake occurred in a setting favorable to intensive media coverage, since the devastating 1939 earthquake that has become part of the folk memory took place in the same location. Although the media initially grossly overestimated the casualties and the damage, the earthquake was reported quickly and completely over a long period of time.

The Erzincan earthquake underscored the importance of preparedness for such a disaster. Emergency plans and scenarios were ready at the government level, but they were not rehearsed or practiced, and this resulted in ineffective response during the first days after the earthquake. The Provincial Rescue and Relief Committee foreseen in the Disaster Response Plan of the city became effective only after the second day of the earthquake. The rescue and relief efforts in the first days of the earthquake were handled in an ad hoc manner by the military personnel.

Starting immediately after the earthquake, the military personnel, police and the gendarmarie provided security against disturbances and looting. This service was the most successful.

During the first 24 hours after the earthquake, most of the survivors from collapsed buildings were rescued by the military personnel stationed in Erzincan and by the unorganized volunteer action of local people and people from neighboring communities. The lack of flashlights, oxygen torches, wire cutters and other necessary equipment hindered these activities. With the arrival of the national and international rescue teams, the rescue efforts became more professional. The delivery of the emergency assistance to the victims was strongly influenced by the experience and composition of the rescue teams. The Turkish rescue team was a special unit of the Civil Defense Organization of the Ministry of Interior. There were great differences between the international rescue teams. Some of them were governmental, some were non-governmental and some were completely voluntary. The need for coordination between the national and other international teams was apparent for their full utilization. It is also necessary to have well-trained local civil defense teams which can respond quickly to such emergencies. This will definitely improve the morale of the affected population.

The first problem faced after the earthquake was the inadequacy of tents for temporary settlements. A total of 28,000 families in the city and 50,000 families in the adjoining localities demanded tents. Most of the demands did not originate from actual need but from the fear of staying in the houses regardless of their safety condition. The unfounded information disseminated in the first days of the earthquake, especially in regard to the quantity and the distribution of tents, hampered the relief activities. Only about 20,000 tents were brought to the city in one month to face this huge demand. This created anarchy and chaos in tent distribution centers which could only be prevented by the use of military force. Under the circumstances, the Red Crescent Society refused to serve as the distributor of the tents, and the distribution was carried out directly by the office of the provincial governor. The inadequacy of the tent stock of the Red Cross seemed to be the key reason for this problem. If it were not for the extraordinary security, the number of casualties during the tent distribution could have surpassed those that arose from the earthquake itself. Thus in urban areas, after a disaster, the friction, potential competition and the likelihood of disturbances among the urban poor will probably increase, and the delivery of emergency assistance may be hindered.

Damage assessment and reporting were conducted by the General Directorate of Disaster Operations with 18 teams. The damage assessment forms used in the surveys actually refer only to one- or two-story rural dwellings and describe the damage in simple categories. The terms of heavy, medium and light damage are well described for such rural structures, but become rather ambiguous and a source of disparity for multi-story reinforced concrete buildings in the urban areas.

After the earthquake, the power, water, sewage, and telephone systems of the city became inoperative. Telephone and power services were restored after three days, but the water and sewage utilities became operative only after four weeks and with heavy external assistance to the municipality.

The removal of debris was efficiently carried out by about 500 pieces of heavy construction machinery belonging to various government organizations. Most of the work was carried out in coordination with rescue efforts. In two months about 80% of the debris (about 100,000 m^3) had been removed.

A total of ten health teams visited all of the districts and communities, rendered first-aid and dispensed medicines. A prefabricated building was quickly set up to resume the activities of the damaged hospitals. The biggest drawback in the medical services has been the request of the government health personnel for reappointments to cities outside Erzincan.

Due to the closure of nine heavily damaged schools, about 22,000 students were left without any schooling. The remaining schools were repaired in 15 days and were put into service with increased class hours to accommodate these extra students. However, due to the earthquake, 1076 teachers out of a total of 1200 filed requests to the Ministry of Education for relocation to other provinces. The acceptance of this request would have created severe hardships for the educational system. Nevertheless, this indicated the unwillingess of the civil servants to serve in cities exposed to high levels of earthquake hazard.

Apart from the casualties and destruction, the earthquake caused significant economic damage. In the long-run the general economic life is expected to resume its normal character. However, the short-term losses encountered by the small businesses can make individual recoveries a difficult task. The government has provided housing or financial assistance to qualified families in order to reconstruct or repair the damaged houses. The government's tendency has been to rebuild and repair as fast as possible to in order to restore socioeconomic normalcy. Public expenditures to this end are estimated to be about 0.5 billion US dollars. In 1992 and 1993 respectively, 4,750 and 3,000 new houses were built and distributed to the eligible families. About 30 public buildings and 6,500 housing units (2,200 in the city, mostly 4-5 story apartment blocks) that received medium damage from the earthquake were repaired and retrofitted to increase their earthquake performance. The retrofitting was conducted by several private construction companies, under the supervision of university professors and at a cost of about US$ 4,000 per household, paid by the government. This repair and strengthening operation constitutes one of the largest in the world. The tenants of about 8,000 houses that were lightly damaged in the earthquake have received a government assistance of about US$ 700 to be used for repairs. About 1,000 small businesses were eligible for US $ 6,500 financial aid to help in setting up their businesses.

On August 28, 1992 a law (No: 3838) for rehabilitation of Erzincan was enacted. The law aims to return the provision of services to normal in the earthquake-affected areas and to prevent migration from Erzincan to other cities. The law enlarges the coverage of those eligible for post-disaster housing and regulates the financial assistance to those affected, including the government personnel working in the earthquake areas. The law also stipulates the payment of about US $6,000 per death to the immediate relatives of those who lost their lives in the earthquake. The injured people will receive between US $1,250 and 3,750, depending on the severity of their injury.

5. Earthquake Risk Mitigation

5.1 GENERAL OPTIONS FOR RISK MITIGATION

The risk of urban disasters can be reduced on the basis of options which include: hazard modification, reduction of the vulnerability of structures, and improvement of urban settlements. Hazard modification can reduce the proneness of the urban centers through protective measures, such as construction of river-regulating structures and/or through site improvements, such as improvement of drainage and slopes to reduce the flood hazard. These are mostly hazard-specific options, and can be regarded as capital investments and costly compared to other options. The reduction of structural vulnerability can entail retrofitting of existing buildings, other structures, and infrastructures for improved hazard performance. The effectiveness of public expenditures to be utilized for such activities should be compared with the cost of repair after the disaster. Urban settlements can be improved by changing the functional characteristics of the settlements through land-use planning and increasing the redundancy of the infrastructure, such as building an additional bridge at a strategic crossing.

The general mitigation options can also be analyzed under two main headings: Non-structural and structural. Non-structural options might encompass the following: introduction or improvement of the disaster legislation and institutions; introduction of incentives such as lower insurance premiums for buildings with better hazard resistance; and awareness-building, education and training. The structural measures include the modification of the hazard and the reduction of the physical vulnerability.

5.2 GENERAL RISK MITIGATION MEASURES

Reduction of the structural vulnerability, siting and land-use regulations, design and construction regulations, relocation of communities, and public education/awareness programs are viable measures for the mitigation of earthquake risk. Specific measures should also be considered for secondary hazards such as fire, landslides and flooding. The reduction of structural vulnerability entails pre-earthquake retrofitting and strengthening of existing vulnerable structures, regulating the design of new structures, and post-earthquake repair and strengthening of damaged structures. The term structure includes buildings, infrastructure and lifelines. Changing the functional characteristics of settlements through government actions and legislation such as land-use and site-dependent building regulations would be effective for the mitigation of losses. An increase in the redundancy of lifelines (water, sewage, power, telephone, and road systems) would greatly decrease their vulnerability during earthquakes and avoid the total interruption of these vital services.

5.3 COST-EFFECTIVENESS OF THE MITIGATION MEASURES

The cost of urban earthquake disasters can be high in terms of casualties, physical damage to property and infrastructure, and loss of economic activities. Urban disasters undermine the incentives for development and cause particular damage to the nonformal sector where the recovery becomes very difficult. Disasters seriously hamper the development, requiring the transfer of resources from development to relief and rehabilitation.

In every country, disasters are influenced by a variety of country- and disaster-specific factors, such as the level of threat, geographical considerations and the availability of resources. The availability of resources seems to be the most crucial factor for developing countries. Since developing countries mostly concentrate on programs for development, disaster issues and associated problems are given a rather modest priority level. Also disaster relief operations are getting more and more expensive. This fact imposes a large burden on the economies of disaster-prone countries. For example, Turkey allocates annually, on average, about 1.5% of its national budget just for rural housing reconstruction after an earthquake disaster. Today, in addition to this cost, there exists a proliferation of relief and development agencies with a substantial amount of money being devoted to research and applications in several fields related to the disaster preparedness and response both on national and international scales.

To be cost-effective with disaster mitigation measures, the government has to evaluate the effectiveness of the measures in relation to the cost of other alternatives, since the benefits to be realized can be difficult to determine. In the least-cost option, the costs of each mitigation measure to achieve a given set of acceptable risk levels are compared, and the one presenting the lowest cost is selected. In the fixed-cost option, on the other hand, the option providing the lowest risk levels among the given fixed-cost measures is selected.

It should be realized that certain simple measures that have essentially zero cost compared to the benefits deserve to be implemented immediately. For example, the tying-down and the securing of furniture, appliances and electronics (e.g. personal computers) in homes and in offices, and the secure attachment of displayed items to their bases in art galleries and museums can very cheaply and effectively avoid costly damages. For urban centers, promising cost-effective techniques worthy of investigation can include (Anderson, 1992): reversal of urban growth, creation of satellite cities, family planning, decentralization of management, and use of new energy-efficient technologies. It should be mentioned that the return on the mitigation efforts will only be realizable if these efforts are coupled with proper training and education at all levels of society. Given the scarcity of resources in developing countries, comprehensive training for disaster preparedness becomes essential to reduce the human and property losses due to disasters and the waste, delay and confused development in the recovery phase.

5.4 LEGISLATION AND INSTITUTIONAL FRAMEWORK

Disaster legislation constitutes an essential element of mitigation activities. In Turkey the disastrous earthquake sequence of 1939-1944 prompted the enactment of the "Law Concerning the Pre- and Post-Earthquake Measures" in 1944. This law stipulated, for the assessment of hazardous zones, determination of appropriate building types and construction techniques, preparation of emergency aid and rescue programs, and planning of temporary housing. In 1959 the "Law for Natural Disasters" was enacted (Act No. 7269-1051: On Measures and Assistance to be Put Into Effect Regarding Natural Disasters Affecting the Life of the General Public, Date of enacting: May 25, 1959; Amended July 17, 1968). This act specified the mode of government assistance for relief and rehabilitation, including post-disaster settlements, established a disaster fund to finance this assistance, and stipulated the formation of the "Central Natural Disaster Coordination Committee" to oversee the relevant activities as indicated in Figure 25.

At the central government level, the General Directorate of Disaster Operations of the Ministry of Public Works and Settlement has the responsibility for the mitigation of all disasters. Most of the activities of this General Directorate have been in the relief and post-disaster settlement activities. The European Natural Disasters Training Center under the same Ministry conducts training and education on disaster mitigation.

At the provincial level, the governor bears the responsibility and has all the authority for the mitigation of disasters. The governorate prepares their own disaster management plans for different disaster scenarios and submits them to the Ministry of Public Works and Settlement. In reality, these plans are inadequately prepared, are based on deficient scenarios and not updated to reflect the changing conditions and the growth of cities.

At the municipality level, there does not exist any specific agency responsible for the mitigation of urban earthquake risk and disasters. The fire departments and other emergency rescue and

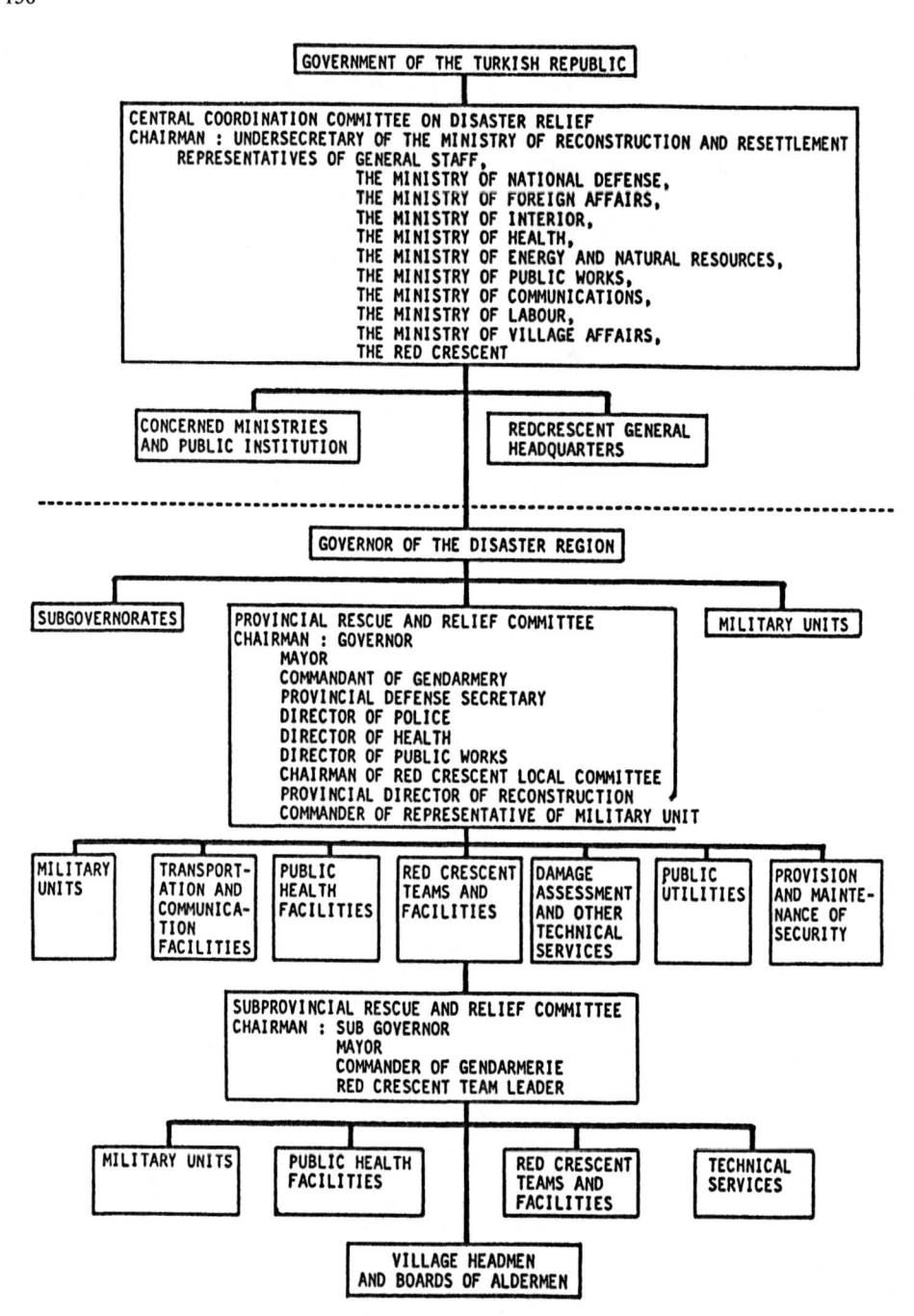

Figure 25. Institutional structure of the "Central Natural Disaster Coordination Committee".

medical teams deal with daily emergencies but are not totally equipped to manage large-scale disasters. In case of emergencies due to fire, heavy rains or snow, that require coordination among the various departments of the Municipality of Istanbul, a so-called "Emergency Crisis Center," composed of various department heads is formed to undertake relevant activities.

6. Mitigation of Urban Earthquake Risk

6.1 GENERAL ISSUES

In technical terms, earthquake risk is the probability of expected earthquake losses (such as lives, injuries, physical and socioeconomic damage). Indirect damage due to the disruption of industry, commerce and services should also be considered losses. The earthquake hazard and vulnerability constitute the key elements of the earthquake risk. Earthquake hazard can only be reduced by either not building in or moving away from the hazardous areas, which in either case is impractical and unrealistic. What remains is then the reduction of vulnerability, in terms of casualties, material losses and socioeconomic losses.

Disasters, regardless of their sources, can be viewed as a break from the normalcy of the socioeconomic system. Thus the people, institutions and the other elements of life which, by nature, function effectively during normal times, cannot necessarily function effectively during the abnormalities created by the disaster. Restoring normalcy necessitates appropriate disaster preparedness before it happens, thereby enabling a community to reduce the recurrence of problems during the abnormal, disaster-induced situation and eventually to serve disaster victims more efficiently—that is, to restore his or her life to what is referred to as normal. In this context, earthquake disaster management becomes the process of anticipating and planning for damage that a major earthquake could create. It is an unbroken chain of concerted actions involving disaster, response, relief, rehabilitation, reconstruction, risk reduction, mitigation, preparedness, and—if possible—warning. Earthquake disaster preparedness and mitigation constitute two of the most important activities of earthquake disaster management.

In general, earthquake preparedness measures encompass: (1) The development of emergency response and recovery plans; (2) The utilization of earthquake hazard and risk assessments; and (3) The dissemination of vulnerability and risk information.

Earthquake mitigation measures encompass: (1) Development of physical and societal impact and loss scenarios; (2) Adoption of seismic zonation and land-use planning; and (3) Reducing the impact of the hazard through earthquake design and construction codes.

For mitigation of urban earthquake disasters, the necessary plans, programs and activities can be listed as follows under the pre-, co- and post-earthquake phases:

6.2 PREPAREDNESS (PRE-EARTHQUAKE) PLANNING AND ACTIVITIES

The pre-earthquake measures that should be implemented in urban centers prone to earthquake risk include:

- Installation of earthquake data acquisition and monitoring stations and services.
- Assessment of earthquake hazard (seismo-tectonic mapping).
- Development of earthquake-resistant design codes and construction standards.

- Pre-disaster planning and management activities and techniques.
- Disaster awareness, public information, education and training.
- Development of methods for retrofitting hazardous buildings and facilities.
- Development of techniques and strengthening of non-engineered, low-strength constructions
- Creation and strengthening of programs and organizations for the prevention of disasters.
- Hazardous material management.
- Legislative and regulatory measures.
- Response readiness.
- Logistical support.
- Resource management and stockpiling.
- Mobile command and communication operations.

6.3 EMERGENCY (MID-DISASTER) PLANNING AND ACTIVITIES

The emergency activities that should be implemented in urban centers right after a disastrous earthquake include:

- Emergency rescue, evacuation, transportation and communication.
- Damage assessment, condemnation, demolition, demarcation of hazardous buildings and zones.
- Debris removal.
- Recovery and disposal of dead bodies.
- Emergency provision of health care, shelter, water, food, and utilities.
- Human response and information management.
- Law enforcement.
- Quick assessment of damage and socioeconomic losses.
- Planning and coordination of disaster assistance.

6.4 POST-EARTHQUAKE PLANNING AND ACTIVITIES

The post-earthquake steps that should be taken in urban centers after earthquake disasters include:

- Detailed surveys regarding repair, restoration and condemnation decisions.
- Assessment of socioeconomic conditions, resources and needs.
- Measures and policies for relief, resettlement, rehabilitation and redevelopment.
- Re-establishment of government services.
- Institutional framework, implementing agencies.
- Hazard abatement.
- Disaster accounting.
- Planning and coordination of rehabilitation and reconstruction assistance.
- Siting of new settlements and communities.
- Retrofit of design codes and construction standards.
- Training and education programs.
- Reconstruction.

6.5 IMPLEMENTATION OF IMPORTANT MITIGATION ACTIVITIES

Japanese cities can be singled out as models for earthquake risk mitigation. The key elements of the policy directions for creation of earthquake disaster-proof living zones are as follows (Nakazawa, 1986):

- In order to prevent spreading of fire after an earthquake, divide urban areas into disaster-proof living zones using roads and rivers as firebreaks.

- Promote disaster prevention measures and supplies in order to facilitate the disaster fighting. Foster awareness for disaster preparedness.

- Improve and construct refuge bases, bridges and roads to protect lives from earthquake disasters. (The refuge bases will have an area of at least 10 Ha with 1 m^2/refugee).

- Improve and establish regulations for earthquake- and fire-proof buildings; improve evacuation facilities and emergency preparedness.

- For gas, electric and water supply facilities, make the conduits earthquake-proof and install safety devices.

- Eliminate the danger of falling objects in homes and from buildings.

Major losses of life in past earthquakes in urban areas were caused by the collapse of buildings with insufficient earthquake resistance or with inappropriate siting considerations. In many developing countries with increasing populations and inadequate housing, the increase in the number of such buildings is bound to aggravate earthquake-related casualties and economic losses in the future. Several developing countries spend about 2% of their Gross National Product (GNP) for post-earthquake reconstruction (Erdik, 1984) and, in some instances, the losses caused by earthquake disasters have completely cancelled out any growth in the GNP (Einhaus, 1988). In this respect, the problems associated with urban planning, infrastructures, lifeline systems, and secondary hazards merit additional considerations. The facilities provided by the metropolitan governments that are essential for the operation of the socioeconomic system (sanitary services, utilities, health services etc.) should be designed with the lowest vulnerability levels. For example, the collapse of the central fire station in the 1972 Managua, Nicaragua earthquake endangered the firefighting, and much-needed ambulances were damaged under a collapsed canopy at Olive View Hospital in the 1971 San Fernando, California earthquake. Transportation facilities (highways, railroads, port and harbors, airports, bridges, etc.) are vital for rescue and recovery efforts. Redundancy in transportation networks is essential for rapid restoration. Metropolitan governments should also be responsible for taking necessary measures to protect the cultural heritage through the maintenance and retrofitting of monuments and museums. Most of these issues can be addressed through proper planning, microzonation and appropriate construction technologies.

In varying degrees, the 1985 Mexico City, 1989 Armenia, 1989 Loma Prieta, and 1992 Erzincan earthquakes provide evidence of the effectiveness of public disaster-preparedness education programs. The 1985 Mexico City earthquake demonstrated the consequences of inadequate preparedness planning (Krimgold and Gelman, 1989). While efforts had been made to develop disaster response planning at the government level, at the time of the earthquake there was no effective response plan in place. Inadequate reconnaissance and damage assessment led to serious delay of full mobilization and application of national and international response capabilities. During the several days in which the ad hoc management of the earthquake response was assembled, valuable time was lost in the search and rescue operation. In regard to the 1989 Armenian earthquake, Wylie and Filson (1989) report that there was no evidence of any pre-disaster public education about the earthquake hazard or about what to do before, during and after the earthquake. Few members of the public, if any, knew of appropriate actions to take during and after the earthquake. There was also no evidence that any pre-disaster earthquake response plans for organizations and communities existed, and in fact, officials reported than none existed. The lack of earthquake disaster response plans likely constrained both the efficiency and effectiveness of initial disaster response and certainly negatively affected the coordination of initial response to the earthquake. As stated by Lee (1990), the 1989 Loma Prieta earthquake and its aftermath underscored the importance of regional planning and preparedness. There have been delays in communication, priority allocation, planning, service restoration, and in the execution of mutual aid agreements and emergency protocols. The study suggests research activities in the following areas: Analysis of local and regional preparedness plans and their implementation following the earthquake; identification and analysis of non-cooperation between agencies; role of planning and redevelopment of agencies on local and regional preparedness plans and reevaluation of the agencies' performance in the aftermath; and an investigation of recovery plans, including those that existed prior to the earthquake and those prepared after the event. Problems similar to those experienced in these earthquakes were also observed in the 1992 Erzincan earthquake.

The pre-earthquake strengthening and retrofitting of critical urban infrastructure, facilities, and vulnerable and hazardous buildings is one of the important physical measures for earthquake risk mitigation. In this respect, the action taken by the city of Los Angeles should serve as a model. Noting that "the pre-1934 unreinforced masonry buildings represent the greatest single threat to the life and limb in Los Angeles in the event of a major earthquake," in 1981 the Los Angeles City Council passed an ordinance (No:154,807) requiring the strengthening or removal of pre-1934 buildings that have bearing walls of unreinforced masonry.

For any earthquake disaster management program, the building of public awareness, information dissemination and the training of personnel constitute the fundamental ingredient of success (Erdik, 1987). For public education programs in earthquake disaster mitigation, the target audience consists of specific sections of the community exposed to earthquakes. The training encompasses public awareness programs to develop and maintain a desired level of awareness and earthquake disaster preparedness. There are several areas which could be used as objectives for program planning in public education (Nigg, 1983): to heighten awareness of the earthquake threat; to inform about possible pre-earthquake preparedness measures; to inform about adaptive behaviors during an earthquake; to inform about adaptive post-earthquake behaviors; and to encourage the implementation of personal or organizational preparedness plans and actions. From the psychological point of view (Hartsough, 1983), any attempts to orient the general public toward appropriate actions before an earthquake must overcome the palliative defenses such as denial and avoidance (i.e.: the earthquake will not happen, and even if it happens, the individual actions are ineffective).

7. Concluding Remarks

7.1 GENERAL REQUIREMENTS FOR EFFECTIVE ADMINISTRATION OF RISK MITIGATION STRATEGIES

Tools for the effective administration of risk mitigation strategies include: (1) legislation, (2) economic aid (i.e. subsidies, infrastructure investments), and (3) training. Organizational and institutional strength are essential for the application of these tools. In light of the particular problems and inadequacies associated with the application of disaster mitigation programs in Turkey, the following general requirements can be stipulated for the effective application of risk mitigation strategies in an urban context:

1. A clear and prioritized urban disaster mitigation policy, well-coordinated and in concert with the national disaster mitigation policies.
2. Effective and continuous assessment of hazards and vulnerabilities for proper identification of the needed measures.
3. An adequate institutional setup, well-coordinated with other municipal and government agencies. The institutional setup should not change with changing local governments.
4. Recognition that disaster mitigation is an integral part of the national development and the responsibility of every sector, from the top levels of the government down to families.
5. Well-organized and continuous public awareness, training, and education programs for a variety of target audiences and conveyed through all possible media.

7.2 APPLICATION OF THE EARTHQUAKE DISASTER MASTER PLAN DEVELOPMENT PROCESS TO OTHER CITIES IN TURKEY AND ABROAD

The earthquake disaster master plan development approach elaborated in this report should be applicable to other urban centers in its generic format. To be effective, however, the other urban centers should not adopt it as a reaction to a given disaster but rather should ensure its continuous implementation. The disaster master plan for an urban center in Turkey or in any other country should be developed in conformity with the existing national level of disaster legislation and should complement it. The focus of this paper has been primarily on earthquakes due to its circumstantial importance both for the city of Istanbul and the author. The master plan process described does not involve specialized economic, social and cultural issues. These issues can be very important considerations for certain urban centers.

7.3 PRIORITIES FOR REDUCING THE EARTHQUAKE VULNERABILITY IN ISTANBUL

The following list of priorities for investments aimed at reducing urban vulnerability in Turkey, and particularly in Istanbul, can be suggested:

1. Development of programs for public awareness-building, information dissemination and the training of personnel for measures that aim at the reduction of urban vulnerability.
2. Development of an institutional base and a mechanism for the control and regulation of the building design and construction sector.
3. Development or modification of land-use planning regulations to reflect, as a priority, the

measures that contribute to the reduction of urban vulnerability and the development of effective enforcement mechanisms.
4. Repair, strengthening, and increasing of the redundancy of urban facilities that are essential for the operation of the socioeconomic system (sanitary services, utilities and health services etc.).
5. Institution of simple measures to protect household and office appliances and furniture from the effects of earthquakes and to eliminate danger from falling objects. Similar measures also apply to exhibited items at museums.
6. Protection of the cultural heritage through maintenance, repair and retrofitting of monuments and museums.
7. Development of methods for retrofitting hazardous buildings and facilities.
8. Development of appropriate techniques for repair and strengthening of non-engineered low-strength constructions in squatter settlements.

References

Akkas, N. and M. Erdik (1984), Consideration on Assessment of Earthquake Resistance of Existing Buildings, Int. Jour. for Housing Science and Applications, v.8, pp.49-66.

Algermissen,S.T., K.V.Stinbrugge, and H.J.Lagorio (1978), "Estimation of Earthquake Losses to Buildings", USGS, Open File Report No: 78-441.

Ambraseys,N.N., A.Zatopek, M.Tasdemiroglu and A.Aytun (1968), The Mudurnu Valley (West Anatolia) Earthquake, Mission Report No.622, UNESCO, Paris.

Ambraseys, N.N. and J.A.Jackson (1981), Earthquake Hazard and Vulnerability in the Northeastern Mediterranean: the Corinth Earthquake Sequence of February-March, 1981, Disasters, Vol. 5,
No.4, pp 355,368.

Ambraseys, N.N. and C.F.Finkel (1991), Long-term Seismicity of Istanbul and of the Marmara Sea Region, Terra Nova, 3, 527-539.

Anderson, M.B. (1992), Metropolitan Areas and Disaster Vulnerability: A Consideration for Developing Countries, in Environmental Management and Vulnerability, World Bank, Washington, D.C.

Barka, A.A. and M. N. Toksöz (1989), Seismic Gaps along the North Anatolian Fault. Abstract, IASPEI General Assembly, Istanbul, S9-1.

Bayindirlik ve Iskan Bakanligi, Afet Isleri Genel Müdürlügü, Afetlere Iliskin Acil Yardim Teskilati ve Planlama Esaslarina Dair Yönetmelik, Ankara, 1991.

Bayülke, N. (1982), Building Types in Bolu Turkey and Their Predicted Earthquake Damages, in Sismic Risk Assessment and Development of Model Code for Seismic Design, UNDP/UNESCO Project

RER/79/014, Sofia, Bulgaria.

Boore, D.M, W.B.Joyner and T.E. Fumal (1993), Estimation of Response Spectra and Peak Accelerations From Western Nort American Earthquakes: An Interim Report, USGS Open File Report 93-509, Menlo Park, CA.

Campbell, K.W. (1981). Near-source Attenuation of the Peak Horizontal Acceleration, Bull. Seism. Soc. Am., 68, 828-843.

Coburn, A. and R. Spence (1992), Earthquake Protection, Wiley, 355pp.

EERI (1986), Reducing Earthquake Hazards: Lessons Learned From Earthquakes, EERI Publication No: 86-02, San Francisco, California.

ESG (1993), European Macroseismic Scale 1992, European Seismological Commission, Luxembourg, 1993.

Einhaus, H. (1988), Emergency Planning and Management for Disaster Mitigation, Regional Development Dialogue, v.9, No.1, pp.1-13, United Nations Center for Regional Development, Japan.

Erdik, M., V. Doyuran, P. Gülkan and G. Akçora (1983), Probabilistic Assessment of the Seismic Intensity in Turkey for the Siting of Nuclear Power Plants, Proc., 2nd CSNI Specialist Meeting on Probabilistic Methods in Seismic Risk Assessment for Nuclear Power Plants, Lawrence Livermore National Laboratory, Livermore, CA, USA.

Erdik, M. (1984), Seismic Hazard in Turkey, in Proc. Conf. on Science and Technology in Socio-Economic Development, Technische Universitat Berlin, Massachusetts Institute of Technology, Berlin, 1984.

Erdik, M., V. Doyuran, N. Akkas, P. Gülkan (1985), A Probabilistic Assessment of the Seismic Hazard in Turkey, Tectonophysics, 117, pp.295-344.

Erdik, M. (1987), Training and Education for Disaster Preparedness, Regional Development Dialogue, v.9, No.1, pp.36-48, United Nations Center for Regional Development, Nagoya, Japan.

Erdik, M., V. Doyuran (1987), Seismic Hazard in the Eastern Mediterranean, in Ground Motion and Engineering Seismology, Ed. by: A.Çakmak, Elsevier, Amsterdam, 1987

Erdik, M. (1987), Site Response Analysis, in Engineering Aspects of Earthquake Phenomena, Ed. by: A.Koridze, Omega Scientific, London.

Erdik, M. (1989), Earthquake Response and the Performance of Rural Masonry Structures in Turkey, in Engineering Aspects of Earthquake Phenomena Vol.3, Ed. by: A.Koridze, Omega Scientific, London.

Erdik, M. (1991), Urban Earthquake Hazards, Risk and Mitigation, Invited Paper, Proc, 5th International Conference on Soil Dynamics and Earthquake Engineering, September 23-26, University of Karlsruhe, Germany.

Erdik, M. (1991), Seismicity, Earthquake Disasters, Hazard, Vulnerability and Earthquake Risk Reduction in South Asia, Invited Report, Proc. ESCAP/UNDRO Regional Symposium on Natural Disaster Reduction Bangkok, 11-15 February 1991

Erdik, M., A. Barka and B. Üçer (1991), Seismic Zonation Studies in Turkey: an Overview, Proc. Fourth International Conference on Seismic Zonation, August 26-29, 1991, San Francisco, CA.

Erdik, M. and A. Barka (1992), A Pilot Microzonation Study for Istanbul, Proc. 10th World Conference For Earthquake Engineering, Madrid.

Erdik, M. and A. Barka (1992), Disaster Management Education for Earthquakes, Proc. 10th World Conference For Earthquake Engineering, Madrid.

Erdik, M., O. Yüzügüllü, C. Yilmaz, and N. Akkas (1992), 13 March 1992 (Ms=6.8) Erzincan Earthquake: A Preliminary Reconaissance Report, Soil Dynamics and Earthquake Engineering, v.11, pp.279-310.

Ergünay, O. and M. Erdik (1984), Disaster Mitigation Program in Turkey, Proc. International Conference on Disaster Mitigation Program Implementation, Ocho Rios, Jamaica, Nov. 12-16, 1984, Center for International Development Planning and Building, College of Architecture and Urban Studies, Virginia Polytechnic Institute and State University.

Everenden, J. F., R. R. Hibbard and J. F. Schneider (1973), Interpretation of Seismic Intensity Data, Bull. Seism. Soc. Am., v.63, pp.399-422.

Evernden, J.F. and J.M. Thomson (1985), Predicting Seismic Intensities, in Evaluating Earthquake Hazards in the Los Angeles Region, pp. 151-202, USGS Professional Paper No:1360, US Government Printing Office, Washington D.C.

Fukushima, Y., T. Tanaka, and S. Kataoka (1988), A New Attenuation Relationship for Peak Ground Acceleration Derived from Strong Motion Accelerograms, Proc. of the 9th World Conf. on Earthquake Engineering, Tokyo, Japan.

Hartsough, D.M. (1983), A Targeted Program for Public Education: A Psychological Perspective, Proc., Workshop on "Continuing Actions to Reduce Losses from Earthquakes in the Mississippi Valley Area," USGS Open File Report No. 83-157, Reston, Virginia.

Jones, G. J. (1992), Population Growth, Urbanization, and Disaster Risk and Vulnerability in Metropolitan Areas: A Conceptual Framework, in Environmental Management and Vulnerability, World Bank, Washington, D.C.

Joyner, W. B., and D. M. Boore (1981). Peak Horizontal Acceleration and Velocity from Strong-Motion Records Including Records from the 1979 Imperial Valley, California, Earthquake, Bull. Seism. Soc. Am., 73, 1479-1480.

Joyner, W. B. and T. E. Fumal (1985), Predictive Mapping of Earthquake Ground Motion, in Evaluating Earthquake Hazards in the Los Angeles Region, pp.203-220, USGS Professional Paper No:1360, US Government Printing Office, Washington.

Kanai, K., and T. Tanaka (1961), On Microtremors- VIII, Bull. Earthq. Res. Inst., v.39, pp.97-114, Tokyo.

Kanai, K., T. Tanaka, K. Osada and T. Suzuki (1966), On Microtremors- X, Bull. Earthq. Res. Inst., v.44, pp.645-696, Tokyo, Japan.

Kandilli (1992), March 13, 1992 (Ms=6.8) Erzincan Earthquake: A Preliminary Reconnaissance Report, Kandilli Observatory and Earthquake Research Institute, May, 1992, Istanbul, Turkey.

Kockelman, W. J. (1985), Using Earth-Science Information for Earthquake Hazard Reduction, in Evaluating Earthquake hazards in the Los Angeles Region, USGS Professional Paper No:1360

Krimgold, F. and O. Gelman (1989), Working Group Conclusions on Lifelines and Disaster Response/Mitigation in Lessons Learned from the 1985 Mexico Earthquake, Ed. by V.V.Bertero, Earthquake Engineering Research Institute, 89-02.

Lee, B. (Technical Editor) (1990), Loma Prieta Earthquake Reconnaisance Report- Supplement to Earthquake Spectra v.6, Earthquake Engineering Research Institute, 90-01.

Medvedev, S. S. (1965), Engineering Seismology, Translated from Russian, Israel Program for Scientific Translations, Jerusalem, 1965.

Nakazawa, M. (1986), Prevention of Urban Disasters in Japan, in Planning for Crisis Relief v.4, United Nations Center for Regional Development, Nagoya, Japan.

Newmark, N. M. (1965), Effects of Earthquakes on Dams and Embankments, Geotechnique, v.15.

Nigg, J. M.(1983), A Targeted Planning Approach for Public Education Programs, Proc., Workshop on "Continuing Actions to Reduce Losses from Earthquakes in the Missisipi Valley Area", USGS Open File Report No. 83-157, Reston, Virginia.

Sengezer, B. S. (1992), 13 Mart 1992 Erzincan Depreminde Meydana Gelen Hasarin Mahalelere Göre Irdelenmesi, Proc. 2nd. National Earthquake Engineering Conference, 10-13 Mart, 1993, Istanbul.

Sugimura, Y., I. Ohkawa and K. Sugita (1982), A Seismic Microzonation Map of Tokyo, Proc. 3rd. Int. Earthquake Microzonation Conference, Seattle, Washington, USA

Tokida, K. (1990), Earthquake Disaster and Approach to Damage Reduction, Proc. ESCAP/UNDRO Regional Symposium on the International Decade for Natural Disaster Reduction, Bangkok.

UNDRO (1990), Mitigating Natural Disasters: Phenomena, Effects and Options, A Manual for Policy Makers and Planners, Office of the United Nations disaster Reklief Co-Ordinator, United Nations,
New York.

U. S. National Academy of Sciences (1987), Proposal for the International Decade for Natural Hazards Reduction, Washington, D.C., May, 1987.

Wylie, L. A. and J. R. Filson (1989), Societal Impact and Emnergency Response, in Armenia Earthquake Reconnaissance Report- Special Supplement to Earthquake Spectra, Earthquake Engineering Research Institute, 89-01.

Yazicioglu, R. (1992), Report of H.E. Mr. Recep Yazicioglu (in Turkish), Governor of Erzincan, 26 May, 1992.

Youd, L. T., J. C. Tinsley, D. M. Perkins, E. J. King and R. F. Preston (1979), Liquefaction Potential Map of San Fernando, California, in Progress on Seismic Zonation in the San Francisco Bay Region,
USGS Circular No. 807.

SEISMIC RISK ON THE TERRITORY OF THE CITY OF YEREVAN, ARMENIA

S. BALASSANIAN
A. MANUKIAN
National Survey of Seismic Protection (NSSP)
Under the Government of the Republic of Armenia

1. Introduction

As determined by both objective and subjective factors, the seismic risk of the territory of Armenia has, at the present time, reached its highest level over the entire historical period.

Practically all of the territory of Armenia is situated in a seismically-active zone. The size of earthquakes ranges up to M=7.1 (according to instrumental recordings) and M=7.5 (according to historical and paleoseismic estimations). Focal depth is, on average, 10 km. All sources are located on the active faults, with an average slip rate of about 1 cm/year. The duration of destructive earthquakes may reach one minute under adverse ground conditions. The average recurrence interval of large earthquakes (M≥5.5) comprises 30-40 years.

A large earthquake is particularly hazardous in the region of Yerevan, the capital of Armenia, because of the following factors:

- High population density—106 people per square km under extremely uneven distribution of the population—more than 40 per cent of the republican population reside in the city of Yerevan;

- Seismically non-resistant housing, structures, communications;

- High concentration of hazardous facilities: nuclear power plant, chemical plants, dams, etc., all having a low level of earthquake resistance;

- Toxic-material and hazardous plants are located within the limits or in the vicinity of densely-populated areas;

- Cities and large communities built on poor grounds as, for example, Giumri (Leninakan), the second largest city in Armenia;

- Sharp deficiency in instrumental observations, prohibiting a reliable assessment of the long-term and current seismic hazard;

- Insufficient preparedness of the Government for undertaking decisive actions immediately before, during and after an earthquake;

- People untrained in the rules of behavior before, during, and after an earthquake;

- Absence of appropriate material and technical resources in the republic for prompt evacuation, relief and recovery, and;

- Hard economical problems caused by the period of transition to another social order.

2. Demography of the city of Yerevan

1.26 million people reside in the 210 km² territory of the city of Yerevan. A high rate of population growth is one of the distinctive features of the city. From 1917 to 1988, the population of Yerevan increased by a factor of 30, with a consequent increase in seismic risk, as determined by the following factors:

- The rapid and cheap construction of houses and city lifelines (including communication, electric power supply, and water supply) in response to population demands, led to an inadequate quality of construction.

- Districts of multi-story buildings appeared whose design level of earthquake resistance corresponded to a VII-VIII intensity on the XII-intensity MSK scale; whereas the level of possible seismic hazard was greater than or equal to IX intensity.

- Within the city limits, the construction of toxic-material and hazardous facilities was begun. Since these plants required a large construction area, a deficiency of space in the center of the city resulted, and the construction of underground structures designed with regard to VIII-intensity seismic hazard was needed. This was considered to solve not only the problem of useful space but also the problem of separating the vehicle and pedestrian traffic over two different surfaces (levels).

The population density is distributed extremely unevenly over the city and has, as a result, created an imbalance in the operation of communication systems, with parts of them not used in full measure and other parts severely overloaded. Moreover, within the city limits, population growth forced areas of many differing altitudes to be inhabited. This severely hampered the development of effective lifeline systems and limited their capacity.

Various migration processes formed the base of Yerevan's population, creating a distinctive way of life for the whole city. According to their places of origin, emigrants could be divided into four categories: rural emigrants, town-dwellers from Armenia, town-dwellers from other republics of the former USSR, and repatriates. In addition, from 1988, the city began receiving refugees (whose numbers totalled 100,000 people) and residents of Northern Armenia who were displaced by the Spitak earthquake.

The catastrophic Spitak earthquake of 1988 occured at a distance of 100km from Yerevan and resulted in the death of 25,000 people. Moreover, 50,000 people were injured severely and 500,000 people were made homeless. This disaster gave rise to a difficult economical situation in the Republic, followed by unemployment; the period of transition from one social order to another was felt keenly. The people experienced deep psychological stress, a sharp drop in living standards, noticable emigration from the Republic, and a total change of the way of living and psychology.

The abovementioned conditions were accompanied by a depletion of Yerevan's material resources and an aggravation of municipal services, condition of lifelines and housing; therefore the seismic risk in Yerevan has reached a very high level.

3. Characteristics of the seismic hazard zone for Yerevan

We have defined the seismic hazard zone of Yerevan based upon the location of active faults and sources of potentially hazardous historical earthquakes.

The part of Yerevan defined by latitudes of 30.5 and 40.5 degrees north and longitudes of 43.5 and 45.3 degrees east has been exposed to destructive earthquakes in the past. Previously, this territory was selected and special seismological investigations were carried out by Karapetian (1986).

3.1 GEOGRAPHIC POSITION

Yerevan's seismic hazard zone is situated in the Ararat valley and outlined by the southern spurs of the Tsakhkunian range: Aragats (4000 m high) and Arailer (2500 m high) mountains to the north; Greater Ararat (5,165 m high) and Lesser Ararat (3,925 m high) mountains to the south; Arghidag range on the southwest; and Ghegam range and Sevan lake on the east.

The average altitude of Yerevan is 1,100 m. The city relief is sharp, with elevation changes reaching 400 m within the city limits.

3.2 GEOLOGICAL AND ENGINEERING GEOLOGICAL CHARACTERISTICS

The Yerevan seismic hazard zone consists of 4-5 km of volcanogenic sedimentary rock underlain by a metamorphosed paleozoic crystalline basement. Volcanogenic sedimentary rock mass is represented therein from bottom to top as follows: chalky porphyrites, sandstones, clays, limestones in alternation; paleogene sand, tuff, clays, andesites; neogene alternation of gypsum and clays, andesite-basalts, andesites; Quaternary tuffs, basalts, and alluvial-diluvial deposits. Metamorphosed rocks of the crystalline basement involve methamorphic schists, quartzites.

The Yerevan city territory, in its different parts, is composed by rocky and large-debris grounds, rocky weathered grounds, coarse-grained sands, clays, sandrocks, and clayey ground.

Ground water level in the region of Yerevan city is on average at 7-8 meters depth.

3.3 SEISMICITY

The chronicle of large earthquakes resulting in human fatalities and the destruction of cities and villages in the territory of Yerevan's seismic hazard zone starts in 550 B.C. The entire historical period from 550 B.C. to 1993 is naturally divided into two parts: historical (550 B.C.-1932) and instrumental (1932-1993). Altogether, 223 earthquakes of different intensity have been revealed and recorded according to macroseismic (till 1932) and instrumental (since 1932) data within the time interval 550 B.C. to 1993. Derived from those data, the statistics of large earthquakes are as follows.

According to the macroseismic data, one large earthquake of M=7.4 and one of M=7.0, five earthquakes of 6<M<7 and ten of 5<M<6 were revealed for the period from 550 B.C. to 1932.

For the instrumental period (1932-1993), one earthquake of M=6.2 and four events of 5<M<6 were recorded. By the term "large earthquake," an earthquake of M=7 is generally meant, but because of Yerevan's seismic risk, we use the term "large earthquake" for those of M=5 or greater, since the occurance of such an event in Yerevan (or any other populated area of Armenia) could cause significant damage, given the dilapidated state of buildings—particularly those in the private sector—and poor-quality construction in the state sector.

When trying to reveal a certain regularity for the occurrence of large seismic events in time within the limits of the considered zone, it is necessary to take into account noticable heterogeneity of the data in the earthquakes' temporal row, for the period from 550 B.C. to 1993.

The overall temporal domain can be divided into three quasi-homogeneous ranges in accordance with the quality of data: I (550 B.C.-1932)—historical, pre-instrumental period; II (1932-1962)—instrumental period when the limited number of stations in the Caucasian regional network was operating; III (1962-1993)—instrumental period when the Caucausian regional seismic observations network operated, including implementation of the Armenian national telemetering seismographic observation network (NSSP network).

Within the I range, seismic activity periods for the events of M=5.5 covering approximately 300 years and passivity periods of the same duration stand out since 736. When this regularity is projected to the future, the next large earthquake of M=5.5 may occur in 2140, i.e. 300 years after the largest Ararat earthquake of 1890 (M=7.4). Within the II and III range periods (instrumental), which may be combined when considering large earthquake statistics, approximately 15-year-long cycles of earthquake (M=5.0) recurrence stand out. When this regularity is projected for the future, the next large earthquake with 5<M<5.5 would be expected by 2007, as far as the last large earthquake of M=5 occurred in 1992 (Martuni earthquake on the territory of Armenia). It should be noted that the 1986 Martuni earthquake was conceptually predicted by Karapetian (1986), who had forecast the 1992 earthquake of M=5.2 in this region on the basis of large earthquake energy emission statistics for the investigated territory. Finally, it should be noted that the recurrence cycles of large earthquakes in the zone of seismic hazard for Yerevan have a very approximate nature. It will be possible to make more precise conclusions after careful study of new data on historical seismicity, re-estimation of old results using new procedures for large earthquake parameters macroseismic assessment, and metrological assessment of the accuracy of all instrumental data are performed. Besides, the recurrence cycles of different intensity earthquakes within small zones, in our opinion, depend on the general regional dynamics of seismic processes. The character of these dynamics is that, along with a certain periodic component, a random component is practically always observed. Reasoning precisely from the above-mentioned factors, the NSSP of Armenia considers current seismic hazard assessment very important. The seismicity of the territory (for the period of 1985-1993) outlined by four main active faults of Armenia is given in Figure 1. From these data it is clear that since 1992, the temporal process of seismic events resembles the period preceding the catastrophic Spitak earthquake.

3.4 ACTIVE FAULTS, SEISMOTECTONICS AND HISTORICAL SEISMICITY

Several large faults of different orientations, each having large earthquake sources, are distinguished by different authors on the territory of seismic hazard for Yerevan. In particular, Karakhanian distinguishes the Garni active fault, the Arax active fault and the Yerevan hypogene

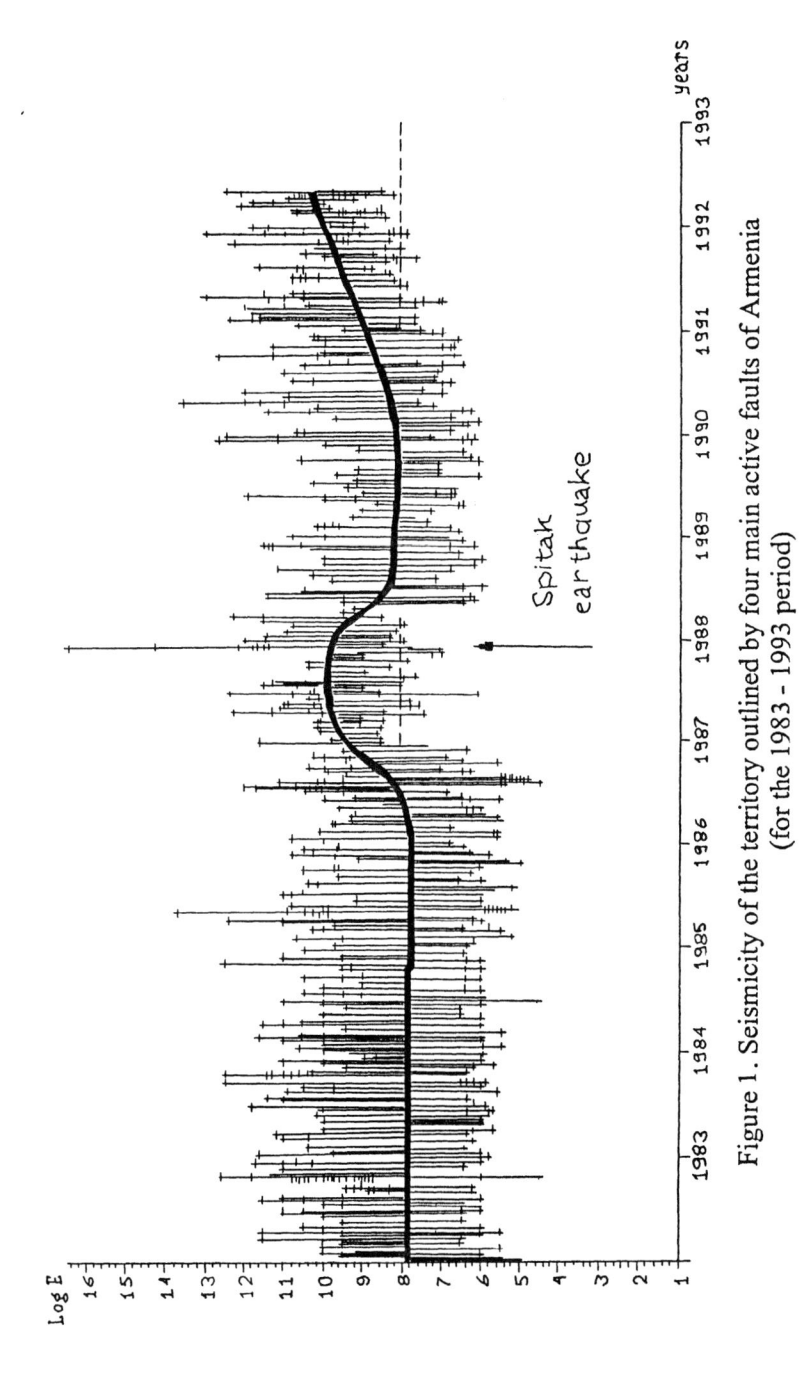

Figure 1. Seismicity of the territory outlined by four main active faults of Armenia (for the 1983 - 1993 period)

fault, which does not manifest noticable current activity but retains the potential hazard of activation [Figure 2].

In addition, Yerevan is exposed to severe hazard from large earthquake sources which destroyed the city structures or caused significant damage to them in the past. Among the farthest from Yerevan is the source of the Ararat earthquake that occurred on 2 June 1840, with M=7.4; the Ararat event was the largest earthquake in the region of Yerevan and also one of the largest in the Armenian Upland. The destructive effect of this earthquake was felt throughout the greater Yerevan area, despite the fact that the epicenter lay 90 km southwards from the city.

In the hazardous vicinity of Yerevan, 20 km southeast from the city, is situated the source of the large Dvin earthquakes. This source manifested especially high activity from 851 to 893. During this rather short interval, five of the large seismic events occured in 851, 858, 863, 869, 893, among them the largest of M=6.0 in Spring, 893. Victims of this earthquake numbered 70,000, and the city was totally ruined.

The Arax active fault, located 20 km southwards from Yerevan, presents significant hazard to the city. This is a dextral strike-slip fault where, according to Karapetian (1986), epicenters of the Igdir group earthquakes (the largest of which took place in 1962 with M5.2) are associated. Large earthquakes along this fault occured in 139 (Ararat mountain, M5.7) and 1841 (Kavdak, M5.7).

The Yerevan hypogene fault revealed by Aslanian (1981) presents an immediate hazard to the city. The Parakar source of large earthquakes associated with it is known for the earthquake sequence known as the Yerevan earthquakes: 1910 (M4.5), 1937 (M4), 1973 (M4), and 1984 (M4). The largest among them was the 1937 earthquake, during which some buildings in Yerevan were damaged due to their dilapidated state, unfavourable ground conditions, and the low quality of construction.

According to the data of Karapetian, if the last three Yerevan earthquakes are related to the third seismically-active period (of the Parakar source) assuming 1973 as its start, then the average recurrence interval for large earthquakes (M4.5-5) epicentered in the Yerevan city region in the 20th century will be 27-36 years.

We consider that the Garni active fault distinguished by Trifonov and Karakhanian is more hazardous than the direct threat of the Yerevan hypogene fault (Trifonov et al., in press). The Garni fault passes 10 km northwest of Yerevan. According to Trifonov et al., this is a right slip with reverse fault component extending in a northwest direction on the territory of Armenia. The length of the fault is 166 km. Amplitude of horizontal displacements for Holocene ranges up to 50-100 m and 200 m for the late Pleistocene. Average horizontal displacement velocity is 3-5 mm/year. This fault is associated with epicenters of large earthquakes.

It has been the activation of the Garni fault which, in Trifonov's and Karakhanian's opinion, primarily led to the catastrophic Spitak earthquake (7 December 1988, M7.1)—the largest seismic event in the territory of Armenia in the entire instrumental observation period.

Among the large earthquakes associated with the Garni fault, that which occurred on 4 June 1679 should be distinguished. It destroyed the city of Yerevan and numerous villages in the Ararat valley. More than 8000 people died in Yerevan, and the whole of population of neighbouring Kanaker village was wiped out (1,228 people). In Garni settlement, 7,600 people died. Aftershocks of the earthquake continued through the end of the year. According to the available historical data, the 1679 Garni earthquake was the most destructive in the history of Yerevan, which was founded in 782 B.C. (Erebouni).

The extreme hazard of the Garni active fault for Yerevan is substantiated by the following: (1) Spatio-temporal migration of large earthquake sources from the southeast towards the

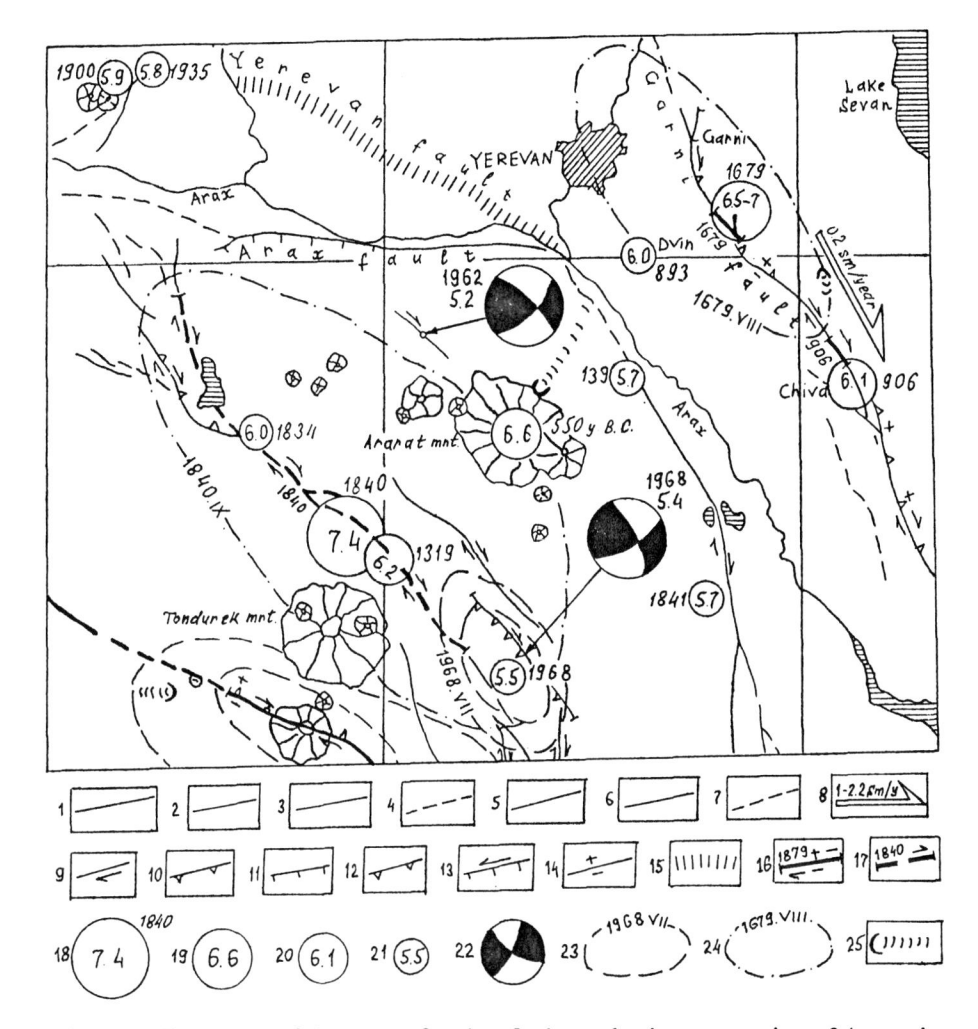

Figure 2. Fragment of the map of active faults and seismotectonics of Armenia
(composed by A. Karakhanian, R. Aroutiunian) - seismic hazard
zone for Yerevan city

1 - the faults active in Holocene (and late Pleistocene); 2 -the faults active in Pleistocene (the last 0.7 million years and only supposed to be active in Hollocene); 3 - proved; 4 - supposed on indirect signs (results of air-space photoes decoding, seismic and fluid gas activity); 5 - transregional (> 0.5 cm/year); 6 - regional (0.5 - 0.1 cm/year); 7 - local (the others); 8 - direction of horizontal displacement along the faults and average intensity; 9 - strike slip faults; 10 - reverse faults and thrusts; 11 - normal faults; 12 - strike slip and reverse faults; 13 - strike slip and normal faults; 14 - vertical displacements; 15 - concealed faults of the basement; 16 - documentally confirmed seismogenic ruptures (seismotectonic dislocations); 17 - seismogenic ruptures supposed on the historical data and fragment field investigation results; 18 - earthquakes (year and magnitude are indicated) of M>7.1; 19 - earthquakes of 6.6<M<7.0; 20 - earthquakes of 6.1<M<6.5; 21 - earthquakes of 5.5<M<6.0; 22 - focal mechanisms of earthquakes; 23 - the highest marked isoseisms on the contemporary data; 24 - the highest marked isoseisms on the historical data and contemporary field studies of historical structures; 25 - Holocene seismogenic landslides.

northwest; 906 (Vaitsdzor, M6.1); 1679 (Garni M6.5-7); 1827 (Tsakhcadzor, M6.7); 1988 (Spitak, M7.1), and (2) by the descending time interval between large seismic events.

In A.Karakhanian's opinion, who had noticed this regularity, it is quite possible that the large Garni earthquake sequence will repeat again, starting from the Vaiotsdzor source. Activation of the Vaiotsdzor source started in 1992 with a swarm of earthquakes (M2-3), anomalous changes in groundwater level, intensive changes of soil gas radon concentration, and other relevant signs.

4. Earthquake resistance of buildings and structures in Yerevan

It is common knowledge that seismic risk depends on the earthquake resistance of buildings and structures within the populated area, more than any other factor. This earthquake resistance of buildings is determined by the building design, quality of building materials, construction quality, and building maintenance.

The most general design construction types in Yerevan are the following: (1) Low-rise stone masonry constructions (private buildings); (2) Stone and complex constructions; (3) Large-panel prefabricated constructions; (4) Framework and frame-and-panel constructions; (5) Buildings constructed by floor-grade method; (6) Frame and braced-frame constructions.

The most general soil types upon which buildings are constructed are: (1) rock and large-debris; (2) weathered rock; and (3) sandrocks.

Eighty percent of the buildings are designed by engineers and architects. A city planning division of 32 employees works within the municipal staff under the Mayor.

Yerevan is being developed in accordance with a masterplan of city development. Before 1960, construction of Yerevan had mainly been done without proper consideration of the territory of Armenia's seismicity. Since 1960, a construction code and rules accounting for the seismicity of Armenian territory began to be used. These code and rules were in effect until 1969, when a more developed code (CH and 11-A, 12-69) was established and enforced until 1981.

In 1981, CH and 11-7-81 code was approved, according to which, when designing buildings and structures for construction in seismic regions, the following requirements had to be met: (1) to use materials, design constructions and design schemes providing adequate values of seismic loading; (2) to approve, as a rule, symmetric design schemes, uniform distribution of construction rigidities and masses (from construction units and loadings upon the slabs); (3) to locate joints out of the maximum stress in prefabricated type buildings, to provide monolithic character and homogenity by using integrated precast units; and (4) to allow for the conditions facilitating plastic deformations development within construction units and their joints, as well as providing general resistance of the structure.

When designing buildings and structures for the construction in seismic regions, it is necessary to take into account both the intensity of seismic impact measured in MSK intensity (seismicity) and the recurrence of seismic impact. For Armenia, the intensity and recurrence had been estimated on seismic zonation maps of the USSR territory. Seismicity estimation for the construction area was made based on the seismic microzonation map. Construction of buildings and structures on the areas with estimated seismicity exceeding IX in MSK scale was forbidden. However, in order to allow for the development of rapid and cheap construction, steps were taken to underestimate, artificially, the seismic hazard of Armenia.

As a result, a seismic zonation map indicating the Spitak earthquake source zone as one of the most undangerous in the territory of Armenia was composed by the Armenian SSR and the USSR

Academies of Sciences, and it was approved by the USSR Gosstroi in 1976. In this map, the rate of hazard was underestimated by 3 to a 7 intensity, instead of 10 (on the 12-intensity MSK scale). Therefore, seismic hazard was not properly considered when designing buildings and structures.

After the 1988 Spitak catastrophe, earthquake-resistant construction codes were reviewed significantly in 1989. The main changes to the codes were to increase the intensity-grading in the territory of Armenia.

In 1992, the National Survey of Seismic Protection (NSSP), under the Government of the Republic of Armenia started composing realistic maps of seismic hazard and seismic risk in different scales, including in them the city of Yerevan. At the present time, new codes of earthquake-resistant construction are being prepared where MSK scale intensities are changed by ground accelerations, and the requirements to the structure's design become more stringent. A revised seismic microzonation map is also being prepared.

5. Preparedness for earthquakes

Before the 1988 Spitak earthquake, neither the leadership of the Republic nor the population had been prepared to protect the Republic against a large earthquake in the territory of Armenia. For a number of reasons this problem was not been considered at all. These reasons included the following: (1) Protection of people against earthquakes was not considered an all-Union problem in the USSR, because only an insignificant part of the Soviet population resided in seismically-active zones; consequently, only the USSR Academy of Sciences was concerned with earthquakes, and they considered them merely as interesting natural phenomena; and (2) the solutions to all problems in the USSR were strongly centralized, and the leadership of the Republic was not able to put forward or solve any large-scale problems requiring great material expenditures and resources.

The period following the Spitak earthquake coincided with a period of social order change in Armenia and its formation as an independent state. This allowed us to advance the problem of protecting the Armenian population against large earthquakes to the foreground of national problems.

According to the project proposed by S.Balassanian to the President, Parliament and Government, the National Survey of Seismic Protection of Armenia (NSSP) was founded under the Government of the Republic of Armenia on 17 July 1991 and furnished with special governmental status and powers. The main task of the NSSP is long-term, mid-term, and short-term, operative assessment of seismic hazard and risk in the territory of Armenia, and the development and implementation of long-term and operative measures for seismic risk reduction in the territory of Armenia.

In order to attain this goal, a special NSSP structure was developed that combined two necessary organizational principles of seismic protection under the conditions of Armenia: (1) joint operation of different centers, covering the entire spectrum of seismic protection tasks, and integrated by the common goal, structure and working programme; (2) vertical subordination of all divisions within the NSSP under its President, who is in turn submitted directly to the Prime Minister. Since the organization of the NSSP in Armenia, the preparedness of the Government and the population for large earthquakes had been improved substantially. The NSSP leadership regularly reports to the Prime Minister about the level of seismic hazard in the Republic and the measures that should be undertaken to reduce the seismic risk.

At the present time, special regulations are being developed by the National Survey for NSSP interaction with other governmental agencies and services under different levels of seismic hazard.

One of the NSSP centers, the Centre of Seismic Knowledge Dissemination, carries on continuous activities for training the population in measures of protection against large earthquakes. This involves the production and demonstration of special training films on television; publication of special information concerning the rules of behavior before, during, and after an earthquake; addresses of the leading NSSP experts in different fields of seismic protection, from seismology to earthquake-resistant construction, on TV, radio and through the press; the NSSP instructors lecturing at schools and enterprises; regular reporting about the current seismic hazard level in the territory of Armenia via radio, newspapers, and television; and preparation of street training drills on seismic protection for the population in cities and other populated areas.

The draft law on the protection of the Armenian population against large earthquakes is now being prepared by the NSSP, and it will be submitted to the Parliament of Armenia in 1994.

One of the principal components of seismic risk reduction involves the training of highly skilled specialists in the fields of seismology, earthquake-resistant construction and other sciences related to earthquake preparedness. 1,600 specialists in the field of civil construction and 48 engineer-geologists are now being educated at Yerevan State University, the University of Architecture and Construction, and American University of Armenia under the supervision of 40 professors. With the financial assistance of the American University of Armenia (founded in 1992), the funds allocated for the education of these students exceed those of other specialities. The faculty of the above-mentioned universities participates directly in the elaboration of construction codes for the Republic, conducts consultations, and takes part in the city construction projects. As of yet, there is no association of engineer-builders or seismological association in Armenia.

6. Prompt response

Emergency preparedness activities in Armenia are coordinated by a special staff under the Government of the Republic. In parallel with this, the Organization of Special Governmental Emergency Situations Board, including the entire Civil Protection System was started. Divisions of militia and fire brigades are submitted to the Ministry of Internal Affairs of Armenia and interact closely with the Civil Protection Staff in case of emergency.

As the Spitak earthquake showed, divisions of militia and fire brigades were of sufficient size but poorly-equipped with specialized means to render timely help to earthquake victims. In this regard, Army divisions usually provide more real help. Yerevan possesses a Prompt Emergency Actions Plan, but a large earthquake has its own distinctive features which are not accounted for in this plan. The gap in this field is filled by the Prompt Actions Plan developed by the NSSP.

The network of medical institutions ready to give urgent help in case of natural disaster is well developed in Yerevan.

The city provides the following alternative communication means in case of a disruption in telephone communications: radio, radio relay, and satellite communication.

When developing a plan of prompt and efficient actions of the Government and people in case of a large earthquake in Yerevan, the hard social and economic condition of Armenia becomes the main problem to be solved.

7. Some quantitative assessments of seismic risk for the city of Yerevan.

This report presents the first attempt to assess quantitatively—and in the most general and rough way—the seismic risk in Yerevan, based on the data collected by the NSSP. Three factors are used as the basis in the first stage of this difficult problem: (1) seismic hazard level in Yerevan, estimated on the entire temporal variation of seismic events; (2) earthquake resistance of buildings and structures within the city limits, taking into account soil conditions and design/construction types; (3) the number of building and structure occupants.

From the data presented in the seismotectonic map of Armenia and the seismic hazard zone for Yerevan, it is seen that the city is situated in the area outlined by the VIII-intensity isoseismal of the 1679 Garni catastrophic earthquake. Furthermore, this isoseismal line divides the city (NW-SE) into two unequal parts, the greater of them located inside the VIII-intensity seismic hazard area.

Figure 4 presents a sketch map of the distribution of the three main soil types within the city. If the VIII-intensity isoseismal line of the 1679 Garni event is placed on this map, the seismic impact on the buildings of Yerevan and the risk the city residents are exposed to becomes clear. A rough estimation of the intensity of seismic impact on buildings and structures under different soil conditions are as follows: rocky ground weakens it by one unit of intensity, weathered rocky soils does not change the intensity of impact, and sandrock formations increase the intensity by one unit.

In 1679, when the Garni earthquake occurred, Yerevan was settled only in its present-day central and northeastern parts—i.e., the buildings were built mainly on the rock—so it is not difficult to suppose that the initial seismic impact in 1679 corresponded to an intensity of MSK IX. Therefore, the seismic hazard intensity on the territory of Yerevan within the VIII-intensity isoseism is assumed to equal IX and, outside of it, VIII.

In order to assess the earthquake resistance of buildings and structures, we have divided the city into 13 parts of approximately equal area and, for each of them, indicated with a symbol the relationship between different design construction types used [Figure 4]. In accordance with their designed earthquake resistance, all buildings are classified as follows: (1) low stone buildings (private constructions), designed in large part for VII intensity, without accounting for soil conditions; and (2) all other types of design constructions (state construction), designed for VII intensity on rocky soils and VIII intensity for other soil types.

When assessing earthquake resistance of buildings, the Spitak earthquake showed that all of them may be divided into three classes, the Most Reliable, Reliable and Unreliable. Each class may be assigned a fractional intensity ($\Delta I_{constr} = +0.5$ intensity), which is not conventional in the practice of construction but can be useful when assessing seismic risk for Yerevan. The Most Reliable include large-panel buildings, framework and braced-frame buildings; it may be considered that these types of construction increase a building's earthquake resistance by $\Delta I_{constr} = +0.5$ intensity. Reliable structures include stone and complex constructions ($\Delta I_{constr} = 0$ intensity), and Unreliable refers to framework, framed-and-panelled, and those constructed by a floor grading method ($\Delta I_{constr} = -0.5$ intensity).

It is very important to distinguish between the designed earthquake resistance (I_{dn}) and real earthquake resistance (I_{rl}). The latter is determined by the construction type, quality of building materials, quality of construction, and correctness of building maintenance.

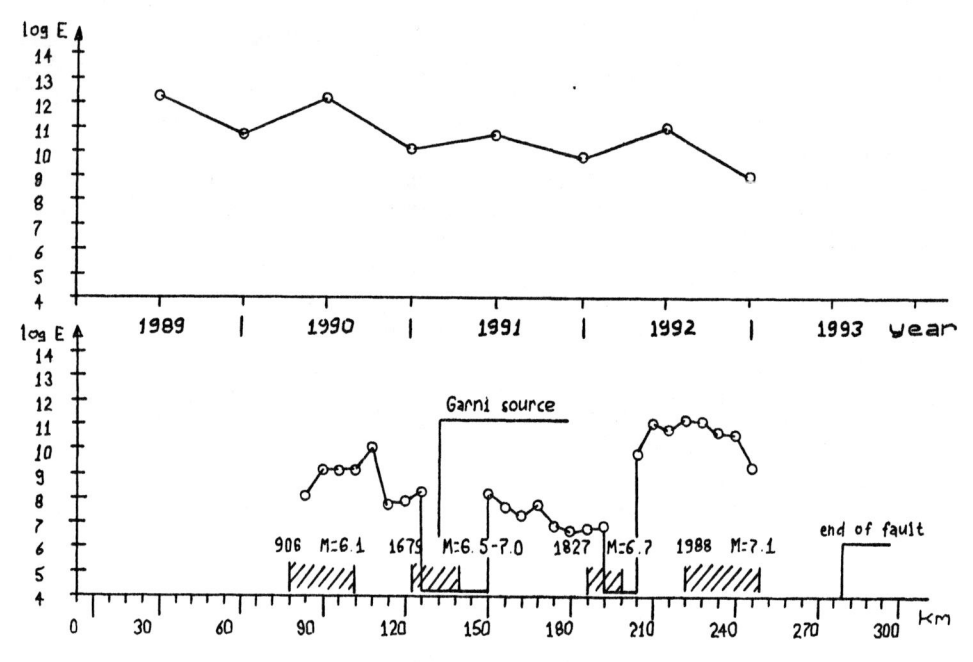

Figure 3. Seismic activity along the Garni fault from 1989 to 1993
(after the Spitak earthquake, 1988)

▨ - source of the large historical earthquakes

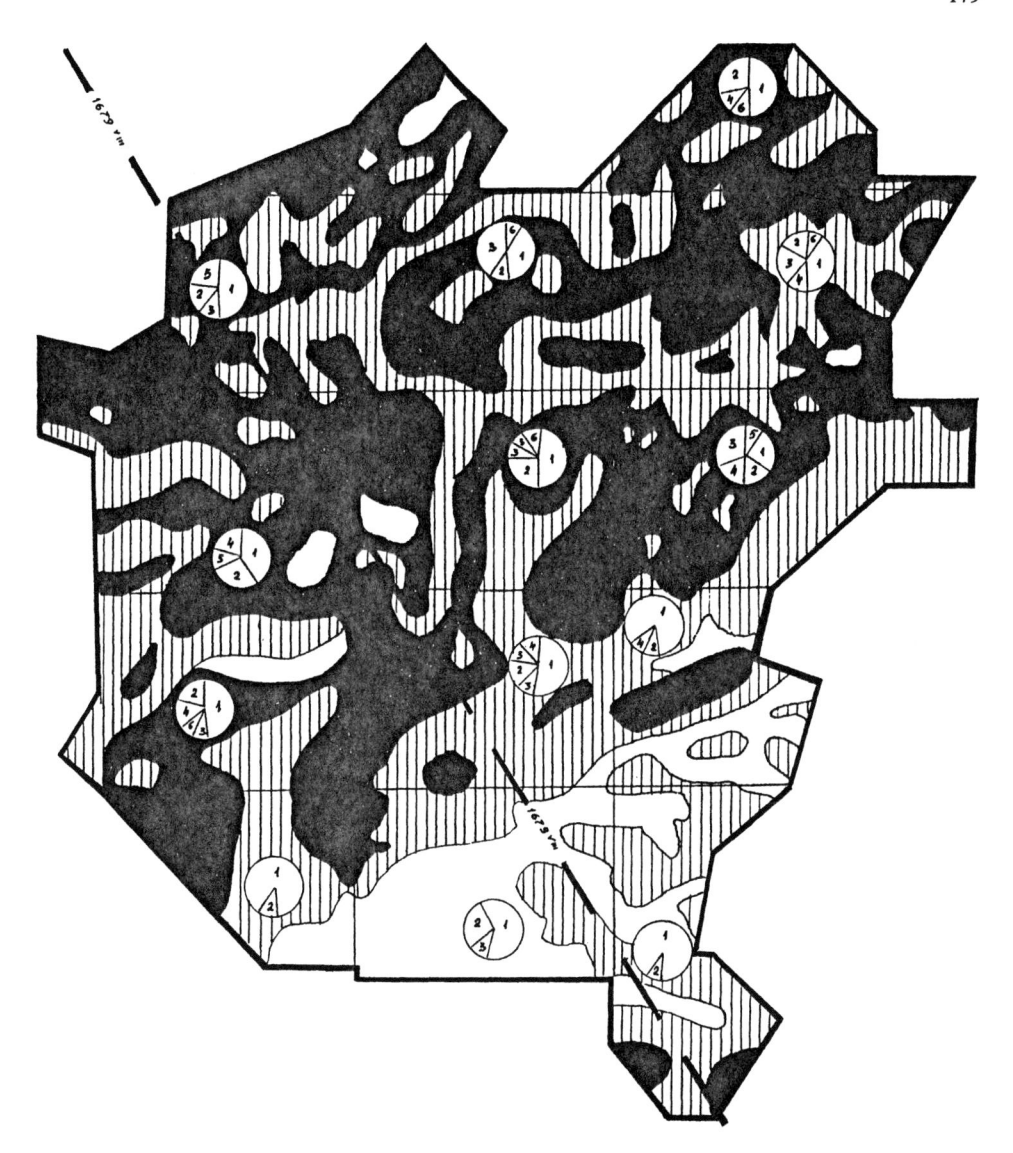

Figure 4. Scheme of Yerevan city indicating prevail ground according
to seismic properties and constructive schemes of buildings

The category of grounding according to the seismic properties:
- ● 1. Rocky and large disintegrated rocks;
- ◐ 2. Rocky weathering; coarse sand; clayey with J>0.5;
- ○ 3. Saddy unconsolidated; clayey with J>0.5.

Relation between the structural schemes of the buildings within the limits of the specified zone in the city of Yerevan

— 1679 VIII — Isoseism

The constructive schemes of buildings:
1. Low-storeyed stone houses (private);
2. Stone and complex constructions;
3. Large-panel;
4. Framework and frame-and-panelled;
5. Constructed by floor-grade method;
6. Frame and frame-brased.

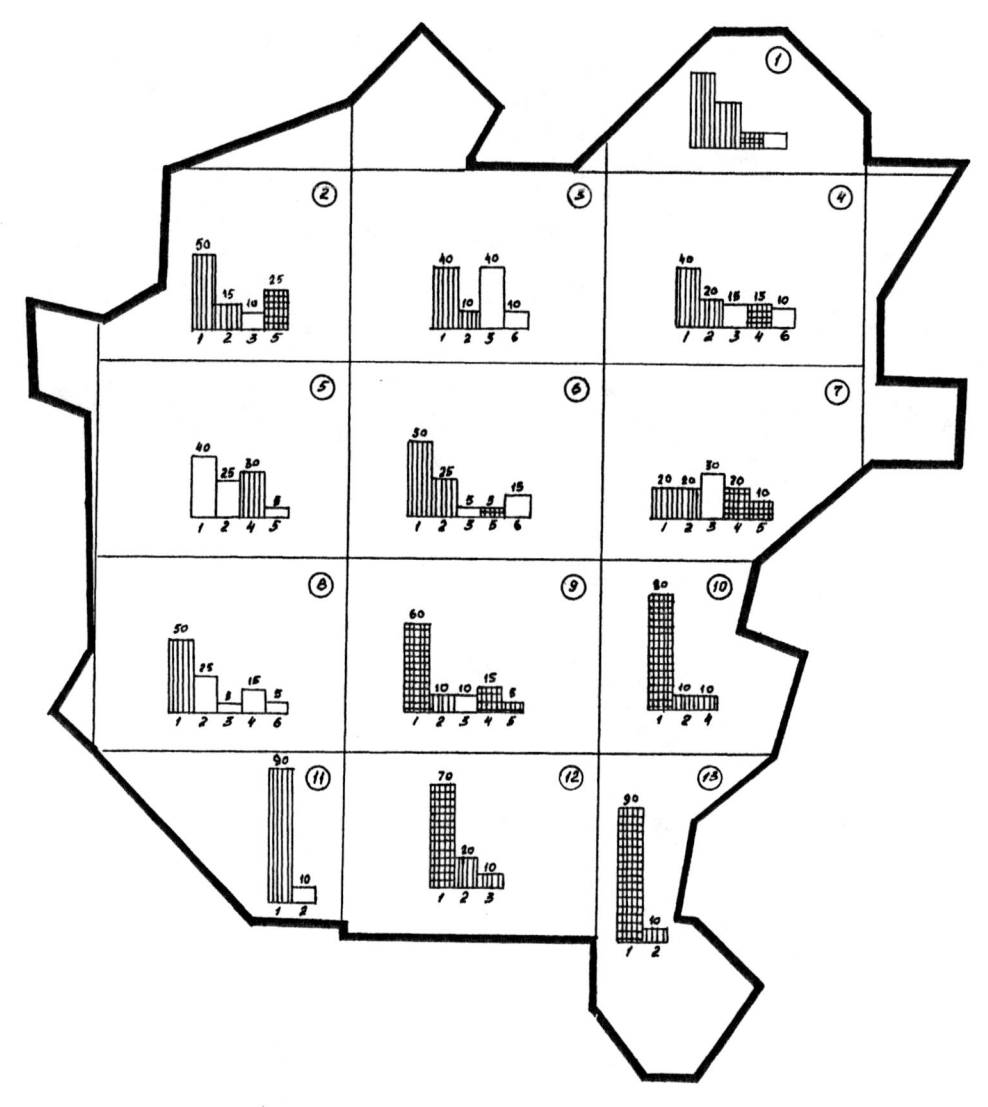

Figure 5. Map of the risk of different contruction type buildings in Yerevan city (design construction types 1,2,..., are indicated in percentage of the total number ob buildings in each square 1,2,..)

1.   - no risk practically (K<1.06);
2.   - moderate risk (1.05<K<1.1);
3.   - high risk (K>1.1)

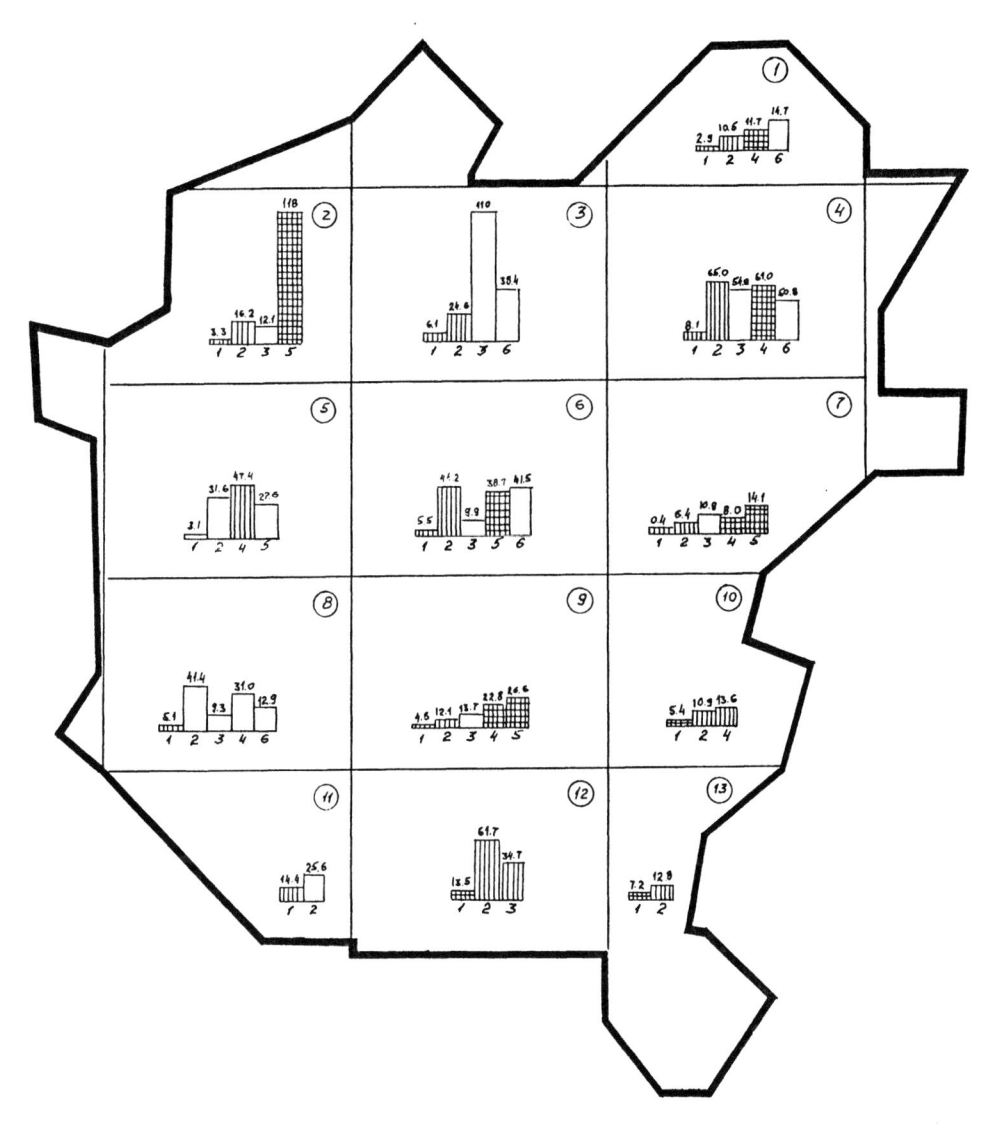

Figure 6. Map of the risk of people's death in Yerevan city
(number of people is indicated in the thousands of people
in eath design construction type 1, 2,..., in each square 1, 2, ...)

1. ▭ - no risk practically;
2. ▥ - moderate risk;
3. ▦ - high risk.

To assess the risk of building collapse, the following criterion may be used: $K = I_{haz} / I_{rl}$ (1), where I_{haz} is seismic hazard estimated in MSK scale; $I_{haz} = I$ in. haz $+ \Delta I_{gd}$, where I in. haz refers to initial hazard in MSK scale, and ΔI_{gd} to the intensity-grading increment on the different soil types; $I_{rl} = I_{dn} + \Delta I_{constr}$, where I_{rl} is the real earthquake resistance of the building. Criterion K' shows that if seismic hazard exceeds the earthquake resistance of the building, i.e. if $I_{haz} > I_{rl}$, then there is a risk of building collapse. And conversely, if $I_{haz} <$ or $= I_{rl}$, then there is no risk of collapse.

Figure 5 presents a map of collapse risk for buildings and structures in an earthquake of IX intensity within the city limits, outlined by the VIII- intensity isoseism in the remainder of the city.

Figure 6 maps estimated fatalities for different parts of Yerevan city. When an integrated analysis of maps 5 and 6 is made, it is seen that the greatest destruction in the mapped territory, covering areas at IX intensity and VIII intensity is expected in the southern part of the city (zones 9, 10, 12, 13), if an earthquake similar to the 1679 Garni earthquake occurs. The number of victims is not expected to be very large when low buildings of private construction are exposed to the risk of collapse. Rather, the greatest number of victims is expected in the northwestern part of the city, Zone 2, in a relatively small destroyed area.

The above-mentioned conclusions and many others which may be made based on maps 5 and 6 allow us to elaborate effective measures for seismic risk reduction in the city of Yerevan. One of these measures is to retrofit buildings and structures within the zones of high risk for collapse.

References

Aslanian, A., "Main Features of the Geological Structure of Armenian SSR" Izvestia of Arm. SSR Academy of Sciences, "Nauki o zemlie", vol.3, 1981, p.3-21

Karapetian, N., "Origination Mechanism of the Earthquakes of the Armenian Upland", Yerevan, Arm. SSK Academy of Sciences, 1986, p. 227.

Trifonov, V., Karakhanian, A., and Assatrian, A., "Active Tectonics of Arabian and Eurasian Plates Interaction Zone", International Geological Review (in press), p.89.

STATUS OF SEISMIC HAZARD AND RISK MANAGEMENT IN KATHMANDU VALLEY, NEPAL

R.P. YADAV[1]
P.L. SINGH[2]
A.M. DIXIT[3]
R.D. SHARPE[4]
[1]*Honourable Member, National Planning Commission, Nepal*
[2]*Mayor, Kathmandu Municipality, Nepal*
[3]*SILT Consultants (P) Ltd., Nepal*
National Team Leader, Building Code Development Project, UNDP/UNCHS/MHPP
[4]*Beca Worley International Consultants, Ltd., New Zealand*
Team Leader, Building Code Development Project, UNDP/UNCHS/MHPP, Nepal

1. Introduction

1.1 GENERAL

This paper begins with a description of the Kathmandu Valley in the context of Nepal. It then outlines the demographics of Kathmandu Valley before describing in some depth the geological and seismological aspects of the seismic hazard existing there. After a summary of the emergency response capabilities of the existing infrastructure, it moves to the current level of seismic awareness, current design practices, and the possible resources that could be mobilised to improve these.

1.2 LOCATION

Nepal is located on the southern hill slope of the Himalayas, in between India to the south, east and west, and Tibet to the north. With an area of 147,000 km^2, the topography of the country varies dramatically from the Terai situated on the Ganges plains in the south to the highest Himals in the world bordering the high Tibetan plateau in the North.

Kathmandu, the capital of the country, is located almost in the middle of the country in a valley amidst the mountains which rise up to an average elevation of about 2,100 m, but with some higher peaks reaching 2,765 m. The valley's roughly elliptical outline is interrupted by rock spurs which encroach from the surrounding mountains. Its maximum width is approximately 25 km from east to west, and 20 km from north to south, and it has an average base elevation of about 1,300 m. The drainage area of Kathmandu Valley is about 585 km^2, and the valley floor area and surrounding terraces are about 400 km^2 [Figure 1].

Three of the 75 administrative districts of the country (ie., Kathmandu, Lalitpur and Bhaktapur) are located within the Valley. The administrative centres of these districts are the three cities which have the same names. These cities are renowned for their historical and cultural richness. The cities of Kathmandu and Lalitpur (Patan) are separated by the Bagmati River and, due to the expansion of the urban areas in both the cities, the boundary between the two now has only historical, political and administrative importance. Likewise, the spatial separation between Bhaktapur and the other two cities is rapidly becoming less and less conspicuous.

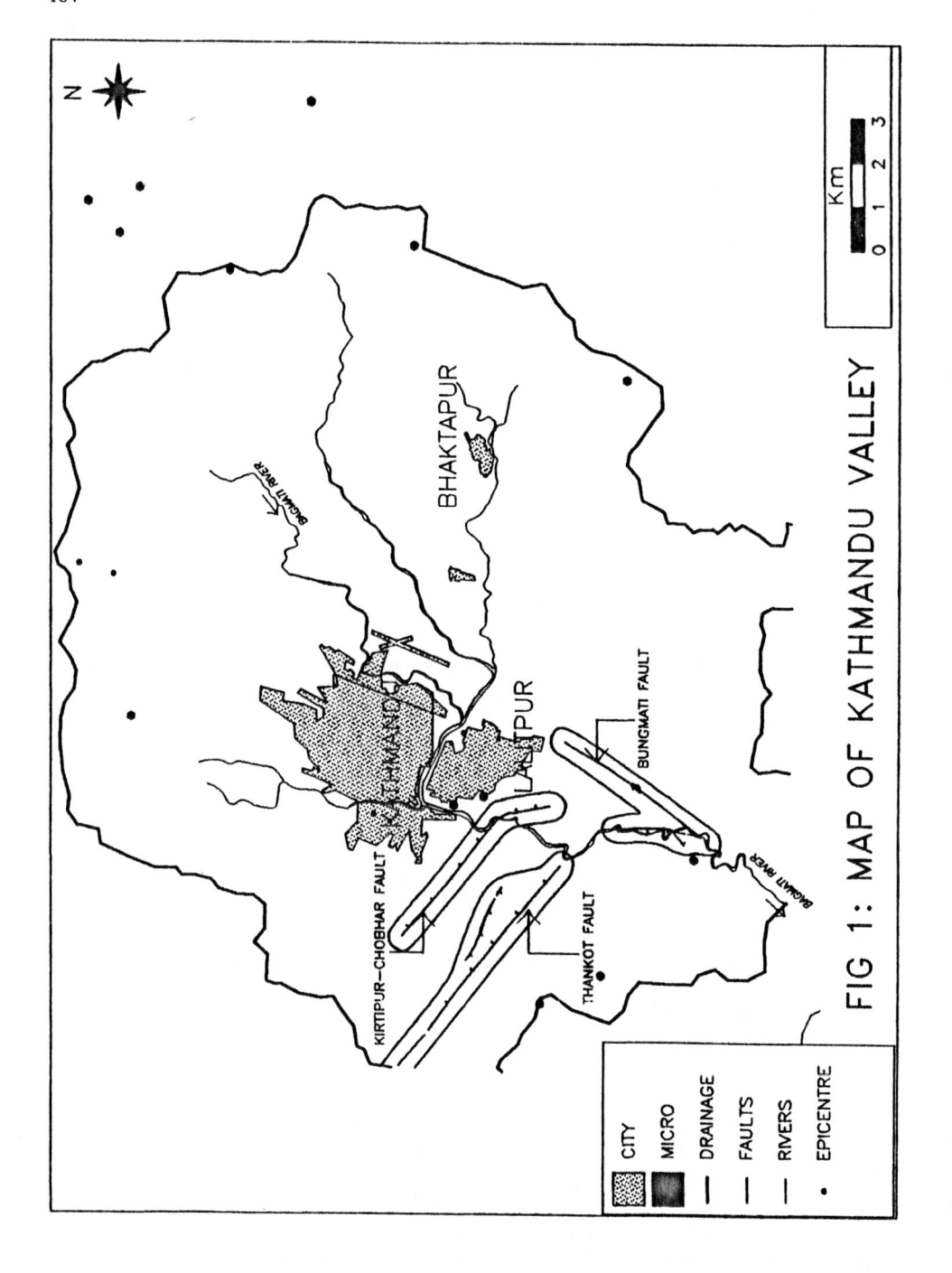

FIG 1: MAP OF KATHMANDU VALLEY

Therefore, the concept of Greater Kathmandu has been introduced lately for preparing the plans for water supply. It includes the municipal areas of Kathmandu and Lalitpur and covers an area of approximately 70 km^2.

Topographic maps and aerial photographs are available for the Valley in different scales. However, detailed maps of existing utility services are not generally available, although work is in progress for their production.

1.3 HISTORICAL EARTHQUAKES IN THE VALLEY

Kathmandu Valley has experienced a number of devastating earthquakes within living memory, such as those in 1934 and 1988. There was a significant earthquake in 1833, and the earliest recorded event in the most comprehensive earthquake catalogue to date occurred in 1255. The damage and casualties due to these events have been great. Bhaktapur has been reported to be typically the centre of damage during several earthquakes. Nepal continues to face a high level of earthquake hazard and risk.

1.4 THE PEOPLE AND POLITICS

Nepal has a population of approximately 19 million people of whom about 90% live in remote rural areas which are mostly inaccessible by road.

A very recent, relatively peaceful democracy movement has seen a progression to a constitutional monarchy. Following this, the country is presently undergoing rapid changes in different walks of national life. The social philosophies, aims, scopes and commitments are being re-defined in this process of national reconstruction. There is presently much local and central government institutional development taking place. A substantial social debate is underway as to whether the Kathmandu Municipality should be administered as an autonomous metropolis. Environmental consciousness of the people is growing.

While there has existed for a long time a full range of government ministries at the national and district level, urban organizations such as municipal councils have been institutionally very weak. Many international aid agencies are very active in Nepal, which has one of the lowest per capita incomes in the world.

In short, it is the right time for the introduction of any rational plan, programme, philosophy, etc., including the conducting of earthquake hazard awareness-raising activities. It is believed that the development of an earthquake damage scenario for the Valley would have a tremendous impact and could bring about a marked change in the attitude of the urban elite to the risk of damage to their houses from earthquakes.

1.5 PREPAREDNESS

Nepal was relatively late in recognising the necessity of understanding the seismic phenomena within and close to its boundaries. Similarly, it has only recently appreciated the necessity of a program of earthquake hazard mitigation and preparedness planning.

There are no regulations presently enforced in Nepal to govern the design for strength of either public or private structures. While there are some competently designed and constructed seismic-resistant buildings in Nepal, the majority of structures are, for a variety of reasons, vulnerable to severe damage from relatively minor earthquakes.

The Ministry of Housing and Physical Planning (MHPP) of His Majesty's Government of Nepal (HMGN) is being assisted by UNDP/UNCHS(Habitat) at the moment to define the seismic hazard within Nepal; to develop a national building code; to plan for its implementation: and to recommend alternative technologies and materials. This work is being carried out by a team of national and international consultants. The project forms part of a larger effort of reconstruction and rehabilitation which was started after the 1988 earthquake.

2. DEMOGRAPHICS

2.1 POPULATION

The populations of the three cities of Kathmandu, Lalitpur and Bhaktapur are as follows:

Cities	Population, 1971 Census	Population, 1981 Census	Total Households (1981)	Population 1991 Census	Avg Household (1991)
Kathmandu	150, 402	235,160	81,139	421,258	5.2
Lalitpur	59,049	79,875	20,630	115,865	5.6
Bhaktapur	40,112	48,472	9,187	61,405	6.7
Total				598,528	

Source: Nepal Central Bureau of Statistics, 1991 Census

It has been predicted that the Valley will have a total population of 1.86 million by the year 2011.

2.2 POPULATION GROWTH AND DENSITY

The consensus of recent population studies for road and water supply planning is :

a) Urban development in the Valley has already exceeded the physical limits of the municipalities and will extend rapidly to the urban expansion areas.
b) Central Kathmandu City has already reached its ultimate population capacity.
c) There will be a gradual increase in the population density in the outer areas (wards) of the municipalities.

The old historic city areas have high average population densities (850 persons per hectare) whereas the peripheral areas with recent development have an average population density of 88 p/ha. The population of Greater Kathmandu is predicted to be 923,900 in the year 2011.

2.3 IMPLICATIONS

Higher population densities and high growth rates in the core areas have made the land very expensive and of high commercial importance. This has resulted, on the one hand, in an adverse impact on the historic-cultural heritage due to indiscriminate reconstruction and, on the other, in vertical growth of the buildings —most often without consideration of the structural strength of the lower parts. The major consideration of owners has been to try to maximize the total floor area. There is a tendency to increase the effective building area by introducing cantilevered projections above the first floor. Under such conditions, the streets in parts of the built-up areas may have become potentially more dangerous during earthquakes than inside the weak buildings.

Resources have become scarce and inadequate. Shortages of water and electricity have become acute even in normally good seasons.

Higher population pressures have created different problems which have diverted the immediate attention of the municipalities to other issues, and earthquake safety has not received any priority.

But things are changing. The municipalities are now embarking upon several new initiatives directed towards raising the quality of life of the city-dwellers. Awareness programs are being launched to keep the population informed about the possibility of improving living conditions. Although earthquake safety has not come to the fore in these, conditions are right for the introduction of earthquake awareness programs.

3. Knowledge of the Seismic Hazard

3.1 GEOLOGY

3.1.1 *Rock/Soil Types.* Kathmandu Valley is an infilled lake basin that appears to have been formed during the Quaternary period, primarily by the rising of the Mahabharat Lekh, the mountain range that forms the southern boundary of the Valley. This uplift is believed to have resulted in the damming of the southward-flowing valley drainage, causing the formation of a lake and the subsequent deposition of layers of river and lake sediments which are at least 450 m thick in total. The lake subsequently drained out via a gorge cut by the Bagmati River.

The topography of the Kathmandu Valley rim is rugged. It is characterized by steep valleys and ridges. The present land surface over most of the basin is characterized by flat-topped terraces at different levels. These have resulted from the various historic levels of the Bagmati River system.

The unconsolidated sediments of the Kathmandu Valley vary widely in composition. In general, the sediments in the northern half of the Valley consist of typical riverbed layered deposits of clays, silts, sands, and gravels, while those in the southern half are mainly lakebed deposits of clays and silts, usually sandwiched between thin layers of coarser sediments. The detailed picture is far more complex, with considerable local variation over short distances.

Such a composition of the valley subsurface makes it susceptible to promoting damage during an earthquake. Many observers have likened the soil conditions of Kathmandu Valley to those of Mexico City which is notorious for its susceptibility to damage during an earthquake.

3.1.2 *Active Faults in the Valley.* Ongoing neotectonic activity within the Kathmandu Valley is demonstrated by the presence of active faults that have displaced sediments of the Pleistocene to Holocene periods. Five such faults have been identified during the seismic hazard mapping study undertaken by the National Building Code Development Project.

The Thankot fault is a well-defined dip-slip (normal) fault that strikes north-northwest and truncates the toe of an alluvial fan (downward to the northeast) on the southern slopes of the Chandragiri Range. The total length of this fault is 8 km. An apparently conjugate fault (Basigaon Fault), with an opposite down-throw but similar strike and a length of 3 km, is located nearby. This is about 5 km from central Lalitpur and about 8 km from the centre of Kathmandu City.

Another dip-slip fault was identified in the flood plain of the Bagmati River in the Pharping area to the south. The western side of this north-striking fault is down-thrown.

A fault that runs along the northern foothills of the Kirtipur-Chobhar Ridge, about 5 km from the centre of Kathmandu, is believed to have been re-activated after the mid-Pleistocene, tilting the otherwise almost horizontal mid-Pleistocene lakebed sediments southward. This is interpreted to be a dip-slip fault with the northeastern side down-thrown.

The Bungamati fault is also within 5 km of Lalitpur. It strikes north-northeast from the channel of the Bagmati River to the village of Sunakothi and has an interpreted length of 26.1 km. Although the sense of displacement is not clear, it appears to be a right lateral strike-slip fault.

These faults are believed to be capable of generating a maximum of a 6.6 Richter magnitude event. However, there has not been any detailed study carried out to quantify these faults as to their activity.

Other active faults have been identified outside the Valley but within distances of 20 to 50 km from Kathmandu. The Kalphu Khola Fault to the northwest of the Valley is interpreted as being capable of generating an earthquake with a maximum possible Richter magnitude of 6.9, while the Kulikhani Fault, lying to the south, has been assigned a maximum earthquake magnitude of 6.9. Similarly, the Sunkosi-Rosi Fault, which lies about 35 km southeast of Kathmandu, has been assigned a maximum earthquake magnitude of 7.1. Any major earthquake along these faults is sure to affect the Valley.

3.2 SEISMICITY

Geologically, Nepal sits astride the boundary between the Indian and the Tibetan plates. The Indian plate is continuously moving (subducting) under the Tibetan plate at a rate thought to be about 3 cm per year. The existence of the Himalayan Range with the world's highest peaks is evidence of the continued tectonic activities beneath the country. As a result, Nepal is very active seismically.

The most recent earthquake(s) felt in Kathmandu was a swarm of 3.9 to 4.2 Richter magnitude events which awakened the city-dwellers on 6 July 1993 between 3 and 4 a.m. The epicenters were located about 40 km north of Kathmandu. Although these were small events, they created a panic among the population as many believed them to be precursors to a larger event. A house collapsed about eight hours after the event, killing one person and burying others. The building collapse was linked to the shaking although the building was fairly old and weak.

The great Bihar-Nepal earthquake occurred in 1934 with a magnitude of 8.4. The epicenter for this event has been located in the eastern Nepal about 240 km away from Kathmandu. The Valley experienced intensities of IX-X on the Modified Mercalli scale. The majority of the reported destruction from the earthquake occurred in Kathmandu Valley and along the eastern Terai (the plains bordering northern India). A total of 8,519 persons were reported dead and more than 80,000 houses damaged within Nepal. Kathmandu Valley's share of this was 4,296 persons dead and 12,397 houses totally destroyed. Seventy percent of the then-standing houses in the Valley were affected. This event confirmed that, because of its geological characteristics, Kathmandu Valley is susceptible to enhanced damage during an earthquake event.

The 1988 (magnitude M_s 6.6) Udayapur earthquake occurred in the vicinity of the Terai and Siwalik Hills 165 km to the southeast of Kathmandu. The damage was concentrated in the plains of the Terai in a pattern similar to the 1934 earthquake. The Kathmandu and Bhaktapur districts were declared earthquake-affected areas. A few very old buildings were damaged in Kathmandu while several houses collapsed in Bhaktapur. Several people were injured, mainly because of the resulting panic. Some deaths were also reported in the Valley.

Earlier major earthquakes were reportedly felt in the Kathmandu Valley in 1255, 1408, 1681 and 1810, with intensities (Modified Mercalli) of approximately IX to X.

3.3 POTENTIAL SECONDARY HAZARDS IN THE KATHMANDU VALLEY

The identification of active faults in the Kathmandu Valley indicates that there are both primary and secondary earthquake hazards in the Valley, since these features can be the sources of strong ground motion as well as surface rupture. The following three main types of likely secondary seismic hazards identified for Kathmandu Valley are considered in detail: surface fault rupture, soil liquefaction and earthquake-induced landslide.

Maj. General B.S. Rana described in his 1935 book on the 1934 earthquake effects which suggest widespread liquefaction occurred in Kathmandu Valley. He described the heavy development of fissures in roads and paddy fields, with water being ejected from them. At some places the height of the fountain of ejected water reportedly reached 3-4 m. It supports the recent assessment that the valley sediments are susceptible to liquefaction during a major earthquake.

Soil amplification is generally regarded as being the cause of some problems during earthquakes in the Valley. It is apparent that areas underlain by thick successions of Kalimati clay are likely to suffer the most damage.

3.4 ORGANIZATIONS STUDYING SEISMIC HAZARD

The following is a list of different national agencies involved in seismic hazard study :

- <u>HMGN Department of Mines and Geology</u>

The Department of Mines and Geology has been running a five-station microseismic network in Central Nepal under a French assistance program since 1978. Within a year, a network of seventeen seismograph stations will cover the whole country. The information on earthquake events is exchanged with international agencies. There are no strong-motion recording instruments in Nepal.

- <u>Ministry of Housing & Physical Planning</u>

The Ministry has been undertaking, since the 1988 earthquake, an Earthquake Area Reconstruction and Rehabilitation Project with the assistance of UNDP/UNCHS(Habitat) and World Bank funding. Reconstruction of seismic-resistant domestic and school buildings has been their major concern.

The MHPP is also implementing a wider program of Policy and Technical Support to the Urban Sector with Habitat's assistance. The National Building Code Development Project is one of the three subcontracts of the program. This UNDP/UNCHS(Habitat)-funded subcontract is underway at present and will define the seismic hazard within Nepal, develop a national building code, plan for its implementation, and recommend alternative technologies and materials.

4. Emergency Response

4.1 HMGN HOME MINISTRY

The Home Ministry is the lead government agency for disaster management in Nepal.

A UNDP-financed project of Institutional Support to Disaster Preparedness Plan (NEP/85/002) was implemented during 1989-1992 and carried out the:

- Preparation of a Comprehensive Disaster Management Plan

- Strengthening of the Special Disaster Unit of the Home Ministry.

- Development of a core group of trained manpower.

The Comprehensive Disaster Management Plan was prepared and discussed in a national seminar in Kathmandu during August 1991. However, it is yet to be implemented.

The Special Disaster Unit was mobilized for the severe floods in July 1993.

4.2 NEPAL RED CROSS SOCIETY

The Nepal Red Cross Society has branches in all 75 administrative districts of the country and it is one of the very active participants in the District Disaster Relief Committees. The Society organizes training and awareness programs regularly and responds immediately and effectively to a disaster situation.

4.3 MUNICIPALITIES

Being the capital city, Kathmandu enjoys access to all the available resources within the country. Likewise, the other two municipalities of Lalitpur and Bhaktapur are also regarded as the best organised and the most resourceful of the municipalities in the country.

Traditionally in Nepal, the municipalities developed as the local political unit as well as the local government. Hence Nepalese municipalities have been somewhat unusual in that the various services, including the lifelines, have continued to be looked after by the various line agencies of the government, and the municipalities have had almost no say in the planning and management of such services as water supply, sewage, electricity. However, following the recent democratic changes in the political system, which have given greater authority to the local government, the scope of municipal works is now being redefined. Kathmandu Municipality has decided recently to form a City Defence Force consisting of 151 full-time personnel. The main activity of the workers is said to be the promotion of cleanliness, security and social mobilization within the city. A committee has been set up to establish coordination between the City Defence Force and the Police. Municipalities, however, have yet to develop strong technical departments which could be mobilized in the event of a disaster.

There is one central fire station in each of the three cities. This paramilitary type of organization belongs to the Ministry of Home.

The water supply and sewage systems in the three cities suffer from the problems of leakage and are susceptible to cross-contamination. Maps of these lifelines to a suitable scale and accuracy are not yet available.

The cities do not have any general emergency response plans, nor any specific plan for an earthquake emergency. The responsibilities for actions (e.g., communication, response, evacuation, and relief) have not been defined, let alone exercised or drilled. This fact was very conspicuously seen at the site of a building believed to have collapsed due to the effect of an earthquake in July this year. Although the police and the fire-brigade were in attendance, together with a commandeered front-end loader, there appeared to be no leadership or coordination of services and it appeared that the rescue operation was conducted on the basis of ad hoc decisions.

The tasks facing the municipalities in bringing about improvements in earthquake safety are very large. Many changes must be introduced into the management of existing organizations before they will be able to monitor and control the implementation of the planning by-laws and the building code. The manpower within the organizations needs to be trained in aspects of seismic safety. The earthquake risk faced by the cities in the Valley needs to be discussed among the politicians, business community and representatives of the service sector, because their participation and contribution are essential for the effectiveness of any program geared towards a reduction in earthquake risk in the Valley.

4.4 POLICE DEPARTMENT

Traditionally, the police do respond to disaster situations with manpower and are equipped with a local mobile radio telephone network and an independent countrywide radio network. They are governed by the Home Ministry.

The police department is implementing a Community Police program under which fifteen policemen are to be posted in each of the 34 wards in Kathmandu. Ward Security Committees are being formed, with participation from social organizations, for solving ward-level problems.

4.5 ROYAL NEPAL ARMY

The Royal Nepal Army is a well-trained and -equipped organization with independent transport and communication networks. It also has an Engineering Corps, helicopters and medium-size transport STOL planes. Traditionally, the Army responds very quickly and substantially to an emergency. The Army was one of the main participants in rescue of people and in the transportation of relief materials during both the 1988 earthquake and the 1993 flood disaster.

4.6 HOSPITALS

There are, in total, eight government hospitals in the three cities. This includes the Tribhuvan University Teaching Hospital, the Army and the police hospitals. The total number of available beds is 896. There are several private nursing homes which are equipped for in-house surgery as well as out-patient medical services. The medical doctors working on part-time bases in these nursing homes are largely the personnel from the government hospitals. Similarly, there are several public and private clinics run by social organizations and by groups of doctors. Mobile clinics are also organized from time to time. However, the available facilities and supply of medicines are generally regarded as inadequate to cater to an increased number of patients after a disaster.

4.7 COMMUNICATIONS

The cities have a telephone network of both underground and overhead wiring. It is very common for the electricity poles to carry the telephone wires as well. There have been isolated cases of the telephone network being accidentally connected to high voltage electricity wires. The telephone network is as vulnerable as any other infrastructure. Recently, the government has announced its intention to allow the private sector to develop a network of cellular telephones, and there has been interest expressed in such a system. Once operative, such a cellular telephone system may serve as an alternative communication system during disaster management, albeit with limitations. Main national and international links out of Kathmandu are now by microwave systems.

There is a separate telephone network belonging to the Army. It is much smaller than the civil telephone system. Although this system facilitates communication within the Army, it also appears to be as vulnerable as the civil network. Therefore, it should not be relied on as an alternative means of communication during disasters.

4.8 ELECTRICITY

The electricity network is thought to be very vulnerable to a major earthquake and could not be relied on after such an event.

5. Current Seismic Design Practices

5.1 BUILDING REGULATIONS

There are no regulations presently enforced in Nepal to govern the design for strength of structures. Where clients require design against earthquake, it is common for the engineer to use the code of the country in which he or she was trained.

New planning by-laws for municipalities in the Kathmandu Valley are being introduced.

Building permits within the Kathmandu Valley are presently given on the deposit of simple plans showing the external architectural features of the proposed building. The masons appear to control the design of the up to five-story structures, usually comprising a light concrete frame and unreinforced brick infilled panels and internal masonry partitions. Many of these buildings will exhibit soft-story collapse of the ground floor because of the open spaces required for shop fronts. The detailing of reinforcing steel does not follow the accepted practices of other highly-seismic countries. Many old buildings are showing obvious signs of distress under static loads alone and collapses during the monsoon season are common.

5.2 STANDARD DESIGN PRACTICE

The standard engineering practice of designing structures is to follow the Indian Standard Code IS:1893-1984 which contains a seismic design coefficient for Kathmandu. Although this code does not take into account the specifics of the Nepalese seismic conditions, it has continued to be the standard guideline for Nepalese designers. However, the consideration of seismicity in design is not mandatory and depends upon the individual initiatives of clients and designers. Very few Nepalese professionals give the necessary level of attention to this problem and the non-implementation of design standards leads to a very poor scenario. Much depends also on the availability of funds, which generally are scarce. This applies not only to the private or governmental buildings and structures. Even several of the structures funded by international aid and development agencies lack consideration of seismicity in design and construction.

In summary, the vast majority of the modern buildings in the Kathmandu Valley have no rational design at all.

It is thought that there are not any consistent anti-seismic measures applied to the design of bridges.

There is evidence in the streets that the electricity authorities do not practice a consistent approach to seismic hazard mitigation with respect to their lifeline. A majority of pole-mounted distribution transformers in Kathmandu are not adequately fixed to their mounts.

5.3 INSURANCE INDUSTRY

The concept of insurance against earthquake damage in Nepal is in an embryonic stage. There are cultural and religious reasons why it is not popular. This means that there are not the incentives at play that work elsewhere to improve seismic resistance.

6. Current Awareness Level

6.1 GENERAL

Recent years have witnessed the initiation of many activities directed towards raising the awareness of the Nepalese planners, decision makers, technicians and the public regarding the possibility of reducing the impacts of the natural disasters faced by the country. Many international agencies such as UNDP, UNDRO and ADPC have been generous in helping Nepal in these initiatives. Professional, as well as social, organizations are also mobilizing their efforts by organizing scientific forums, training programs and media releases. There is a positive and gradual shift in the attitude of Nepalese society towards the way they handle natural disasters. The recent flood disaster has awakened the affected public to the potential threats and the necessity to undertake mitigative measures.

However, the general awareness level of the Nepalese continues to be very low, be it with respect to the threat of an earthquake or to the threat of any other natural hazard the country is facing. Although many people remember the last earthquakes in their lives, there is no understanding of the necessity to undertake concrete action. It is common for them to believe that no action within their reach could increase their safety from earthquake. This attitude continues to be held by both government officials, the business community and people from different all walks of life.

Awareness-raising activities have not yet been institutionalized. All of the activities undertaken during the last couple of years were guided primarily not by the recognition of the necessity and the possibility of disaster reduction, but in response to the availability of funds, individual initiatives and enthusiasm. Emphasis continues to be placed on post-disaster relief/rescue and rehabilitation aspects.

Whatever awareness regarding the earthquake threat to the city or the country has been created, it lies in a dormant condition and receives the least priority of action. There is no program of drills conducted in schools, hospital or any other places of public gathering. There has not been any central organization created that is responsible for undertaking or facilitating such activities. There is no legislation to regulate for earthquake preparedness.

6.2 CURRENT INITIATIVES

A very successful training program on disaster management in Nepal was organized in Kathmandu during May 1993 by the Home Ministry of HMGN in cooperation with the United Nations Department for Humanitarian Affairs, the Asian and Pacific Program for Development Training and Communication Planning, the Water-Induced Disaster Prevention Training Centre, and UNDP.

Forty participants in the training program were drawn from different line agencies of the government, NGOs, and other public and private institutions representing all the five development regions of the country. The participants included Chief District Officers, Local Development Officers, District Superintendents of Police, senior officers from the Royal Nepal Army, medical doctors, engineers from the HMGN's Departments of Roads and Irrigation, and representatives from the Nepal Red Cross Society, Nepal Scouts, the Nepal Geological Society and other institutions.

The Ministry and the UNDP have plans to organize a workshop on disaster management in the near future. Participants will be the high-level decision makers in the government. Three case studies on the status of natural disasters and disaster management are to be presented and discussed in the proposed workshop. Earthquakes will be the topic for one case study. The others will cover deforestation and erosion, social disruption, and the refugee problem.

MHPP, through its UNDP/UNCHS(Habitat)-supported Earthquake Affected Areas Reconstruction and Rehabilitation Project and National Building Code Development Project, is formulating information dissemination strategies and plans for implementation. The relevant dissemination materials are being prepared under the ongoing subcontracts. The strategies seek to out reach to a cross-section of the population, including the rural mass.

With support from Habitat, the World Bank and GTZ, MHPP is implementing also a Town Development Fund Board Project which aims at institutional development of the municipalities in the country.

The newly-formed National Society for Earthquake Technology - Nepal (NSET/NEPAL) is planning to organise in November 1993 a High Level Meeting between the policy-level personnel of Nepal (politicians and bureaucrats) and experts of international repute in the fields of earthquake technology, in cooperation with the International Association for Earthquake Engineering (IAEE) and their World Seismic Safety Initiative (WSSI). The international experts—senior professionals and high officials of IAEE Executive Committee—are expected to meet also with representatives from the professions, the business community, and influential private citizens.

On a long-term basis, NSET/NEPAL has plans to organize and coordinate possible international assistance in training Nepalese professionals in the field of earthquake engineering and technology. The Society also wants to start a modest earthquake hazard and safety awareness program targeted at the common citizen.

These programs, no doubt, can contribute much in raising the awareness level in the country. However, it must be recognized that the task ahead is immense and a systematic approach is required, including institutionalization and the definition of responsibilities. The lack of a central agency to develop the philosophical approach and to coordinate all aspects of disaster management, including disaster mitigation, is very acutely felt.

7. Current Resources and Deficiencies

7.1 MANPOWER

On paper there are many academically-qualified geologists, geophysicists, earthquake engineers, architects and engineers available to form a core group of specialists in seismic safety. Many of these have undergone specialist training in earthquake engineering and hazard mitigation, etc., in various well-known institutions worldwide. The Asian Disaster Preparedness Centre at the Asian Institute of Technology in Bangkok has provided training in Disaster Management for Nepalese nationals ever since its establishment. In contrast, however, there are almost no earthquake engineers implementing acceptable seismic design in Nepal and many of the trained professionals have been unable to implement the knowledge gained. The reason for this situation are complex and are bound up with cultural practices and the slow development of professionalism.

7.2   INFORMATION

7.2.1 General. Information in libraries and archives with complete and accessible records of past studies are virtually nonexistent.

7.2.2 Geographical Information Systems (GIS). The Danish International Development Agency (DANIDA) is currently assisting the Nepal Telecommunication Corporation to assemble computer-based maps of the Kathmandu Valley. There appears to have been as yet no attempt at coordinating the basic parameters such as accuracy, projection and accepted digital versions of common information such as national and district boundaries among the growing number of GIS users in Nepal.

7.2.3 Bibliography. The National Building Code Development Project is currently preparing a computer-based bibliography of local and international documents relevant to their study. This will be particularly strong with respect to the seismic hazard mapping of Nepal.

7.3 MANAGEMENT RESOURCES FOR POST-DISASTER MANAGEMENT

The institutional set-up for the management part of post-disaster activities appears to be well-organized at the highest government levels. However, frequent transfers of the officials from the Special Disaster Unit (SDU) in the Home Ministry mean there is no continuity of special knowledge. The present SDU consists of officials who were recently transferred from general administration positions. It is anticipated that this practice will continue in the foreseeable future.

This dilutes, to a large extent, any effort to establish permanently the core group of specialized and trained personnel so necessary for the successful implementation of the preparedness and post-disaster mitigation plans.

The administrative district organizations suffer from similar problems—probably on a larger scale than in the head office.

The recent experience in flood disaster management showed that there remains much to be desired and improved in the national capability in disaster management. Lack of strong leadership at various levels and lack of coordination was conspicuous. UNDP was asked to coordinate donors. The country was totally unprepared for the disaster. Kathmandu Valley was inaccessible by road from the rest of the country for weeks because of washed-out bridges, resulting in acute shortages of food, commodities and petroleum products. The police, the Army and international donors provided a remarkable service in re-opening the road system.

7.4 POSSIBLE INSTITUTIONAL RESOURCES FOR PRE-DISASTER HAZARD MITIGATION

7.4.1 *IDNDR National Committee.* A National Committee for the International Decade of Natural Disaster Reduction (IDNDR) has the Home Minister as chairman. Its activities are linked to those of the Special Disaster Unit and it is not thought to be active in the scientific and technological aspects of hazard mitigation.

7.4.2 *National Society for Earthquake Technology.* This new Society, as mentioned previously, is enthusiastically aiming to foster the advancement of the science and practice of earthquake engineering and technology. It plans to do this by providing a professional forum for the coordination of the intellectual resources from both government and private organizations.

7.4.3 *Tribhuvan University.* While the Institute of Engineering and the Department of Geology within the university have well-trained staff in structural design, architecture, geology and seismology, there is little evidence of their contribution outside the campus to the mitigation process. The lack of strong graduate research schools undoubtedly is hindering their progress.

7.4.4 *National Bureau of Standards and Metrology* The National Bureau of Standards, under the Ministry of Industry, is a member of the International Standards Organization and has published a relatively small number of its own standards. A few of these are relevant to the construction industry. Its role in maintaining building standards is expected to rise dramatically in the near future.

7.4.5 *Asian Disaster Preparedness Centre (ADPC).* ADPC, based in Thailand, has consistently been very generous in providing training to Nepalese nationals in the field of disaster reduction. This has resulted in a number of individuals receiving advanced training, but the expected strengthening of national institutions is less obvious.

7.5 NATIONAL AND INTERNATIONAL NETWORKING

There are a number of strong relationships between individual overseas organizations and their Nepalese counterparts in fields relevant to seismic safety. Many of these are driven by very specific projects which are part of the overseas organization's research agenda. Others are part of intercountry aid and development programs.

7.6 PROFESSIONAL DEVELOPMENT

7.6.1 *Advanced Training.* Advanced on-the-job training for graduates with geological, engineering and architectural degrees is almost nonexistent. This is a serious deficiency which needs to be addressed both by government regulation and the professional organizations.

7.6.2 *Self-Regulation & Ethics.* The Nepal Engineers Association, an umbrella organization for both architects and engineers from a number of subdisciplines, is reported to have had institutional difficulties in recent years and is not seen as a professional leader. Two recently-formed societies have attempted to fill some of the gaps :

The Society of Consulting Architectural and Engineering Firms (SCAEF) has approximately 25 Kathmandu-based members and has subcommittees investigating professional ethics and advanced training.

The Society of Nepalese Architects (SONA) has recently been formed and is accepting the chance to represent the profession on government committee investigating by-laws.

None of the above institutions has yet been able to introduce a system of peer review leading to an accepted professional qualification such as is common in most more-developed countries.

There is therefore no existing infrastructure that would ensure that all design professionals would be obliged to make their clients aware of the importance of anti-seismic features. Similarly, there is no acceptable means by which either the government or the public can judge the competence of design professionals.

8. Summary

Nepal is, right at this time, taking many of the required steps for improving seismic safety of her people. The country is very fortunate in that her efforts are being assisted by international agencies like UNDP, Habitat, and WSSI.

However, the task lying ahead is immense. At the root of the required action is the need to raise the awareness of all sectors of the community to the seismic risk with which they are living.

Considering the perceived vulnerability of Kathmandu, which will be the source of all assistance immediately after such a disaster, a worthwhile first step would be the participation of as wide as possible section of the Kathmandu Valley community in an earthquake scenario exercise. This would very much complement the progressive and planned introduction of the strength-based building code being developed at the moment.

EARTHQUAKE DISASTER COUNTERMEASURES IN SAITAMA PREFECTURE, JAPAN

FUMIO KANEKO
OYO Corporation
2-2-19 Daitakubo, Urawa
Saitama 336 Japan

1. Introduction

Located just north of Tokyo, in one of the most active seismic zones in Japan, Saitama prefecture has been hit by several destructive earthquakes in the past. Though there has not been any major seismic activity in the last sixty years, the prefecture has developed and become highly urbanized. It has been recognized that there are some weaknesses in newly-developed areas and facilities and that they are vulnerable to future natural disasters, especially earthquakes. Therefore, one of the most important and urgent needs of the prefecture is to have a plan for mitigating earthquake damage that includes measures to be taken before and after disasters.

2. Demographics

Saitama prefecture covers an area of approximately 3,800 km^2. It consists of 92 cities, towns and villages. As of 1992, there were approximately two million households and a total population of 6.45 million. Several cities in the southeastern part of the prefecture have populations of 300,000 to 400,000, with population densities of 4,000 to 9,500 persons per square kilometer. There are also several sparsely- populated towns and villages in the western part of the prefecture with population densities of 6 to 150 persons per square kilometer. The average population density in the prefecture is about 1,700. The growth rate of the population has recently slowed to 2%, one-third of which is due to natural growth and the rest to people moving from neighboring areas. Figures 1 and 2 show the geomorphologic characteristics and the distribution of the population in Saitama prefecture.

While the western part of the prefecture is mountainous, with hard soil and rocks, the eastern part consists of densely-populated lowlands, with soft soils found along the rivers. The central part is hilly, with volcanic ashes covering stiff soil. Recently, the population of this region has been increasing rapidly.

3. Seismicity

The 1855 Ansei-Edo earthquake (M=6.9, epicenter about 10 km south of the prefecture), the 1923 Great Kanto earthquake (M=7.9, epicenter about 30 km south of the prefecture), and the 1931 Nishi-Saitama earthquake (M=6.9, epicenter in the northern part of the prefecture) are the most recent disastrous earthquakes in the history of Saitama. According to the damage reports, 316 people were killed and over 10,000 houses were destroyed by the 1923 Kanto earthquake.

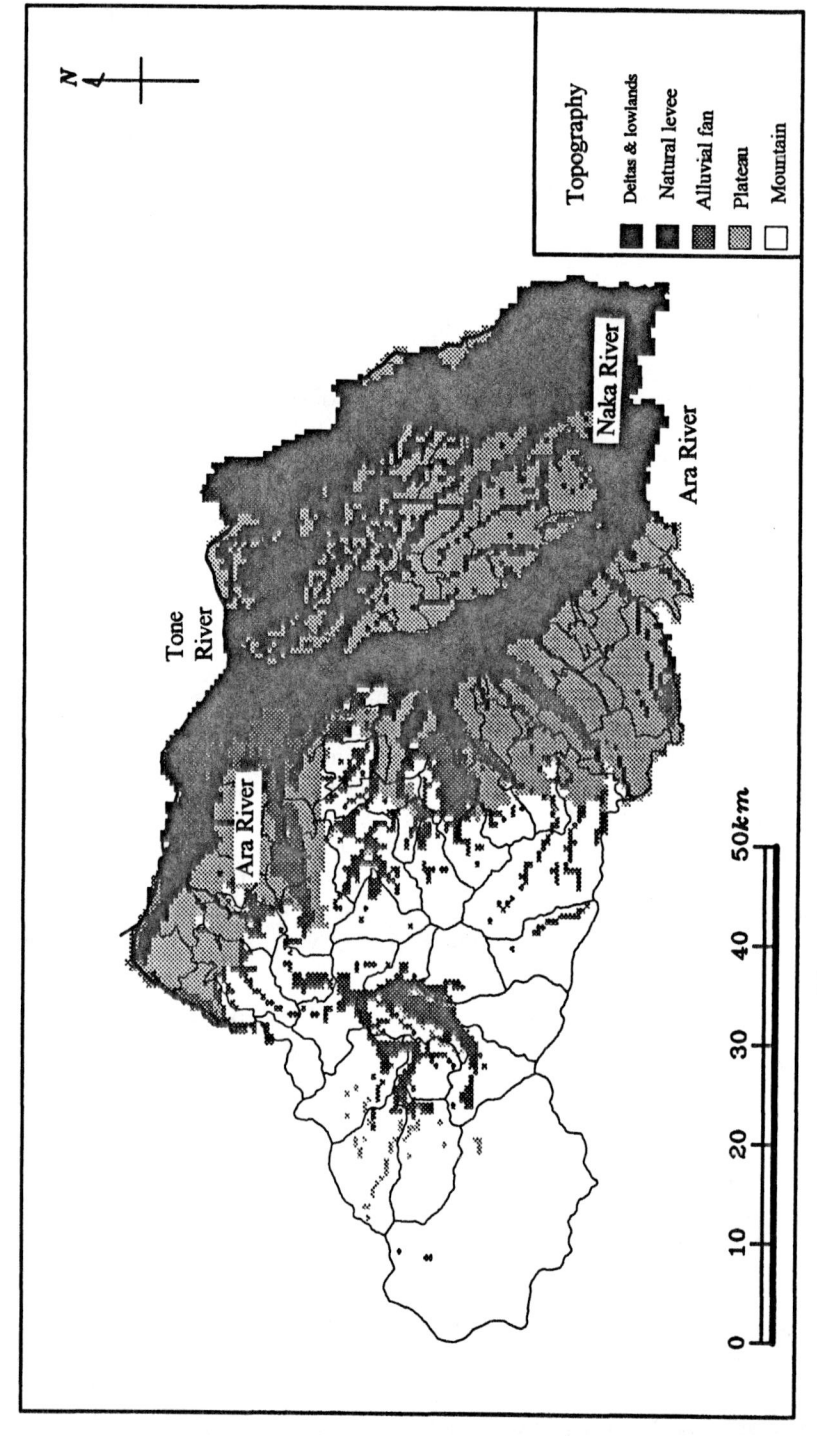

Figure 1. Geomorphologic characteristics of Saitama prefecture

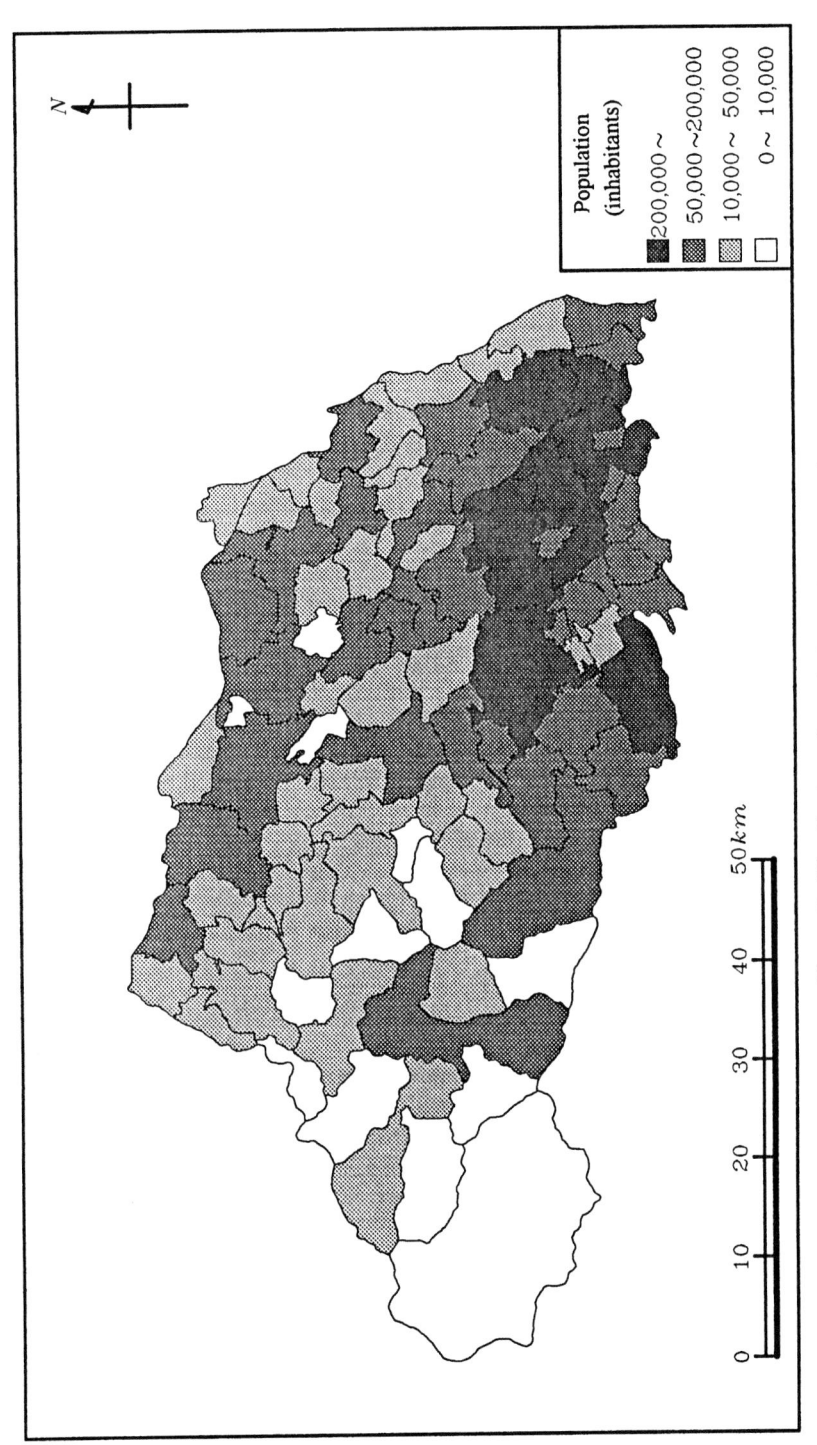

Figure 2. Distribution of population in Saitama prefecture

The seismic intensity is estimated to have been V to VI by the JMA (the Japan Meteorological Agency) scale [ed. note—approximately VII to X MMI]. The 1931 earthquake killed eleven people and destroyed around 200 houses. Usually more than twenty earthquakes are felt every year in the prefecture.

The National Land Agency of Japan, in its report of August 1992, estimated that a recurrence of the Kanto earthquake with a magnitude of around 8 would not occur in the next 100 years. On the other hand, the probability of occurrence of an earthquake of magnitude 7 in an area including Saitama, caused by interaction between the Philippine Sea plate and the Eurasian plate, is higher than that for the recurrence of the Kanto earthquake. It is expected such an earthquake will occur within the next 20 to 30 years.

4. Construction

Of the two million houses and buildings in the prefecture, the majority are wooden houses of one or two stories or reinforced concrete buildings having three to five stories. In the central part of the major cities, many tall RC and SRC buildings can be found. Figure 3 shows the distribution of wooden houses.

The total number of buildings is growing by about 3% a year, mainly due to new housing construction. The most rapid growth takes place in the central part of the prefecture. In the eastern part, due to the high population density, many new homes are built on soft sediments or other poor soils. This is a typical problem of housing construction in Japan.

Earthquake-prone Japan, including the Saitama prefecture, has strict building codes, especially concerning aseismic design. Also, land-use regulations exist for the most populous areas which comprise around 20% of the prefecture's total area. Almost 100% of the houses, buildings and various kinds of structures and facilities are built under these codes or regulations. The basic value of seismic force for aseismic design in Japan is 0.2 G, and various coefficients are used to account for seismicity, soil conditions, and structural types.

The codes and regulations for the aseismic design of buildings and houses in Japan were greatly improved in the revisions of 1950 and 1975. Also, some small changes were made in 1992 in land-use regulations and in design codes for three-story wooden houses.

5. Emergency Response and Earthquake Awareness

According to an investigation conducted by the prefecture, 65% of the people are conscious of the threat of a major earthquake. This is one of the reasons why the prefecture has been so active in preparing against disasters. There is a Prefectural Disaster Prevention Council, chaired by the governor and including representatives from all the prefectural departments, as well as from the private sector. This council has discussed and promoted disaster control programs. When a disaster occurs, the council becomes the headquarters for emergency response activities. Also, the council coordinates research on disaster reduction problems with experts from universities, research organizations, and consultant companies.

Saitama administers other disaster prevention programs through its fire and police departments. They include a wireless communication system covering the whole prefecture, a helicopter for the various disaster control activities, and five Disaster Emergency Centers (two have been already

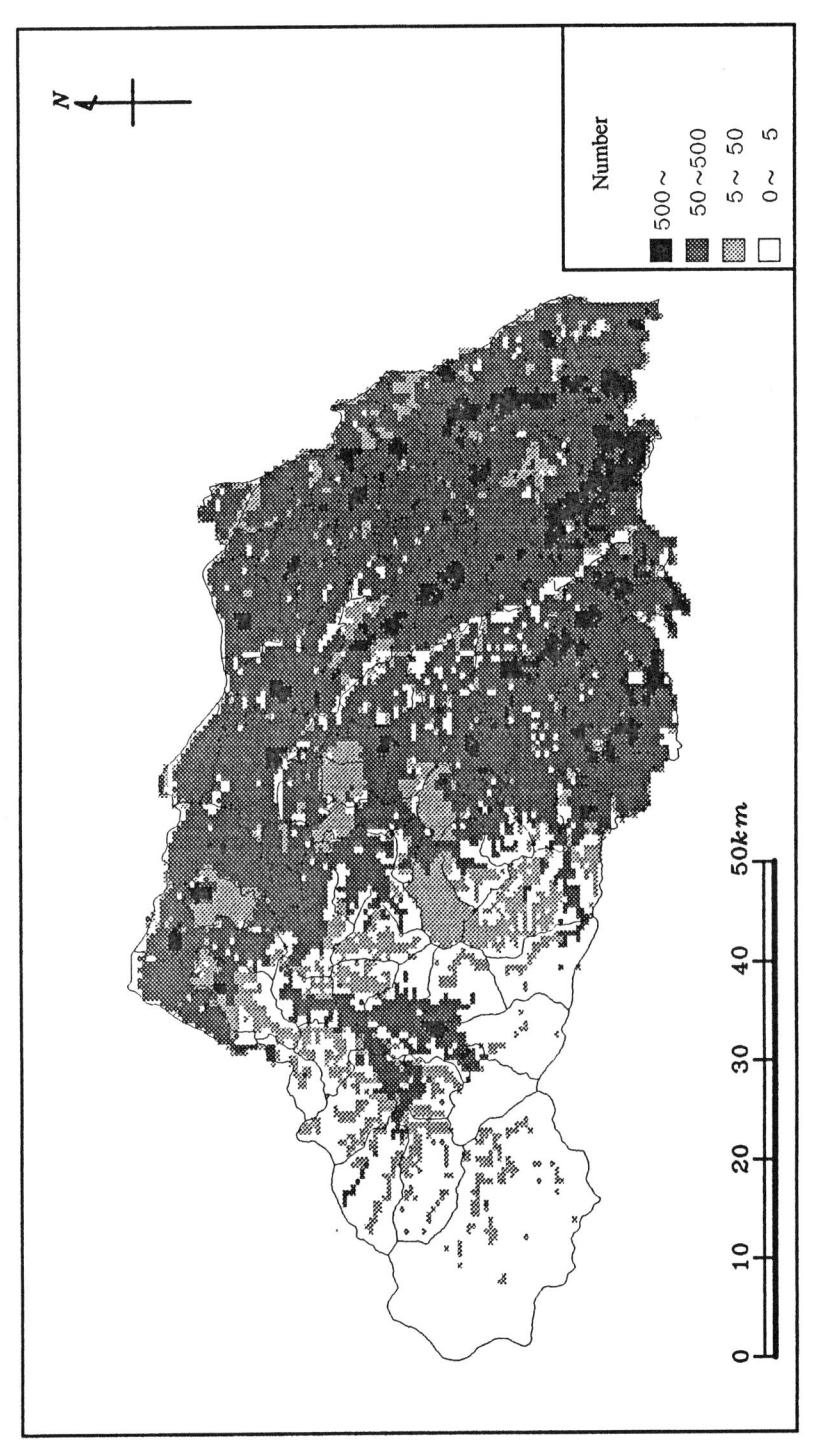

Figure 3. Distribution of wooden houses in Saitama prefecture

completed) with heliports and warehouses for storage of vast amount of food, water, blankets and medical supplies.

To heighten public awareness of disaster control, a Disaster Control Education Center is now being built, and educational leaflets and videos have been published and distributed. Training and simulations, with citizens' participation, are carried out every year on September 1, the anniversary of the 1923 Great Kanto earthquake.

About 30% of Saitama residents are organized into local disaster control units. The main activity of these units is to study ways individual families can best control damage from disasters. When disasters occur, they are prepared to respond properly and act as the people of San Francisco did after the Loma Prieta earthquake of 1989.

6. Earthquake Damage Scenario and Damage Estimation System

6.1 EARTHQUAKE DAMAGE SCENARIO

In order to establish adequate and effective earthquake disaster countermeasures, earthquake damage scenarios or seismic microzoning, which estimate seismic intensity and various kinds of damage caused by earthquakes, are very helpful. Saitama prefecture commissioned earthquake damage scenarios in 1980, assuming major earthquakes like the Kanto earthquake and the Tokai earthquake. These scenarios have been used to improve the prefecture's disaster countermeasures. In 1989, the prefecture started a second earthquake damage scenario project to provide more precise estimates about local earthquakes, following warnings by the National Land Agency of Japan.

The following is an outline of the results obtained from the study:

6.1.1 *Contents and Procedure of the Analysis.* The scenario analysis consists of several steps in which natural sciences, social sciences and government have roles. The main procedure of the analysis is as follows:
- Identification of disaster prevention problems in the objective area.
- Postulation of the kinds of earthquakes that may affect the area, mapping the distribution of their seismic intensities and assessing the probable effects of their seismic motion.
- Estimation of damage to structures, lifeline facilities, etc.
- Estimation of fires, casualties and time to restoration of normal conditions.
- Summarization of findings and formulation of strategy.

6.1.2 *Seismic Intensity Distribution (SID) Estimation.* SID is a basic part of the earthquake damage scenario.

First, four earthquakes were hypothesized for the scenario. Historical earthquakes, seismicity, active faults and seismo-tectonics in and around the prefecture were taken into consideration. Table 1 and Figure 4 show these hypothetical earthquakes, which may endanger the prefecture in the future.

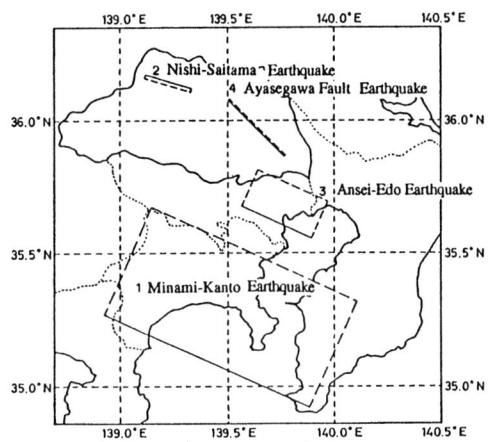

Fig. 4 Epicenters of hypothetical earthquakes

No.	Name	Description	Fault Length (km)	Presumed Magnitude
1	Minami-Kanto Earthquake	Recurrence of 1923 Kanto Earthquake	85	7.9
2	Nishi-Saitama Earthquake	Recurrence of 1931 Nishi-Saitama Earthquake	20	6.9
3	Ansei-Edo Earthquake	Recurrence of 1855 Ansei-Edo Earthquake	30	6.9
4	Ayasegawa Fault Earthquake	Though not recorded so far, it could cause severe damage	35	7.4

Secondly, SID was calculated on the basis of geological analysis and seismic motion analysis. Since it is well known that seismic ground motion varies according to geological features, geological analysis is very important.

In this case, geological information, including distribution of soil types, was compiled and analyzed. As a result, geological conditions in the prefecture were classified into more than 300 representative ground types, and the amplification characteristics of seismic motion were calculated for each ground type. Geological structures and physical properties of the soil layers were taken into account. Then, for the convenience of the following analysis, the area was divided into about 15,000 squares, 500 m on a side. Each square unit was given a corresponding ground type.

Next, using a semi-empirical method that considers a fault model, ground motion at the basement that would be caused by each of the hypothetical earthquakes was calculated for each square unit.

Finally, surface ground seismic motion for each square unit was calculated using ground amplification and basement motion. Figure 5 shows one example of a SID map for the Minami Kanto earthquake (recurrence of the 1923 Kanto earthquake). Seismic intensity values were converted from calculated surface motion. After estimating SID, such effects of seismic motion as liquefaction and slope failure were assessed. One example is shown in Figure 6.

6.1.3 *Damage to Structures.* Damage to structures is estimated by assessing seismic intensity, vulnerability functions (the relationship between damage rate and seismic intensity) and number and types of structures. Vulnerability functions, sometimes called damage matrices, are mainly deduced from data on past earthquakes or by theoretical analysis. They are used to estimate the number of damaged structures and damage rate to each structural type. As an example, damage to wooden houses caused by the Minami Kanto earthquake is shown in Figure 7.

6.1.4 *Damage to Lifeline Facilities.* Lifeline facilities include water supply, sewage pipes, gas supply, power supply, roads, and railways. The procedure for estimating damage to lifeline facilities is similar to that of structural damage estimation, that is, it is deduced from data on past earthquakes. As an example, damage to railways caused by the Minami Kanto earthquake is shown in Figure 8.

6.1.5 *Other Estimates.* Other types of damage assessment concern casualties, monetary loss, and loss due to fire. Time of restoration of lifeline facilities to normal functioning was also estimated.

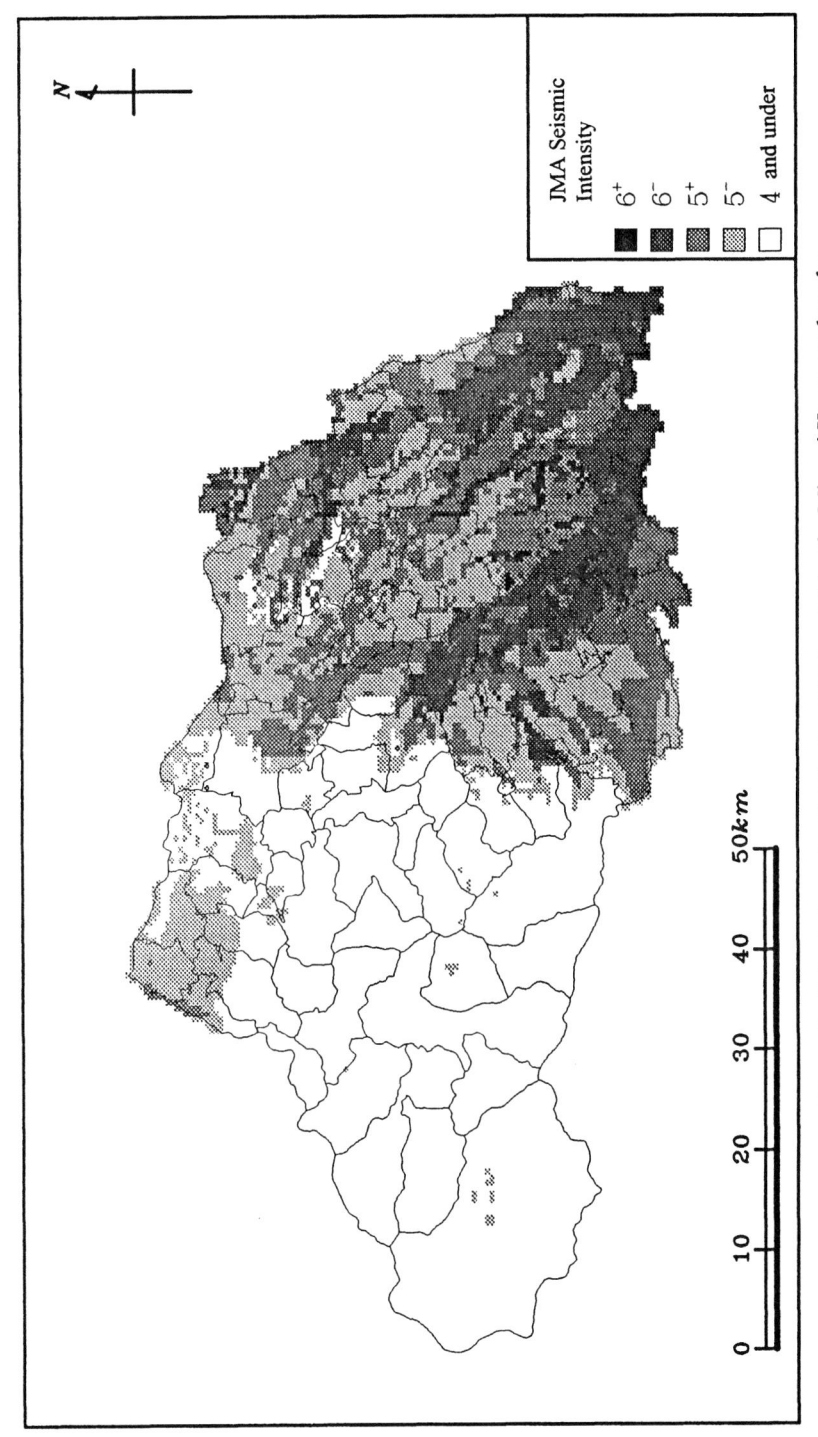

Figure 5. Seismic intensity distribution (JMA scale) by the Minami Kanto earthquake

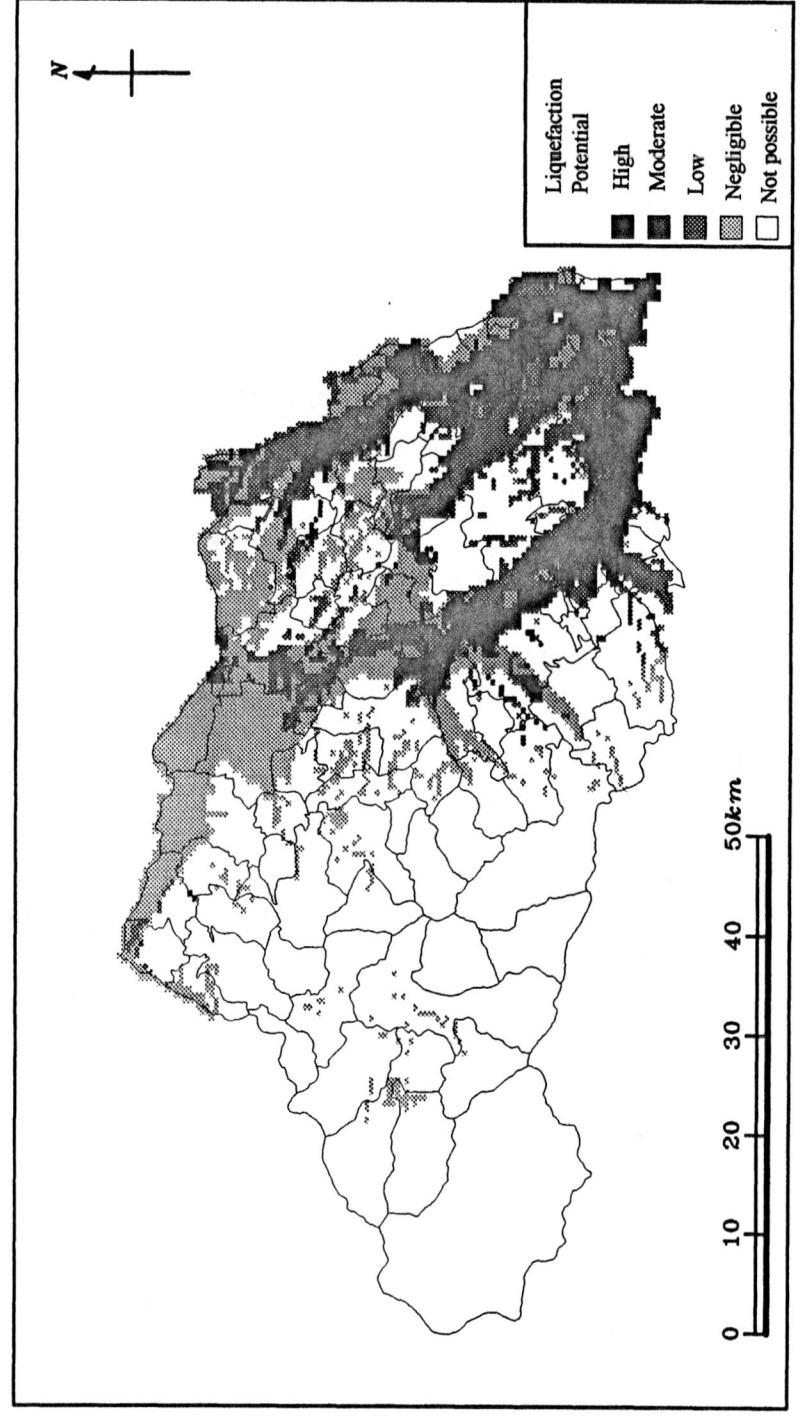

Figure 6. Liquefaction caused by the Minami Kanto earthquake

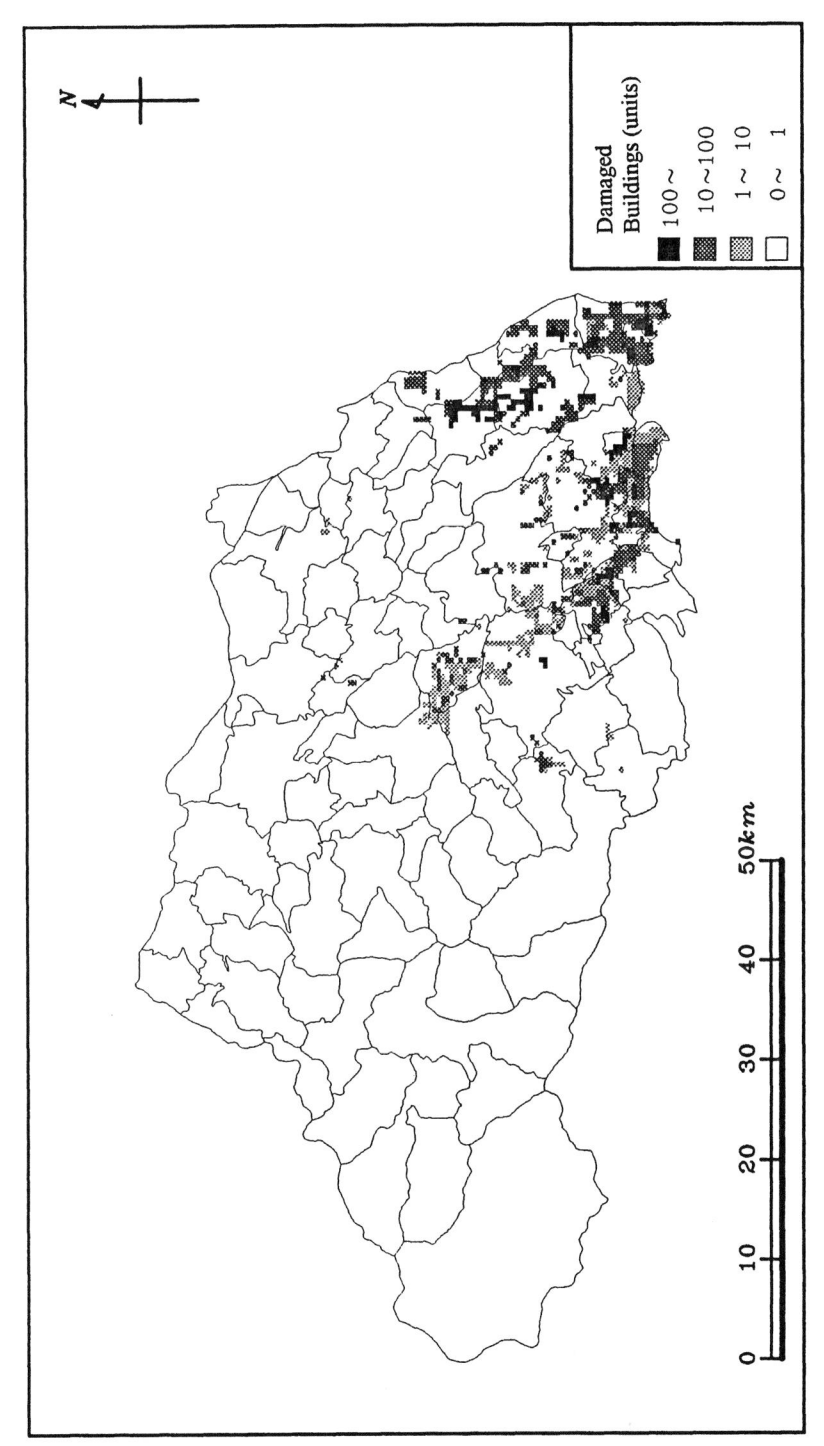

Figure 7. Damage to wooden houses caused by the Minami Kanto earthquake

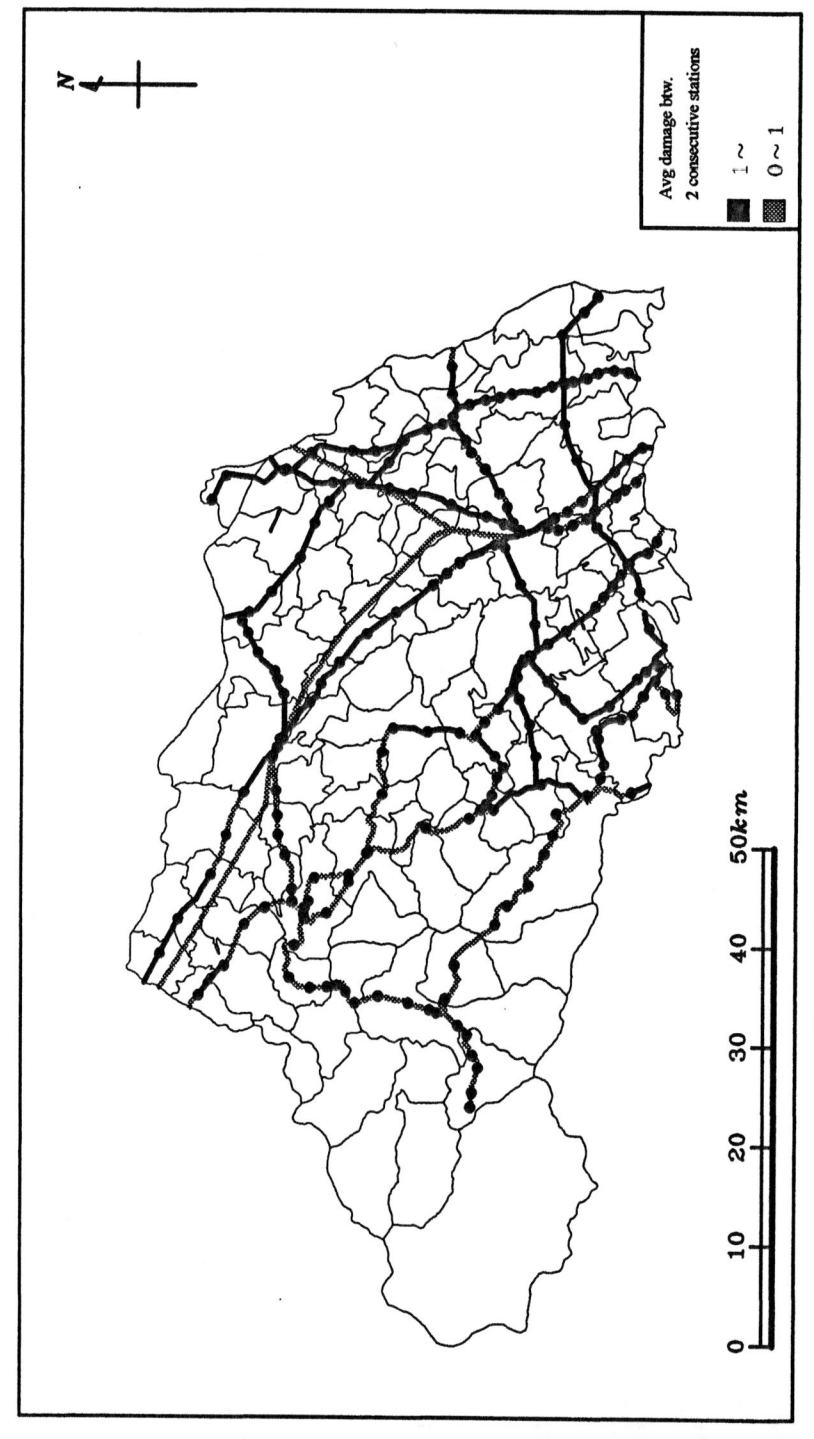

Figure 8. Damage to railways caused by the Minami Kanto earthquake

For these estimates, simulations or empirical methodologies were used.

6.1.6 *Comparison of Damage from the Four Hypothetical Earthquakes.* Different earthquakes have different kinds of effects. When planning earthquake countermeasures, characteristics of damage and the probability of a given kind of earthquake should be taken into consideration. For this purpose, results of damage due to the hypothetical earthquakes were compared. Table 2 shows this comparison.

For Saitama prefecture, the Minami Kanto and the Ayasegawa-fault models cause greater damage, but their probability of occurrence is lower than that of the other models.

6.1.7 *Public Information.* A scenario booklet, videos and leaflets were distributed throughout the prefecture to disseminate information about earthquakes and their effects.

6.2 SAITAMA EARTHQUAKE DAMAGE ESTIMATION AND INFORMATION SYSTEM (SEDEIS)

Small changes in the various factors would lead to large variations in the results and would also increase costs and time required for the future projects. Therefore, in order to utilize the results, the methodologies, and the vast amount of data compiled by the project, a system for estimation of damage was developed for use on a personal computer.

The SEDEIS (Saitama Earthquake Damage Estimation & Information System) consists of the following three sub-systems:
- The Data Managing System deals with natural and social data on disaster prevention in Saitama prefecture. It provides a function of arrangement for data variation according to time.

- The Precise Estimation System can make a precise damage estimate before an earthquake occurs. Such factors as location of epicenter, season of the year and wind velocity can be programmed in.

- The Urgent Estimation System can make a rough estimate of damage in a short time. It is meant to be used immediately after an earthquake occurs, when only epicenter and magnitude are known, for the purpose of quick response and decision-making in an emergency.

These systems should be a powerful tool for establishing an earthquake disaster countermeasure plan for Saitama prefecture.

```
                        ┌──── Data Managing System
        SEDEIS  ────────┼──── Precise Estimation System
                        └──── Urgent Estimation System
```

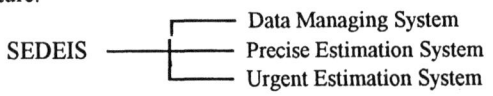

Figure 9. Main Features of SEDEIS

7. Final Remarks

A recent example of a seismic microzoning project in Japan has been introduced. Since Japan is an earthquake-prone country that has urbanized rapidly, more precise and useful seismic microzoning studies are in demand. Technical advances have increased the effectiveness of these studies and the speed with which they can be conducted. In the past twenty years, quite a number of seismic microzoning projects were conducted in high-seismicity regions in Japan. Their

Table 2 Comparison of damage from the four hypothetical earthquakes

	Item		Current Condition	Damage for Each Hypothetical Earthquake				Remarks
				Minami-Kanto	Nishi-Saitama	Ansei-Edo	Ayase-River	
	Magnitude		—	7.9	6.9	6.9	7.4	
Damage description	Structures	Wooden buildings	1,936,180 u.	41,707(1.4%)	3,751(0.1%)	9,493(0.3%)	62,938(1.9%)	Damaged units = No. of severely damaged + moderately damaged units. Figure in parentheses: Damage rate = Rate of severe damage + 1/2 rate of moderate damage
		RC buildings	80,881 u.	1,026(1.1%)	1,435(1.0%)	474(0.5%)	2,351(1.7%)	
		SF buildings	96,884 u.	1,644(1.5%)	840(0.7%)	308(0.3%)	2,983(2.3%)	
		Brick walls	678,000 u.	22,326(3.3%)	2,229(0.3%)	10,037(1.5%)	14,505(2.1%)	Include stone walls
		Falling objects	117,765 u.	5,988(3.4%)	1,907(1.1%)	4,736(2.7%)	7,852(4.4%)	From 3-or-more-stories-high buildings
	Fires	Outbreak	—	83 cases	6	2	264	at 18:00 in winter (max.)
		Spreading	1,936,180 u.	49,794(2.6%)	77(0.0%)	4(0.0%)	49,740(2.6%)	Wind velocity at 8 m/s
	Lifeline facilities	Water Supply Region	19,494 km	10,881 locations	423	2,238	5,372	Water supply pipelines
		Water Supply Prefecture	511 km	6 locations	0	1	2	Prefectural network
		Sewerage Region	9,125 km	1,441 locations	9	295	365	Objects: Duct lines
		Sewerage Prefecture	303 km	8 locations	0	2	6	Prefectural main collectors
		Gas	819 km	18 locations	0	4	5	Medium-pressure pipelines
		Electric power Electric poles	835,781 u.	22,708 u.	135	714	22,021	Incl. damage due to fires
		Electric power Service wires	4,790,749 u.	127,394 u.	451	1,961	126,843	Incl. damage due to fires
		Telephones	2,477,900 u.	8,860 u.	1,288	5,405	5,063	
	Transportation facilities	Roads All roads	45,038 km	2,809 locations	1,252	1,969	2,936	Main and secondary roads
		Roads Main roads	600 km	55 locations	21	39	58	
		Bridges	596 u.	8 u.	0	0	3	Bridges over main roads
		Pedestrian bridges	348 u.	4 u.	2	2	0	
		Railways	675 km	338 locations	152	266	320	
	Others	Cliffs	734 locations	70 locations	53	39	62	
		Reclaimed lands	52 locations	0 location	0	0	0	
		River banks	318 km	194 km	53	100	133	Flood-prevention sections
		Reservoirs	580 locations	0 location	0	0	0	
Casualties and social impact		Deaths		1,580 pers.	50	150	1,540	
		Injuries	6,319,639 pers.	15,520 pers.	1,440	3,030	16,790	
		Homeless		162,930 pers.	1,610	10,840	150,340	
		Refugees		203,800 pers.	—	—	169,090	
		Water-suspended households	2,028,201 u.	662,427 u.	21,738	220,267	340,781	Immediately after earthquake
		Electric power-suspended households		595,396 u.	33,712	255,621	260,531	Immediately after earthquake
Losses		Buildings	—	¥1,346.5 bil.	108.5	165.4	1,644.3	Incl. damage due to fires
		Water supply	—	¥29.8 bil.	1.0	5.9	14.2	
		Sewerage	—	¥12.0 bil.	0.1	2.7	3.0	
		Gas	—	¥12.0 bil.	1.2	11.1	11.7	
		Electric power	—	¥62.7 bil.	3.6	29.4	36.5	Incl. damage due to fires
		Roads	—	¥147.5 bil.	65.7	103.2	154.3	
		Bridges	—	¥5.2 bil.	0	0	1.9	
		Railways	—	¥2.1 bil.	0.9	1.6	2.0	
		Total	—	¥1,617.8 bil.	181.0	319.3	1,867.9	Prefectural budget for 1990 was ¥1,212.6 billion

* This amount considers only direct costs necessary for reconstruction, restoration, etc., and does not include the indirect losses related to economic activities, etc.

findings have been applied to measures for earthquake disaster reduction programs. These efforts in Japan have improved the state of preparedness for earthquake disasters.

Finally, I would like to thank the advisory committee, chaired by Professor Tsuneo Katayama of the Tokyo University, for giving technical advice for the project. Also, I express my gratitude to Saitama prefecture for allowing me to present this paper.

ARGUMENTS FOR EARTHQUAKE HAZARD MITIGATION IN THE KAMCHATKA REGION

MARK A. KLYACHKO
Kamchatka Center on Earthquake Engineering
& Natural Disaster Reduction (KamCENDR)
9 Pobeda Avenue
Petropavlovsk Kamchatsky, 683006 RUSSIA

1. Introduction

The Kamchatka Peninsula is unique in Russia and the entire world in terms of its concentration of various extreme natural phenomena, such as snowfalls, hurricanes, tides and tsunamis, avalanches, mudflows, landslides, rockfalls, typhoons, floods, volcanoes and, lastly, frequent and devastating earthquakes.

Petropavlovsk (the town of St. Peter and St. Paul) is the major town in Kamchatka. Founded by Vitus Bering in 1740, it is the oldest Russian seaport on the Pacific coast of Russia's Far East, situated both on the coast and on a steep slope of the Avacha Bay and embracing the latter. The population exceeds 300,000.

The seismic hazard for Kamchatka has already been investigated in detail for almost 30 years, and the regional maps of seismic activity and town seismomicrozoning, made in 1978, have become the basis for seismic hazard information.

The Kamchatka Peninsula is located within the Pacific Seismic Belt, in the vicinity of the seismically active Kurilla-Kamchatka abyssal trough, which explains the regional seismic intensity (IX on MSK scale) and recurrence interval (200 years). The average recurrence of VII shocks is 15 years and that of VIII shocks is approximately 60 years. The most recent damaging earthquakes of VII-VIII MSK in Petropavlovsk were in 1952, 1959 and 1971.

It has been estimated that the probability of an M7 earthquake in the near future is about 50% and that of M8 or greater is 20 %. In 1992-1993 earthquakes of M7.2, 6.3 and 7.4 occurred at 100 km from Petropavlovsk.

2. Secondary Dangers from Earthquakes

Joint work between owners of structures and the KamCENDR staff has led to an assessment of secondary hazards, both their consequences and how corresponding damage can be reduced. Three years were needed for this study.

2.1 MILITARY AND INDUSTRIAL HAZARDS

Hazards caused by fire and explosions in Kamchatka's residential areas pose no major threat because there is no gas pipe network there, and electricity lines are automatically switched off at shocks of MSK VI and greater.

Industrial explosions and fires, however, are considered major hazards, as there are large numbers of various fuel tanks aboveground in residential zones and petroleum product storage facilities on the bay shores. Damage to them may result in serious ecological repercussions. To mitigate the risk from petroleum spills, a method of seismic protection has been developed. This method can be used for seismic protection of fuel tanks, water towers and dams.

Chemical hazards pose another serious secondary threat because of the use of ammonia at refrigerating enterprises and chlorine at water treatment stations. The levels of secondary radiation hazard are still being determined.

2.2 TSUNAMI AND SEICHE WAVES

Tsunami waves near the Khalakhtyrka Beach can reach 15 m. The maximum propagation of a tsunami on the eastern coast is accepted to be approximately 8 km, so a tsunami may reach and cover the urban areas. The coast of the Avach Bay is not exposed to great tsunami hazard, but earthquakes can cause tsunami and seiche waves as high as 4-5 m. The limits of this hazard are clearly defined and a warning system is in operation. Damage of marine hydrotechnical constructions has been estimated, and a tsunami vulnerability map compiled.

2.3 SNOW AVALANCHES

Snow avalanches are hazardous for residential areas located on steep slopes only. A map of snow avalanches has been compiled, and 98 of these points within the city determined. The number of anti-avalanche constructions in the city is insufficient. As for landslides, the slopes of in Petropavlovsk are relatively stable.

2.4 LIQUEFACTION

Liquefaction of soil is a real hazard for trade and fishery port construction for earthquakes of intensity VIII-IX because all such constructions are on reclaimed, saturated, silty marine sands. The analysis of liquefaction potential was based on the experience at the Ishinomaki Port during the Miyagi-Ken-Oki earthquake in 1978 (Iai et al., 1985). Taking into account liquefaction, we determined the possibility of gentry-crane overturning during earthquakes and suggested additional measures to increase the reliability of their operation. Further, hydrotechnical grooved piled walls are not seismically stable.

3. The Construction Data Base (DB)

The building stock of residential Petropavlovsk consists of old wooden houses, masonry, large-block concrete buildings of 20-30 years of age, and modern large-panel concrete buildings. Public buildings are typically frame-reinforced concrete panel structures and old masonry.

In 1988-1990, building owners in Petropavlovsk compiled a database and carried out certification, inspection, and primary analysis of level of earthquake resistivity (LER) for public buildings and municipal residential stock in accordance with the Manual (K.B.DALNIIS, 1987). To oversee the work and draw the final conclusions, a Coordination-Methodical Council, consisting of specialists in earthquake engineering and chaired by the author, was established. The results of the work have revealed the following: (1) a generally unsatisfactory state of buildings in Petropavlovsk, a serious deficiency of LER in masonry apartment buildings, and insufficient LER of large-block fund and public buildings, while large-panel buildings were in fairly good shape; and (2) the necessity of constructing new health care facilities and the need for investments in hospital, school and food industry construction. These results are discussed in more detail in Klyachko and Putinsev (1988); Klyachko, 1990; Klyachko and Sopilnyak, 1989; and Klyachko et al., 1992.

Thus the expert LER evaluation and corresponding estimation of primary vulnerability were based on certification results of the civil and industrial buildings (Klyachko and Sopilnyak, 1989, Klyachko et al., 1992)—especially buildings related to life support (health care, energy supplies, lifelines, and emergency response equipment)—and dedicated to the elimination of an emergency situation and earthquake consequences. Then an assessment of primary earthquake vulnerability for separate urban areas was made.

All modern buildings in the town are designed by professional engineers in accordance with the existing Seismic Code SNIP-7-81. All constructions built before 1970 were designed for the load of VII MSK instead of IX MSK as now. Urban building control exists and is enforced, but it is not adequate, and therefore many buildings show a low quality of construction work.

In terms of materials, local volcanic slag, which has some peculiar features, is used for stone masonry and for light concrete.

4. Emergency Response

Government officials and business leaders are well aware of the earthquake danger; however, practical preparedness for the disaster is hampered by financial problems and leaves much to be desired.

The government organizations responsible for elimination of the earthquake disaster are: the regional and municipal Civil Defense, the Commission on Emergency, and local departments of Ministry of the Interior. At their disposal they have a special Civil Defense batallion, two rescue detachments (Civil Defense & the Ministry of Interior), and eight fire brigades. The police forces are unable to rescue people and have no elementary medical training. In an emergency situation, the presence of numerous military forces is considered as positive.

The population itself has very poor knowledge of the main methods of assistance and survival, and there is not sufficient training to withstand the disaster.

The number of special hospital beds and medical equipment is not sufficient.

In case of telephone communication failure, there is practically no other means of communication available. Other communication means are at the disposal of Civil Defense staff and state officials only.

5. Research

Kamchatka has three scientific organizations which carry out research in the field of seismic safety: the Institute of Volcanology of the Russian Academy of Sciences (seismological aspect); the Centre of Earthquake & NDR (KamCNDR) (seismic microzoning, geological aspects, geotechnical and structural engineering including earthquake and urban planning); and the Far East Association for Construction Reliability and Population Safety (Scientific Body of the State

Committee on Civil Defense in the Russian Far East).

6. Conclusion

Earthquake-prone Petropavlovsk is characterized by insufficient seismic performance of constructions as well as numerous secondary natural and man-made dangers. Such a combination of danger and vulnerability, when added to the isolation of the area, severe climate, and inadequate earthquake preparedness of city services and the population, results in inadmissibly high socioeconomic risk to the regional center. There is a particularly urgent need to evaluate and analyze the vulnerability and risk, and to develop and implement a long-term program of preparedness for earthquake disaster. In the meantime, prompt measures must be taken to prevent human losses and very extreme damage. The best way to mitigate the earthquake disaster and ensure the seismic safety in Petropavlovsk is to fulfill all the works in accordance with Figure 1.

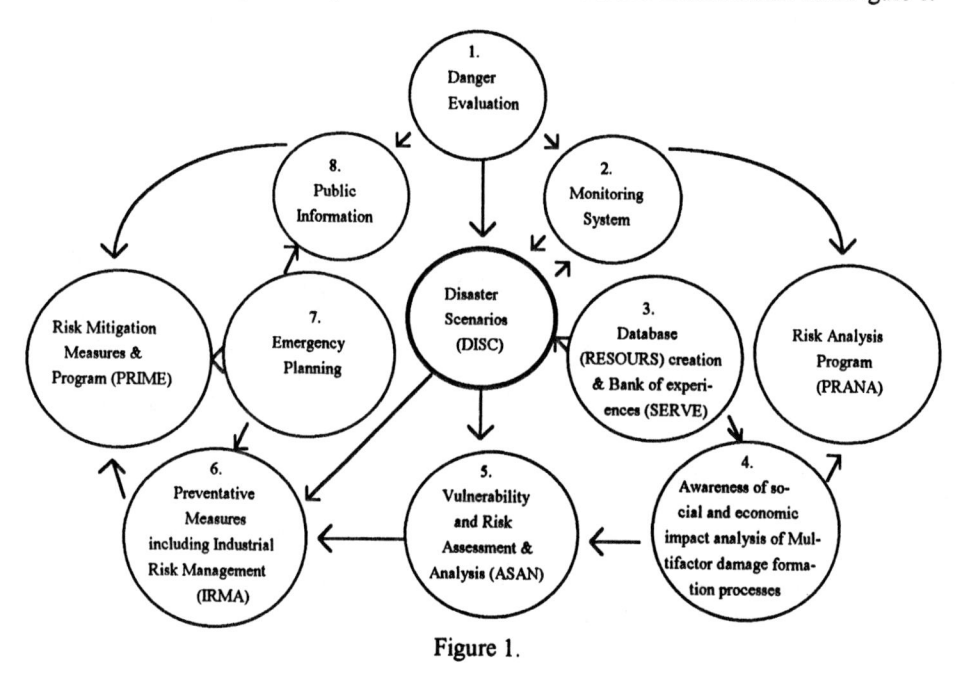

Figure 1.

References

"Methodical Manual for certification of buildings", K.B.DALNIIS, Petropavlovsk-Kamchatsky, 1987.

Iai S., H. Tsuchida, and W. D. Finn, "An effective stress analysis of liquefaction at Ishinomaki Port during the 1978 Miyagi-Ken-Oki Earthquake".- Report of the Port and Harbour Research Institute. Vol.24, No 2, 1985.

Klyachko M., "Certification of buildings in seismically hazardous areas of the Kamchatka region", 9 ECEE, Moscow, 1990.

Klyachko M., U. Peicheva, O. Savinov, A. Uzdin, and V. Uzdin, "Antiseismic pneumo-protection of reservoirs" (pat.right).

Klyachko M., J. Polovinchik, and A. Uzdin, "A Complex Program for Natural Disaster Mitigation on Urban Areas of Russian Far Eastern on Example of Petropavlovsk Kamchatsky", Tehran, May 1992.

Klyachko M.A. and E. O. Putinsev, "Certification and examination of buildings in the seismically hazardous regions of Kamchatka", MSSSS Session, Irkutsk, 1988 (in Russian).

Klyachko M.A.and Yu. Ya. Sopilnyak, "Complex engineering program of preparatory measures on mitigation of damage from the forecasted earthquake in Kamchatka", UNDRO Conference, Moscow, 1989.

SEISMIC RISK AND HAZARD REDUCTION IN VANCOUVER, BRITISH COLUMBIA

CARLOS E. VENTURA
NORMAN D. SCHUSTER
Department of Civil Engineering
The University of British Columbia
Vancouver, British Columbia
Canada

1. Introduction

Several regions in Canada have been subjected to the devastating effects of earthquakes. Some of these earthquakes have caused significant property damage and some loss of human lives. The province of British Columbia, located on the west coast of Canada, has suffered the direct effects of very large earthquakes. Over the past century, several earthquakes with Richter Magnitude 7.0 or greater have affected the provincial territory. The southwest region of British Columbia, where about 70% of the three million inhabitants of British Columbia live, is considered the most active earthquake region in Canada. Because of the geological features of the region, there is a very good chance that British Columbia will experience a large earthquake in the near future. The city of Vancouver and its surrounding communities, known as the Greater Vancouver Regional District, are located in this region.

Considering the size of the population, and mounting evidence of a potentially large and long duration earthquake, media and governments have focused their attention on the seismic risk of the province. Federal and provincial emergency preparedness officials are continuously planning and evaluating measures to deal with the damage and chaos resulting from such an event. To cope with the resulting crisis from such a large earthquake, emergency measures have been planned to ensure rapid mobilization of resources such as medical care, temporary housing, food distribution and financial relief. Emergency planning includes establishment of Emergency Operation Centres by local municipalities, the provincial government (through the planning of the Provincial Emergency Program), and support from across Canada, organized by Emergency Preparedness Canada.

This paper presents an overview of the seismic risk and hazard reduction programs of the Greater Vancouver Regional District. It provides statistical data about the population distribution and growth of the area, and examines the seismicity of the area, including historical earthquake information. The current emergency response programs and earthquake awareness efforts are described, and a general overview of construction practices and current hazard reduction programs are provided. There exist several hazard reduction programs in the province, but the emphasis here is only on building construction. Because of space limitations, existing programs aimed at reducing the seismic hazard in dams, bridges, pipelines, communication systems, etc. will not be discussed here in detail.

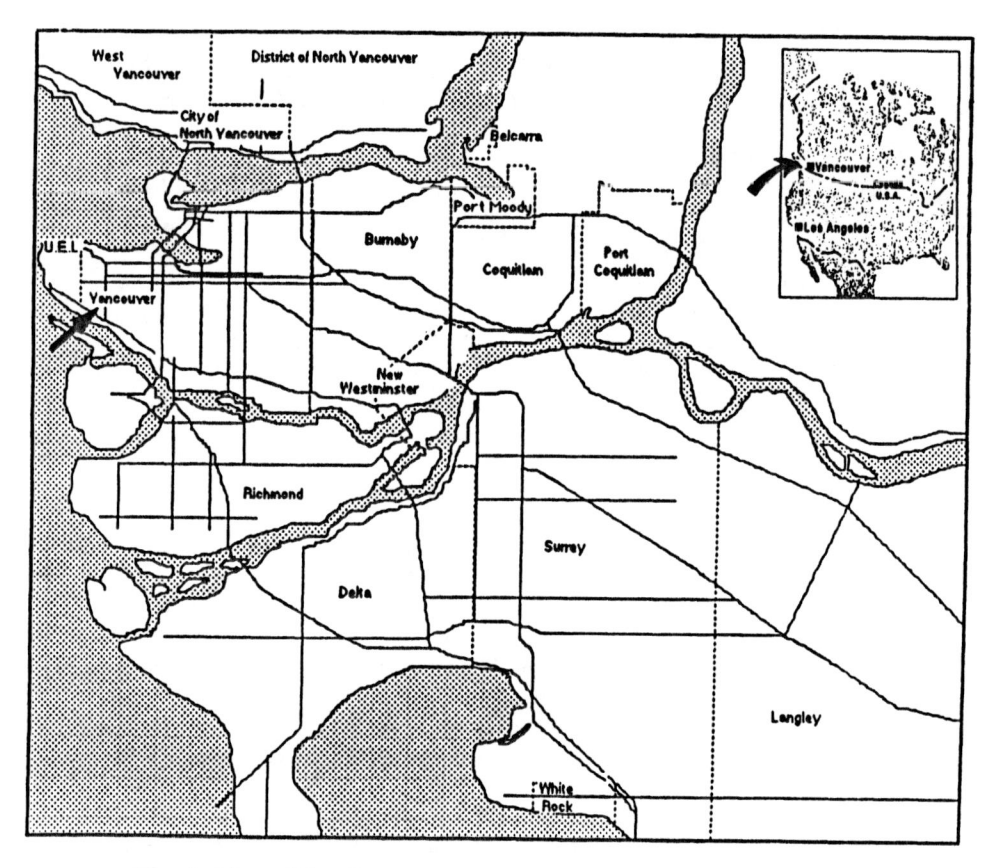

Fig. 1. Map of the Greater Vancouver Regional District (GVRD).

2. Demographics

The Greater Vancouver Regional District (GVRD) is the most densely populated region of the province. The district includes twenty municipalities, three unincorporated areas and all Indian reserves; all of these covering an area of about 293,000 Hectares. A map of the GVRD is shown in Figure 1. Selected statistical information about the district is presented in Table 1 (Greater Vancouver Key Facts, 1993). This table shows that the population of the district exceeded 1.6 million people in 1992. The average annual income for each municipality ranges from about $19,000 to $38,000. A graphical comparison of the population and population density values listed in Table 1 is presented in Figure 2. The city of Vancouver shows the largest population and the highest density as well. North Vancouver City has the second largest population density, but its total population is relatively small. Four of these municipalities, namely Vancouver, Burnaby, Richmond, and Delta, face a high seismic risk because of their proximity to the potential seismic sources as well as their large population and high population densities. Richmond and Delta also face liquefaction and tsunami hazards.

Statistics Canada recently released figures showing that close to 23,000 Canadians relocated to the GVRD last year (The Vancouver Sun, 1993). This number represents about 7% of all the Canadians that moved around the country in 1991-92. Among major population centers in Canada, the GVRD recorded the largest net gain. In addition to this, about 27,500 (12%) of all the foreign immigrants arriving in Canada chose to settle in this area. The net growth of the entire province, including Canadians and foreigners, was over 66,000, which was just exceeded by the growth in the province of Ontario.

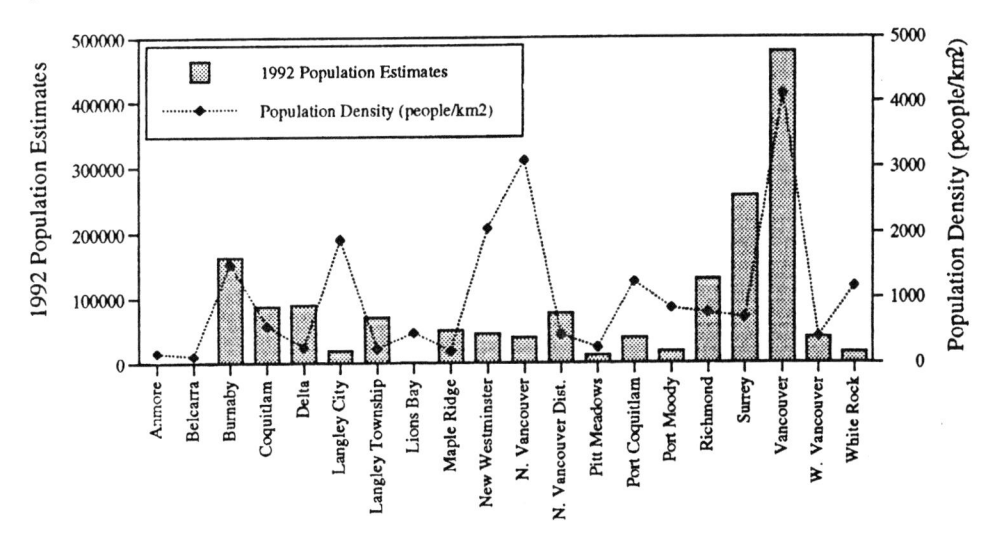

Fig. 2. 1992 Population Distribution of the GVRD.

Table 1. 1992 Demographic Statistics for the GVRD

Municipality	Population Total	Density (persons/km²)	Annual Population Growth (% of previous year's population)	Average Annual Income (US $)
Anmore	796	158.3	N/A	N/A
Belcarra	613	109.2	N/A	N/A
Burnaby	162,817	1,525.4	1.4	20,222
Coquitlam	87,147	571.2	2.7	21,492
Delta	90,171	247.5	1.6	22,986
Langley City	19,311	1,896.6	2.4	21,253
Langley Township	69,616	219.2	3.3	21,253
Lions Bay	1,346	469.5	N/A	N/A
Maple Ridge	50,093	187.5	3.3	N/A
New Westminster	45,480	2,067.1	1.2	19,737
North Vancouver City	39,283	3,099.7	1.2	23,794
North Vancouver District	76,806	431.0	1.4	23,794
Pitt Meadows	11,772	235.2	4.4	N/A
Port Coquitlam	38,556	1,243.1	2.5	21,420
Port Moody	18,304	839.8	1.6	22,267
Richmond	128,773	766.2	2.4	21,789
Surrey	256,452	690.5	4.0	19,355
Vancouver	477,872	4,114.3	1.2	21,642
West Vancouver	39,265	396.9	0.8	37,845
White Rock	16,399	1,170.1	1.7	23,597
Other	12,628	-	-	-
Total	1,643,500			

3. Seismic Risk

The southwest corner of British Columbia is situated over a seismically-active subduction zone. In this geological setting, earthquakes that may represent a hazard to the GVRD region occur in three distinct source regions (Rogers, 1992): crustal earthquakes, deeper earthquakes within the subducted plate, and very large subduction earthquakes on the boundary between two lithospheric plates. Figure 3 shows a tectonic map of the province, where four plates meet and interact. The plates converge, diverge and pass each other at transform boundaries (Information Circular, 1991):

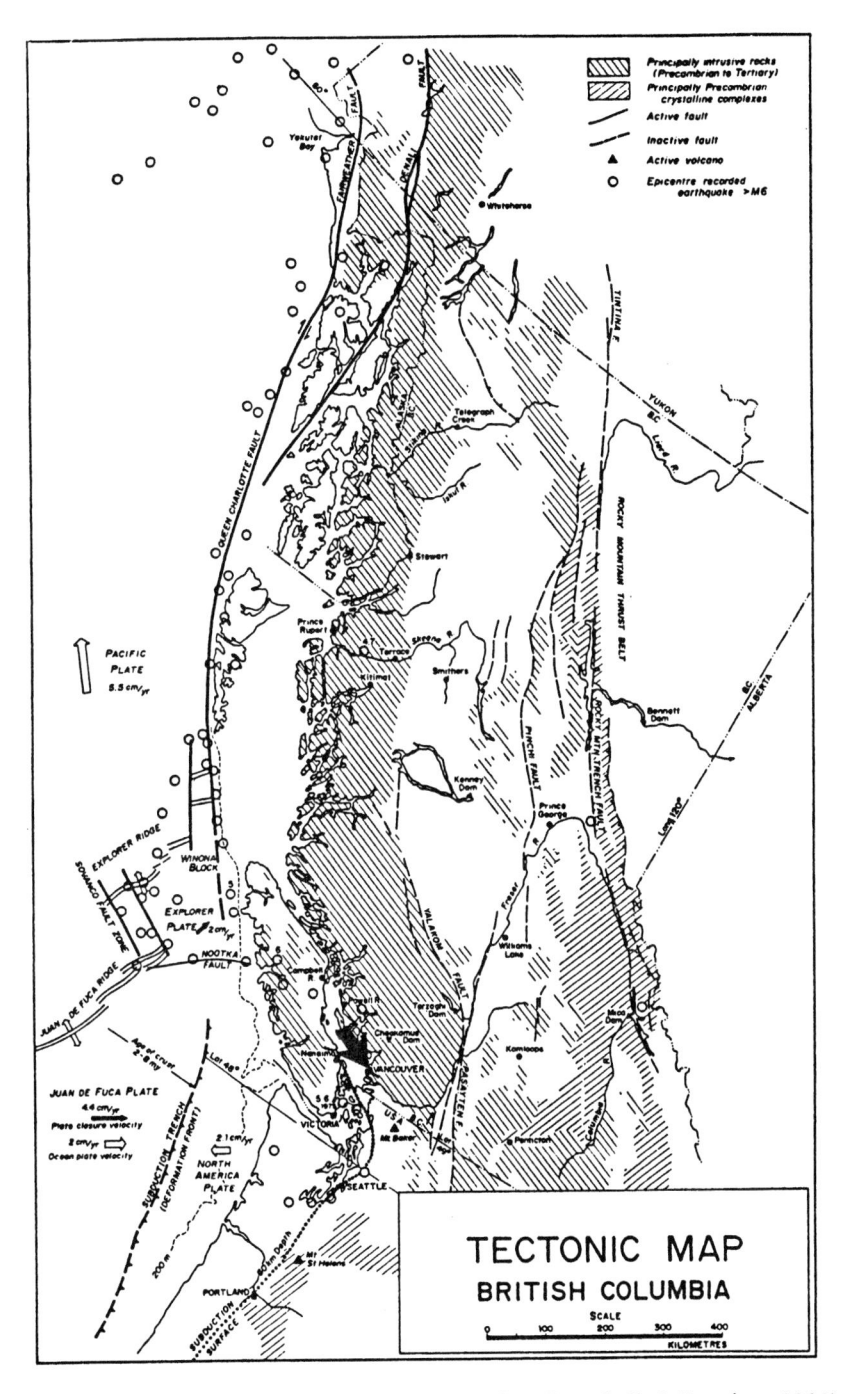

Fig. 3. Tectonic Map of British Columbia (after Campbell & Rotzien, 1992).

- About 200 km off the west coast of Vancouver Island, the Juan de Fuca plate and Pacific plate are diverging or spreading apart along the Juan de Fuca ridge.

- Further east, the Juan de Fuca plate is converging with and sliding beneath the North America plate.

- Another small plate, the Explorer, is also sliding beneath the North America plate, and at the same time the Juan de Fuca plate is sliding past it along the Nootka fault.

- In the north, there is a transform boundary between the Pacific and the North America plates called the Queen Charlotte fault; in 1949, this fault was the site of Canada's largest earthquake.

British Columbia's southwest corner is considered the most active seismic region in Canada. More than 200 earthquakes are recorded each year in the GVRD region and Vancouver Island. As an illustration, a simplified map of British Columbia with the location of earthquakes of magnitude 3 and greater up to the end of 1986 is shown in Figure 4. Although most are too small to be felt, an earthquake capable of producing significant damage can be expected to occur somewhere in the region about once every ten years. On average, three or four earthquakes are felt each year.

Large earthquakes that have been felt or have affected British Columbia during the past century are listed in Table 2 (Information Circular, 1991). Significant recent earthquakes that have affected the GVRD area include a 1965 event (M6.5) in the Puget Sound, Washington, US, and three events in the Vancouver Island region: a M6.0 in 1958, a M5.7 in 1972, and a M4.9 in 1993. The largest and most recent earthquake in southwest British Columbia happened in 1946, when a M7.3 event rocked central Vancouver Island. In 1965 a M6.5 occurred south of the Canada/US border beneath the city of Seattle causing damage to the city and surrounding area.

Table 2. Major Seismic Events that Have Affected British Columbia During the Past Century

Year	Region	Richter Magnitude
1872	Central Washington	7.4
1899	Yukutat Bay, Alaska	8.0
1918	Vancouver Island	7.0
1929	Queen Charlotte Islands	7.0
1946	Vancouver Island	7.3
1949	Puget Sound, Washington	7.0
1949	Queen Charlotte Islands	8.1
1958	Alaska - B.C. Border	7.9
1964	Alaska	9.2
1970	Queen Charlotte Islands	7.4

To date no earthquake has been observed to be centered in the interface between the subducting

Juan de Fuca plate and the overriding North America plate (see Figure 4). There is, however, mounting evidence that a major megathrust earthquake of magnitude 8.5 to 9.1 is possible at this interface (Rogers, 1988). Some investigators, however, suggest that the maximum magnitude of such an earthquake will not exceed M 7.0 (Campbell and Rotzien, 1992) due to the characteristics of the Cascadia subduction zone. Nevertheless, some of the hazards to the GVRD from significant subduction earthquakes are the long duration of strong shaking, and the large areas affected by this shaking.

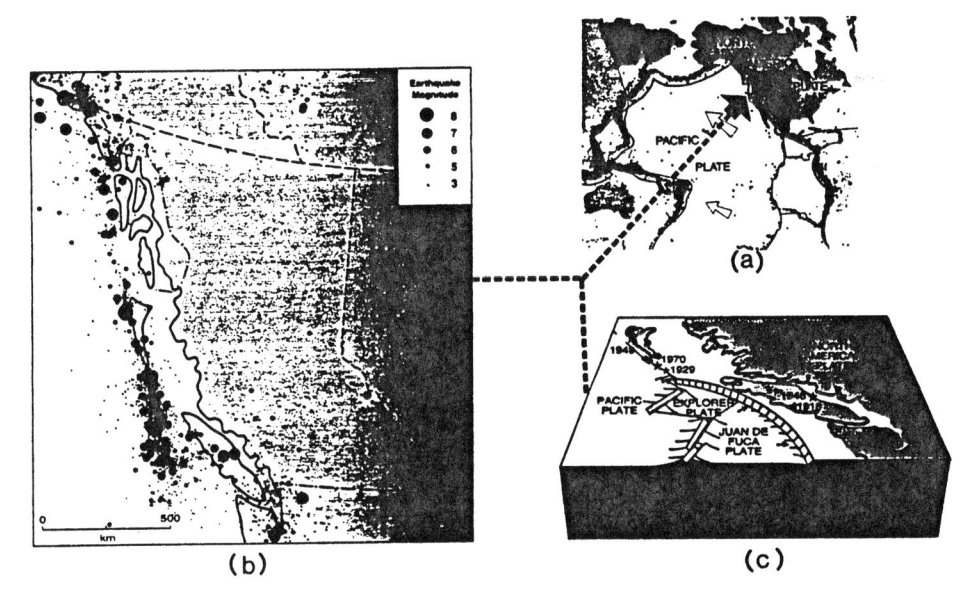

Fig. 4. Seismicity of British Columbia (adapted from Information Circular, 1991): a) global tectonics, b) provincial seismicity, c) local tectonics.

The damage distribution resulting from a significant earthquake affecting the GVRD will not only depend on the size and epicentral distance of the event but also on certain geological characteristics of the areas being affected. In the GVRD region these include:

a) Ground Motion Amplification:

A significant parameter that determines seismic risk is the potential for ground motion amplification due to site conditions. Recent studies conducted by Sy, Henderson, Lo, Siu, Finn, and Heidebrecht (1991) indicate that because of their geology, areas of the GVRD in the Fraser River Delta may be subjected to considerable ground motion amplification. These amplifications occur not only at the fundamental periods of the sites, but also at periods corresponding to higher modes of vibration, and they depend on the ratio between acceleration and velocity of the ground motion. A recent study by Byrne, Yan and Lee (1991) suggests possible amplification of peak base rock accelerations, increasing them from 0.21g to 0.30g.

b) Liquefaction:

Loosely-packed, water-saturated sediments usually found in certain parts of deltas, river channels and uncompacted landfills are likely to liquefy during an earthquake. Such is the case for several communities of the GVRD in the area forming the delta of the Fraser River. The liquefaction susceptibility of the Fraser Delta has been investigated by Watts, Seyers and Stewart (1992) and others. A preliminary map of liquefaction susceptibility for the region is presented in Figure 5. This map shows that the communities of Delta and Richmond have liquefaction probabilities of over 30% in 50 years. The effect of liquefaction in these communities could result in substantial damage to buildings and lifeline structures.

c) Tsunamis:

A big earthquake on the outer coast of British Columbia, or even as far as Alaska, Chile, the Aleutian Islands or Kamchatka, may generate tsunamis that will significantly affect the west coast of British Columbia. Tsunami simulation studies (Dumbar, Le Blond and Murty, 1989) have predicted wave amplitudes in excess of 9 m and current speeds of 3 to 4 m/s. Sheltered areas can expect to see maximum water levels less than 3 m. Low level areas such as Richmond, which are sheltered from the direct effects of tsunamis, will be subjected to threats of flooding if a tsunami arrives at high tide (Murty, 1992).

4. Earthquake Awareness and Emergency Response

It is estimated that about 50% of the people living in the GVRD are aware of the seismic hazards of the region. The public is also aware of recent scientific evidence of the threat of a major earthquake in the region. The 1989 Loma Prieta Earthquake in California served as a catalyst for public concern as well as political response to the seismic threat. In response to this threat, emergency programs exist to cover earthquakes and other types of natural hazards. These programs include a Federal Emergency Program, a Provincial Emergency Program (PEP), and a Municipal Emergency Programs. To keep the public informed, many municipalities and organizations offer public presentations on subjects about natural hazards. Also, regular

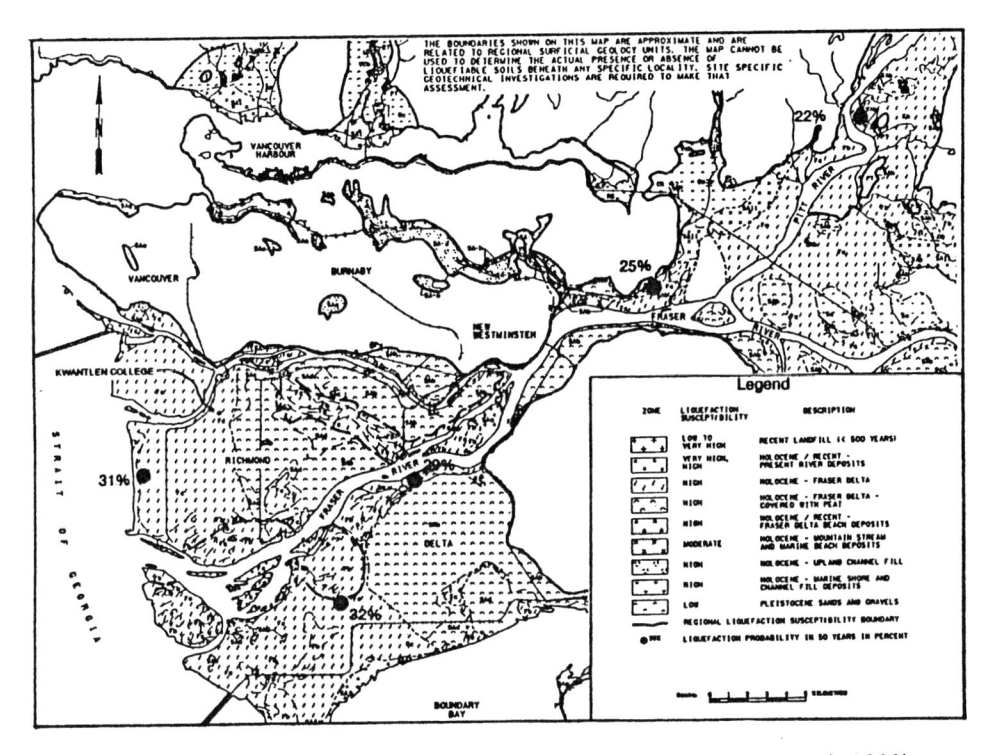

Fig. 5. Liquefaction Susceptibility Zones of GVRD (after Watts et al, 1992).

technical seminars are organized by professional organizations to keep their members abreast of the latest technical developments on seismic design.

The ability of government organizations to respond to an earthquake disaster depends not only on well- developed earthquake preparedness programs but also on their budget and size. Budgets and manpower for several relevant institutions, for the prominent municipalities of the GVRD, appear in Table 3. It is interesting to note that Vancouver city, which has the largest population and highest density, has also the largest institutional manpower. Also worth noting is that twelve hospitals are located in Vancouver city, compared with three in Burnaby, one in Richmond and one in Delta. In the event of a major earthquake directly affecting the city of Vancouver, the ability of the GVRD to provide emergency services may be severely limited if some hospitals in the city are incapacitated. The hospitals in the other areas may not be sufficient to provide adequate service.

Many of the municipalities of the GVRD have active seismic hazard reduction programs. For instance, the city of Vancouver's ten-year budget for earthquake-related projects is about $20 million (all currency in US$). Of this budget, $4 million is devoted to bridge and waterline seismic upgrades, $12 million is for saltwater fire pumping stations, and $4 million is for schools upgrades (Maki, 1993).

School districts have also established seismic hazard mitigation programs. For example, the Vancouver School Board's seismic hazard mitigation program includes: a) emergency preparedness, b) non-structural hazard reduction, and c) structural retrofits. The district-wide emergency preparedness (EP) plan includes formation and training of EP teams in every school, EP first aid kits, drinking water, and supplies and equipment at every school, and emergency communications between the district control center, sub-area offices and the schools. The non-structural hazard reduction effort consists of a detailed survey of each school to identify 47 classes of non-structural hazards and a plan to eliminate or significantly reduce each of them. The total hazard reduction program has an estimated cost of $8 to $10 million. The current available funding is about $0.32 million per year. Because of the funding available, the most serious class of hazards identified in each school are being mitigated first. The ongoing structural retrofit effort includes three stages: risk assessment, analysis and evaluation, and final design and retrofit construction.

Table 3. Institution Statistics for Prominent Districts of the GVRD
(Currency in US Dollars)

	Vancouver[1]	Burnaby[2]	Richmond[3]	Delta[4]
Fire Department				
Firefighters	733	162	179	126
Clerical	61	83	20	-
Firefighters/city population	0.15%	0.10%	0.14%	0.14%
Annual Budget	$70,467,925	$12,873,257	$10,013,172	$8,478,834
Budget/ Total City Budget	13.83%	3.11%	2.42%	2.05%
Police Department (R.C.M.P.)				
Officers	1156	233	147	159
Clerical	254	85	162	-
Officers/city population	0.24%	0.14%	0.11%	0.18%
Annual Budget	$123,917,573	$16,020,436	$10,850,034	$9,052,134
Budget/ Total City Budget	24.32%	3.88%	2.62%	2.19%
Hospitals[5]				
Total Personnel	19450	2345	1022	400
Total beds	4295	1126	281	175
Personnel/city population	4.07%	1.44%	0.79%	0.44%
Number of hospitals	12	3	1	1
Annual Budget	$658,976,645	$65,971,505	$31,782,075	$12,972,960
Budget/ Total City Budget	159.40%	15.96%	7.69%	3.14%
Planning Department				
Total staff	136	45	34	18
Staff/city population	0.03%	0.03%	0.03%	0.02%
Annual Budget	$10,965,123	$2,249,808	$1,709,448	$1,233,102
Budget/ Total City Budget	1.55%	0.54%	0.41%	0.30%
Department of Licenses and Permits				
Total staff	143	57	66	28
Staff/city population	0.03%	0.04%	0.05%	0.03%
Annual Budget	$10,259,258	$2,480,407	$3,177,252	$1,557,660
Budget/ Total City Budget	2.01%	0.60%	0.77%	0.38%

1) Budget Department, City of Vancouver. 2) 1993 Annual Operating Budget, City of Burnaby. 3) Finance Department, City of Richmond. 4) Finance Department, City of Delta. 5) Canadian Hospital Directory, 1992-93.

5. Construction Practice

Building construction practice is a key element for reducing the hazards posed by earthquakes. The design of buildings in British Columbia is normally based on the current edition of the British Columbia Building Code (BCBC) that, in turn, is based on the National Building Code of Canada (NBCC). The NBCC provisions are revised and updated every five years. The latest edition of the code was issued in 1990.

According to the Permits and Licenses Department of the City of Vancouver, about 1,000 new buildings are constructed every year. This represents about 1% annual rate of growth. About 90% of the tall buildings are made of reinforced concrete, and about 90% of the low-rise buildings are made of wood frames. Of all the buildings, 25 to 30% are designed by professional engineers and architects. For seismic evaluation purposes, the City uses the following building classification:

>   CF - concrete skeleton   CSW - concrete shear wall
>   WF - wood frame   SF - steel skeleton
>   PB - timber post and beam RM - reinforced masonry
>   URM - unreinforced masonry OSB - old style brick construction
>   PC - precast concrete

Because of the geological setting of the GVRD, especially the areas within the Fraser River Delta shown in Figure 5, the BCBC requires that the possibility of ground motion amplification, lengthening of predominant period and liquefaction be taken into consideration during the design phase of a structure. The code includes regulations to account for ground motion amplification and period lengthening, but not for liquefaction, which instead requires a special study. Since liquefaction is a major concern for the communities in the Fraser Delta, guidelines for design where liquefaction is a possibility were recently developed by a task force chaired by Drs. Byrne and Anderson of the University of British Columbia (Task Force, 1991).

The NBCC does not require that an existing building be upgraded to meet today's more stringent requirements if found to be seismically deficient. However, the Code does require that modifications and additions to a building meet the new requirements. The City of Vancouver also does not require seismic upgrading of a deficient structure unless there is a change in occupancy, of loading, or when any proposed upgrading exceeds a certain percentage of the assessed value. Following the 1990 edition of the NBCC code, the city now requires that architectural, mechanical and electrical components be seismically braced and that professionals submit letters of assurance that the work has been properly designed and constructed.

Seismic upgrading takes into consideration the level at which the building adheres to the 1990 NBCC as well as the function, population, mobility, cost of upgrade and replacement cost, potential relocation of function to a facility with a lower seismic facility level, danger to passersby, long-term plans for the facility and post-disaster function of the building (City of Vancouver Seismic Report, 1991).

The current NBCC provisions cover both crustal and intraplate subduction seismic events but not interplate subduction events. According to the Code, for the 475-year return period event, it is estimated that a M7 earthquake centred about 65 km from Vancouver could produce peak ground accelerations between 0.16g and 0.27g. An appropriate peak ground motion on rock or firm ground for the region is 0.21g (Task Force, 1991).

5.1 SEISMIC ASSESSMENT OF EXISTING BUILDINGS

Although the seismic hazard has existed for a long time, it is only recently that measures have been taken to prepare for an earthquake. This is due to a lack of a major seismic event in the GVRD which diminishes the perceived risk. In any event, the municipalities are currently assessing the seismic performance of their buildings.

a) Public-Owned Buildings:
As an example, the city of Vancouver performed an assessment of all public buildings in the district (City of Vancouver, Seismic Report, 1991). Each building was categorized as Level 1, Level 2, or Level 3. Level 1 indicates that the structure meets the intent of the current building code (NBCC, 1990) while Levels 2 and 3 indicate that the structure does not meet the intent of the building code. Further, Level 2 structures are capable of withstanding between 50-75% of the seismic loads specified in the building code, while Level 3 structures can only withstand less than 50% of these seismic loads. The highlights of the report include:

- Of the 74 buildings evaluated, the designated levels were 34%, 24%, and 42% for levels 1, 2, and 3 respectively. Some buildings designated as post-disaster facilities, that is City Hall, the police stations, and several firehalls are classified as Level 3 structures and potentially may not be able to fulfill their function.

- Some of the emergency communication systems have inadequacies associated with the buildings within which they are located, that is, they are located in Level 3 facilities. In addition, some components require additional protection from falling debris and from self-damage.

b) Schools:
The current building code designates schools as post-disaster structures. Since schools are generally scattered uniformly throughout the city, they are ideal for emergency shelters. As with the public-owned buildings, it is possible that this requirement of the building code may not be met.

The school boards have also performed seismic assessments of their facilities. The highlights of two of these reports, one obtained from the Delta School board (DELCAN Corp., 1991) and the other obtained from the Vancouver City School Board (Transit Bridge Group Ltd., 1990) are presented here. Unlike the city's categorization scheme, the school boards prepared a ranked list which would prioritize which buildings should be upgraded first. The ranking was based on the highest risk of personal injury.

Each school board used a different approach. The report presented to Delta was based on ATC-21 Rapid Visual Survey (ATC-21, 1988) which provides a rough evaluation of a building based on type of construction, year built, etc. The buildings are ranked on a scale from 0-6, where 0 indicates a poor performance. Accordingly, 38 of the 66 buildings received a score of 2 or less This indicates that 58% of these buildings are possibly hazardous and therefore require a more detailed analysis.

The Vancouver School board used a similar approach when their schools were evaluated. Except that they prepared their own guidelines to assess the schools so that the results of independent assessments would be consistent (Angel, 1992). Accordingly, schools were ranked based on a personal injury/safety index. Scores ranged from 0-3181. Of the 302 schools assessed, 95.7% scored between 0-1000, 3.6% scored between 1000-2000, and 0.6 % scored above 2000. One interesting point is that it will probably take the Vancouver School board 25 years to upgrade all their schools. In that time, the building code will undergo five revisions. And, therefore, it is possible that a building which satisfies the current building code may not satisfy the requirements of future versions of the building code.

c) Historic Buildings:

Vancouver has a Heritage Planner, a program of Heritage Awards, and a Heritage Foundation that look after the preservation of historic buildings. The City of Vancouver uses the code standards for the seismic performance of these types of buildings. These standards have to be complied with, whenever remodelling work is to be undertaken in an historic building. The City is self-insuring and is liable for failures or accidents at buildings in which retrofit or remodelling work has been approved by its inspectors. Because of the potential liability, the City imposes strict standards of compliance for these buildings, which are generally located in the old, poorer or less-developed parts of the city. Many of the building owners in Vancouver's historic districts are caught between trying to meet the city's demands for expensive safety upgrades and fulfilling their vital role of providing affordable space for low-income residents or small businesses.

d) Private-Owned Buildings:

In addition to the assessment of public buildings, the city of Vancouver is currently evaluating 1500 private buildings.

6. Seismic Research

Educational and government organizations of the province are actively involved in seismic research focusing on damage mitigation. Various departments of the University of British Columbia conduct research on seismology, geology, geophysics and earthquake and geotechnical engineering. A 3 meter square shake table in the department of Civil Engineering has been used for more than fifteen years for testing structural models and buildings components.

The Earthquake Studies group at Pacific Geoscience Centre (PGC) in Vancouver Island operates a seismic network to monitor earthquakes in western Canada. About fifty seismograph stations throughout western Canada provide the capability to locate earthquakes. A network of about forty strong motion freefield instruments is operated in British Columbia. PGC also conducts special purpose seismic hazard assessment studies.

Provincial government institutions such as B.C. Hydro and the Ministry of Transportation and Highways have very active programs, including research, for mitigating seismic hazards in dams and bridges, respectively.

7. Conclusions

British Columbia is located in an area where very large earthquakes, although very uncommon, are likely to occur. The population and public officials of the GVRD are aware of the seismic risk of the region. A large earthquake affecting the region at this time could be devastating. However, the level of expected devastation is decreasing as preparations are made through public education and emergency programs. Regulation of construction practice is also helping to reduce seismic hazards in new and existing structures.

Acknowledgements

The information presented here was provided by officials from the cities of Vancouver, Burnaby, Richmond and Delta and the school boards of Delta and Vancouver. In particular, J. Oakley of the Office of Emergency Management, City of Vancouver, T. Angel of the Vancouver School Board and R. Maki of Permits and Licences Department of the City of Vancouver provided very helpful information. The interest of these persons and their organizations is acknowledged with thanks.

References

Angel, B.A., (1992) "Report on Seismic Evaluation and Upgrading of Schools in Vancouver, B.C.," prepared for EERI Conference in San Francisco.

ATC-21, (1988) "Rapid Visual Screening of Buildings for Potential Seismic Hazards," Applied Technology Council, Redwood City, California, 185 pages.

Byrne, P.M., Yan, Li and Lee, M., (1991) "Seismic Response, Liquefaction, and Resulting Earthquake Induced Displacements in the Fraser Delta," Sixth Canadian Conference on Earthquake Engineering, Toronto, June 12-14.

Campbell, D.D. and Rotzien, J.L., (1992) "Deterministic Basis for Seismic Design in B.C.," Geotechnique and Natural Hazards, First Canadian Symposium on Geotechnique and Nat. Hazards, BiTech Publishers, Vancouver, British Columbia, pp. 71-80.

Canadian Hospital Directory, 1992-1993, Vol. 40. Compiled by the Canadian Hospital Association, Ottawa, Ontario, Canada.

City of Vancouver, Seismic Report, (1991), Extracts courtesy of J. Oakley, Director, Office of Emergency Management, City of Vancouver.

DELCAN Corp. (1991), Report on Seismic Assessment of Schools (no title available), prepared for Delta School Board.

Dumbar, D., LeBlond, P.H., and Murty, T.S., (1989) "Maximum Tsunami Amplitudes and Associated Current on the Coast of British Columbia," Sci. Tsunami Hazards, 7(1), pp. 3-44.

Greater Vancouver Key Facts - A Statistical Profile of Greater Vancouver, Canada, prepared by

Strategic Planning Dept., GVRD, 1993.

Information Circular, (1991) "Earthquakes in British Columbia," Circular No. 6, produced by the Geological Survey Branch of the B.C. Ministry of Energy, Mines, and Petroleum Resources in cooperation with the B.C. Ministry of Environment, the B.C. Provincial Emergency Program and with the assistance of the Geological Survey of Canada.

Maki, R., (1993) City of Vancouver, Permits and Licences Department, Personal communication.

Murty, T.S., (1992) "Tsunami Threat to the British Columbia Coast," Geotechnique and Natural Hazards, First Canadian Symposium on Geotechnique and Nat. Hazards, BiTech Publishers, Vancouver, British Columbia, pp. 81-89.

NBCC (1990), National Building Code of Canada, National Research Council of Canada.

Rogers, G.C., (1988) "An Assessment of the Megathrust Earthquake Potential of the Cascadia Subduction Zone," Canadian Journal of Earth Sciences, Vol. 24, pp. 844-852.

Rogers, G.C., (1992) "The Earthquake Threat in Southwest British Columbia," Procs. of First Canadian Symposium on Geotechnique and Nat. Hazards, BiTech Publishers, Vancouver, British Columbia, pp. 63-69.

Sy, A., Henderson, P.W., Lo, R.C., Siu, D.Y., Finn, W.D.L. and Heidebrecht, A.C. (1991) "Ground Motion Response for Fraser Delta, British Columbia," Proceedings of the 4th International Conference on Seismic Zonation, Stanford, California, Vol. II, pp. 343-350.

Task Force Report (1991) "Earthquake Design in the Fraser Delta," Byrne, P.M. and Anderson, D.L., chairmen, Soil Mechanics Series No. 150, Department of Civil Engineering, The University of British Columbia, Vancouver, British Columbia, 42 pages.

Transit Bridge Group, Ltd., (1990),"Seismic Risk Assessment of Vancouver School Board Schools, Final report to VSB," Victoria, B.C.

The Vancouver Sun (1993) "Statistics - B.C. is Canada's favourite destination," August 6, Section B, page 8.

Watts, B.D., Seyers, W.C., Stewart, R.A., (1992) "Liquefaction Susceptibility of Greater Vancouver Area Soils," Proc. of First Canadian Symposium on Geotechnique and Nat. Hazards, BiTech Publishers, Vancouver, British Columbia, pp. 145-157.

EARTHQUAKE HAZARDS IN THE GUATEMALA CITY METROPOLITAN AREA

HÉCTOR MONZÓN-DESPANG[1]
JOSÉ LUIS GÁNDARA G.[2]
[1]*Sismoconsult, Guatemala City*
[2]*Research Centre of Architecture of San Carlos University*

1. Introduction

Founded in 1776, Guatemala City is the largest city in the country of Guatemala and the largest metropolitan area in Central America. It houses 7.5% of the total population of Central America. In 1989 the average population density was 5746 hab/km²; in 2000 it will reach 7600 hab/km².

The Guatemala City Metropolitan Area (GMA) comprises seven municipalities and occupies an area of about 350 km². A simplified map is presented in Figure 1. Currently, there is a population of about two million. That is 50% of the urban population in the entire country, and 20% of the total population of Guatemala. The population growth rate is 4%, compared to 2.9% for the country as a whole. Most of the excess growth rate consists of squatters immigrating from elsewhere in Guatemala and settling in marginal areas.

The GMA is the hub of the industrial and economical activity in Guatemala, of which it represents more than the desirable share, in regard to the country as a whole. It attracts 78% of the non-agricultural private investment, 67% of the country's industry, and 45% of total formal employment.

The metropolitan area is located within the Valle de Guatemala, a 500 km² valley atop the volcanic mountain chain of Central America. Its climate is mild, rarely below 15° C or above 28° C.; humidity ranges from 70-80%; annual rainfall is about 1500 mm. The valley is about 1500 m above sea level and just on the continental divide. At that location, water for two million inhabitants is a somewhat precious commodity.

2. An Overview of the Natural Hazards

Within the metropolitan area, there is no possibility of floods (except smaller localized storm floods). There is no record of winds exceeding 80 km/hr (except very occasional small whirlpools).

Fall of volcanic tephras is always a possibility with two nearby active volcanos. There is evidence of some large ash falls in recent millenia, especially in the south of the valley, but there have been only light historic ash falls. Nevertheless, the very floor of the entire valley is a consequence of several catastrophic ashflows which occurred within the past quarter million years. The top 10 m of the valley floor were deposited by three events within the past 80,000 years. But given the extremely low probabilities of occurrence of such events and their cataclysmic consequences, no practical preparedness applies.

Earthquakes, slope stability and mudflows are hence the main natural hazard concerns in the

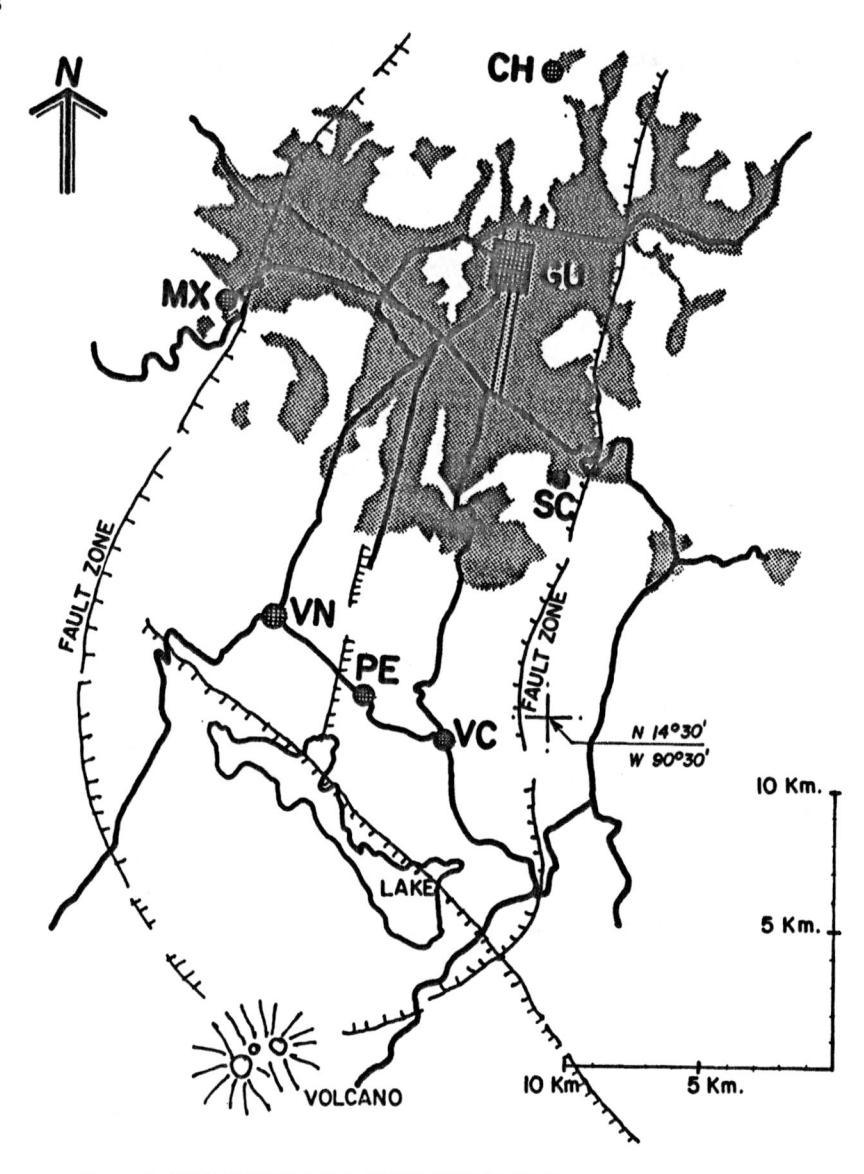

Figure 1. THE GUATEMALA CITY METROPOLITAN AREA
The shaded area shows urban construction in 1985 (after Fernando Masaya); the valley is the region bounded by the geologic faults; the dark square indicates the size of Guatemala City when founded in 1776; dark circles (and the square) indicate the seat of the municipalities: CH - Chinautla; GU - Guatemala City; MX - Mixco; PE - Petapa; SC - Santa Catarina; VC - Villa Canales; VN - Villa Nueva; geologic fault zones are schematic and some segments may be inactive; dark lines are important roads within the city and through the valley.

Guatemala City metropolitan area; the former include earthquakes of both distant and local origin; the latter include landslides and mudflows usually triggered by rain, or by earthquakes when they do occur. There are several other ground-related problems, especially in marginal, topographically-difficult areas that have been occupied in recent decades by poverty-stricken squatters from the rural areas of the country.

3. Geography and Geologic Setting of the Guatemala Valley

The Valle de Guatemala is an active geologic graben. The trench, depicted in Figure 2, is 12-15 km wide and about 30 km long. It literally splits the main Guatemalan mountain chain, whose longitudinal axis is being subjected to tectonic tensile stresses. As the graben has widened, its floor has subsided more than 400 m, relative to the surrounding highlands. However, several gigantic volcanic ashflows, 100-150 m in thickness, filled the bottom of the graben, generating a remarkably flat floor (today attractive for urban development). Yet, there is a hitch in the gentle flatness of the valley floor: water streams have scoured steep-sloped, deep ravines in the volcanic ash fill. Being on the continental divide, two systems of streams have developed leaving a central "island" surrounded by ravines. Each half of the central "island" has long, narrow fingers of flat land generated by adjacent merging streams, and eventually two rivers—always guarded by deep ravines—emerge at each end of the geologic trench.

The Spanish Colonial authorities found it interesting to settle on the ravine-isolated middle portion of the flat valley. A number of small villages developed on the outskirts of the valley, just across the ravines, providing agricultural support. Every town in the valley was both nearby and distant. Today, as shown in Figure 1, the city and its satellite towns are a single metropolitan area, seven densely-populated municipalities, geographically packed together, but physically separated. Unfortunately, nowadays they are still administratively uncoordinated and divorced from each other.

4. Seismicity and Related Hazards

The Guatemala City metropolitan area is under the threat of local, active geologic faults and major, regional active faults. The last very damaging earthquakes occurred in 1917/18 and in 1976.

Local earthquakes are a major source of seismic hazard. They develop high intensities (Mercalli VII and above) generated by small to middle magnitudes (Ms 5.5 to 6.5), and affect small- to middle-sized areas (say, one to three municipalities). Local earthquakes in the valley are accompanied by landslides, passive ground cracking and, very possibly, active fault rupturing.

As shown in Figure 2, the east and west flanks of the valley are fault scarps. At the foot of each flank is a strip, a few kilometers wide, containing numerous parallel normal faults, many of them active; normal faulting is a feature that runs almost continuously for some 30 km on either flank. Moreover, at the center line of the graben seems to be another strip of smaller active faults. Therefore, intense small magnitude local earthquakes may be generated almost anywhere within the valley (e.g., an event such as the San Salvador earthquake of 1986). Additionally, intense middle-magnitude earthquakes may be generated near the flanks of the valley (of which the 1917-18 earthquakes in Guatemala City are a possible example).

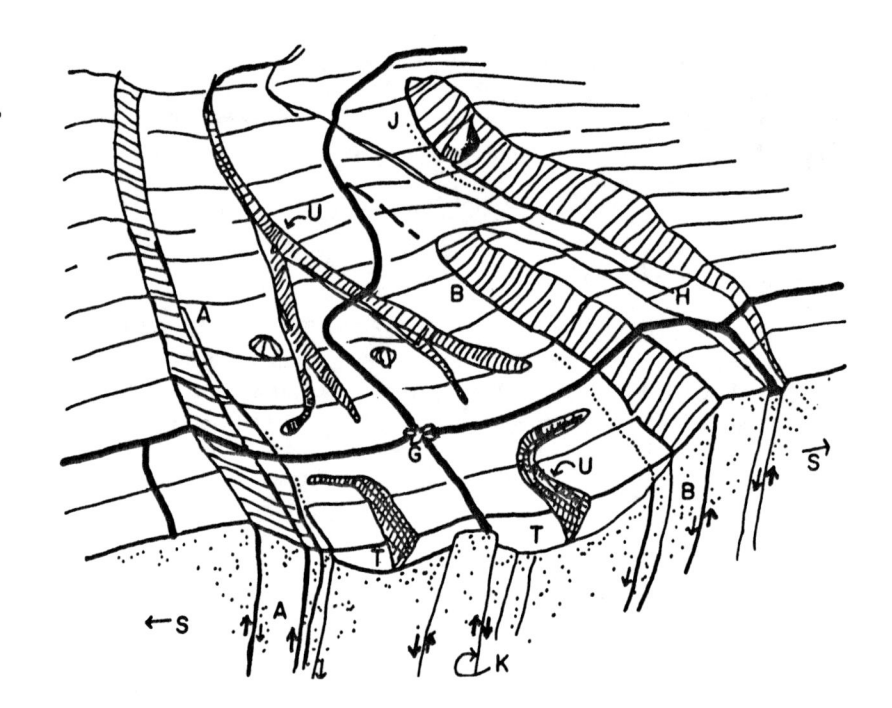

Figure 2. MORPHOLOGY OF THE GUATEMALA VALLEY
Perspective and cross-section of the geologic graben;
A - Mixco fault zone; B - Santa Catarina fault zone;
K - minor faulting within the valley; J - Palencia fault zone;
T - pyroclastic fill, mainly ashflows 100 to 150 m thick;
U - ravines cut by water streams into the pyroclastic fill

Regional earthquakes, generated outside the valley, may develop high intensities (Mercalli VII or even VIII) in the metropolitan area provided they are of a larger magnitude (Ms 6.75 to 7.75). Historic instances have already occurred and remain a future possibility. These earthquakes are less likely to produce ground rupturing and ground cracking within the metropolitan area, but they would trigger many landslides within the metropolitan area.

Regarding regional earthquakes, the Motagua fault system is part of a major tectonic plate boundary and runs just 25 km north of Guatemala City. It produced the large 1976 Guatemalan earthquake (magnitude Ms 7.5) which developed damaging intensities in the northwest of the metropolitan area; numerous housing units and buildings were destroyed or damaged, and deaths reached a thousand in the metropolitan area (23,000 was the total death count). However, another earthquake of this magnitude from the same sector of fault seems unlikely in the near future.

Just south of the southern reach of the Guatemala City metropolitan area is another fault, the Jalpatagua, large enough to generate events Ms 6.0 to 7.0. This fault poses a hazard; a larger event here may affect all of the southern municipalities. It is not known whether the 1885 Amatitlán earthquakes were generated on this system, but the 1913 Cuilapa earthquake probably was. (Amatitlán is 5 km south, Cuilapa is 40 km east of the metropolitan area).

Distant earthquakes: Earthquakes magnitude Ms 7.5 and larger, originating 80 km away in the nearby subduction zone, appear less likely to cause widespread damage to the metropolitan area. Yet they will trigger landslides and cause fear, panic, and sporadic damage.

4.1 SEISMICALLY-RELATED GROUND HAZARDS

The real dimension of seismic ground hazards was only apparent after the 1976 Guatemalan earthquake. The pyroclastic fill that blankets the Valle de Guatemala and other neighboring highland valleys is very stable under gravitational stresses. It is possible to build and load the very edges of the ravines. But when it shakes, the walls of the ravines undergo shallow landslides, usually thinner than 5 m; occasionally a larger portion slides with fatal consequences. In the next large earthquake, many people will be affected, especially in the poorer urban quarters where people live in small plots at the edge and on the slopes of the ravines. More affluent quarters will also be affected among those who live on the beautiful escarpments which dominate the valley. It is difficult to give general zonation rules; it appears better to set rules for the kind of geotechnical investigations that are needed in each zone, leaving the individual decisions to engineers.

In 1976, one of the faults within the valley ruptured in connection with the main seismic event and another ground hazard became evident. Besides the main breakage which caused several vertical ground offsets, a strip of the valley floor about 5 km wide underwent passive cracking. Fissures in the ground ran for hundreds of meters, causing increased damage to buildings on top of it. The problem with this rupturing style is that no specific terrain strips can be easily segregated to ban construction on it. It is simply not practical because the affected bands are kilometers wide.

5. Construction

For a brief description, construction will be divided in three categories: housing, commercial and office buildings, and industrial.

Housing: Most of the dwellings in the Guatemala City Metropolitan Area are single-family units. Only a small percentage of the population live in apartment buildings. Almost half of the single-family units are reasonably reinforced masonry[1]; a quarter are adobe; and the remainder are shacks built from a variety of materials, from wood boards to scrap material.

Engineered housing construction is restricted to a fifth of the total housing being built. It is mostly reinforced masonry. Another portion of masonry construction is executed by empiricists and much of it is rather good. The seismic performance of this type of housing ranks from very safe to reasonably safe; moreover, masonry has performed very well under severe seismic conditions, including ground rupture beneath the house (USGS, 1976 pp 38-51).

The informal shack housing, which is self-constructed, poses intermediate to low seismic risk by itself (except in cases where unreinforced and even non-bonded masonry is used). However, the shacks are frequently built on hazardous terrain subject to landslides during earthquakes. Pressure of squatters for land will not only increase the depletion of important green areas and potential parks, but will also continue mounting seismic risks as more ravine edges and the slopes themselves are occupied.

The adobe dwellings are either self-constructed or built by local masons; the economics of building with adobe reside in extracting the material from the lot itself; hence, there is little hope to improve such a material. Guatemalan adobe is non-cohesive and extremely brittle. The situation worsens when heavy tile roofs are used. The latter are the main cause of casualties during earthquakes. Adobe with a light tin roof fares marginally better.

In the metropolitan area, assuming the seismic hazard from ground shaking intensity is relatively uniform, the highest housing risks correspond to city quarters where adobe prevails. When ground effects are taken into account, the edges and slopes of the ravines and the steep flanks of the graben are additional localized risk microzones, as discussed elsewhere in the paper.

Commercial and office buildings: These buildings are usually reinforced concrete frames. Of those built in the past twenty years that exceed eight stories, about half have shear walls. The tallest buildings are no more than eighteen stories above street level due to local airport regulations. Despite the absence of legally-enforced seismic regulations, most are responsibly designed and built. Use of the US Uniform Building Code 1985 (latest edition 1991) is common.

The most hazardous of this kind of building are the older ones, between 40 and 25 years of age, in which reinforced concrete is brittle and the frames are relatively bare. None of the larger buildings of this type collapsed during the 1976 earthquake, but a few had to be demolished and others underwent major repair. It is likely that in a future event this group will fare much worse.

One conclusion is that retrofitting regulations may currently be more urgent than regulations for new buildings.

Industrial buildings: These are usually steel frames. Their largest vulnerability is not the building itself, but the machinery and equipment, which are seldom prepared to withstand earthquakes.

5.1 PLANNING REGULATIONS

In the metropolitan area, only one municipality besides Guatemala City proper has an urban planning department. The one in Guatemala City is only able to monitor about 40% of the formal

---

[1] Reinforced masonry is usually brick or cement block tied together with wall-thick reinforced concrete bond beams and RC vertical chords; these confining elements are cast while the walls are being erected. Hollow units with integral reinforcement are also used.

construction. It lacks personnel, and the volume of clandestine construction is high. Currently there is no master plan of urban development. There are some urban strategies such as the green belt; the industrial, residential and commercial zonation; and the transportation master plan. These are, however they are uncoordinated efforts. Furthermore, there is no disaster mitigation planning, yet there are encouraging signs that the necessary strategies are going to be implemented soon.

Municipalities are autonomous entities in Guatemala, but they are normally extremely short of funds. The Guatemala City municipality has a total budget of US$16 per capita. Other cities in Central America have comparatively larger budgets.

6. Emergency Response

In Guatemala, there is a theoretical framework to respond to emergencies, as set forth in the 1969 regulations governing the National Emergency Committee. In case of an emergency, the first to react should be the neighbors, followed by the fire department; police should be on alert to keep order. Depending upon the magnitude of the emergency, higher orders are activated. There is the municipal emergency committee, the provincial committee, and the national committee. The last is in charge of coordinating local or international aid in case of larger emergencies or disaster. Only after the emergency situation has ended should the National Reconstruction Committee coordinate the repair and reconstruction effort.

For decades, firefighters have provided effective emergency service in Guatemala City and many other municipalities, including handling daily emergency situations and also major emergencies and disaster situations. There are two parallel firefighting organizations. There is the older "Volunteer Firemen Corps," which has a national level of organization with fire departments in many municipalities. In Guatemala City proper there is also the Municipal Fire Department, which has tried to organize sister departments in other municipalities. Counting the two corps in the metropolitan area, there are about 20 stations, some 200 professional firefighters, and some 400 additional trained volunteers who are on standby; there are about 100 each of fire cars, ladder cars, water trucks, ambulances, and rescue units. Also, they have had several disaster response experiences in neighboring countries.

On the emergency management level, the National Emergency Committee is under the authority of the President of the nation and has a military organization. There is a coordinator-in-chief and a small staff, professionally trained. They give attention to the "chronic" natural emergencies in the country, which include floods and mudflows on the Pacific coast, and the problems caused by three active volcanoes. There are a number of volunteers who join in case of a larger calamity.

There is no specific emergency framework at the level of metropolitan area. Currently, the framework jumps from municipal to provincial. The GMA includes seven municipalities and its province seventeen, and the main problem is one of jurisdiction because municipalities are autonomous entities and the province is dependent on the central government. A metropolitan emergency committee should still function at the autonomous level. As mentioned in the section on research, there are efforts to set up a metropolitan emergency response. On its side, the Municipality of Guatemala City is trying to improve the internal coordination among their administrative departments.

Comments on how the theoretical emergency response mechanisms work in practice are included in the next section.

7. Earthquake Awareness and Preparedness

Hunahpú and Xbalanqué, the two mythological Maya-Quiché heroes, managed to seize and bury underground the nasty and haughty Cabracán—the earthquake—who occasionally moves and shakes the earth. Ever since, every inhabitant of Guatemala is aware that the earth shakes strongly now and then.

Despite such an awareness, everything related to seismic preparedness is typically postponed to the "near future"; earthquakes are usually perceived as being in the past or else in the distant future. Consistent with this attitude, owners demand that new engineered construction be built "earthquake-proof," but when told that existent property is at risk, retrofitting is usually put off. The same attitude prevails toward general preparedness.

Such an attitude also pervades the higher government authorities. A report from CEPREDENAC, 1989, clearly states (pg. 24) that planning agencies have not considered resources to mitigate the future effects of natural hazards. As a result, mitigating actions have been improvised, isolated and uncoordinated, and there is also high dependence on foreign donations. Moreover, the people expect the authorities to be somehow ready to take charge in case of emergency, but they do not perceive their role as active (until they discover during an emergency that they must be).

As described before, there is a specific structure throughout the country to set up municipal, provincial or national emergency committees. The structure has worked in the past when disasters hit, but the response tends to be an improvised problem-solving approach; there has not been any planning in preparation for major emergencies.

During such emergencies there has always been precious time lost while a plan is tailored for the specific calamity. Citizen response has been spontaneous but not necessarily effective, due to lack of disaster education. The response capacity of firemen, the only trained rescuers, is frequently questioned by other institutions who do not have such prominent participation. There is usually some lack of coordination and degree of friction among those trying to attend the emergency, which include rescuers, citizen protection corps (police and army), public health agencies, non-governmental organizations (national and international), and the Emergency Committee. A problem of "too many chefs" has sometimes arisen during a calamity.

Nevertheless, there are serious efforts to improve the emergency awareness and response. Recently, a system of non-governmental organizations (including bilateral, multilateral and scientific ones), has been created to give coordinated support to the National Emergency Committee (CONE). Moreover, CONE with the support of the Office of US Foreign Disaster Assistance has started a natural-hazards education program for teachers and government officials.

8. Applied Research

There are a number of studies and research efforts on the subject of disaster mitigation. Yet to date, the information that has been produced is not readily available: most institutions develop internal reports instead of publishing for a larger target audience. Among the institutions that have undertaken some research are the following:

- INSIVUMEH, the national institute of hydro-meteorology seismology and volcanology, was created after the 1976 Guatemalan earthquake to keep track of natural events and hazards. Its purpose is scientific. It is regarded as important by the authorities but financial support is low. INSIVUMEH has not yet fully developed self-supporting activities.

- CEPREDENAC is a Central American institute to promote and strengthen programs on disaster preparedness. It has been functioning since 1988 with Swedish support. It has emphasized the physical aspect of preparedness. It is little known in Guatemala outside academic and civil defense circles.

- The Universidad de San Carlos (USAC), the national university, has been active in disaster preparedness, emphasizing the social aspects. The school of medicine has worked in simulation. The school of architecture, by means of its research institute, CIFA, has implemented a permanent program in disaster mitigation studies and is currently working on the metropolitan area, identifying vulnerability and proposing mitigation strategies (CIFA/DIGI, 1993).

- The Universidad del Valle de Guatemala has initiated a tentative program in Earth Sciences.

- There are government agencies such as Secretaría de Planificación Económica (socioeconomic planning), Instituto Nacional de Electrificación (power utility), and Instituto Geográfico Militar (mapping agency) that have developed important information which can be very valuable to implement disaster mitigation plans. Unfortunately, most of this information is generally unavailable.

- There are some organizations approaching the topic of disaster mitigation in Guatemala, such as the Organization of American States, United Nations Development Program, US Agency of International Development, and the Pan American Health Organization. Since 1990, these organizations that have been literally pushing the development of an adequate system of disaster mitigation in the country.

A very fruitful task would be to collect and organize the available information, currently scattered throughout many institutions, and put it in a disaster-related information center. This will preclude duplication of tasks and ease the work of researchers and planners. For example, the references listed at the end of this paper in "Selected Bibliography" are valuable documents, of which there are only four or five copies shelved in the private libraries of the institutions that funded the studies.

9. Final Remarks

The Guatemala City Metropolitan Area is a zone of significant seismic hazards. Due to the socioeconomic conditions, the population density and the concentration of resources, the area is vulnerable. It is imperative to be prepared for future earthquakes.

Since 1990 there has been increasing awareness of the need for such preparedness. The National Emergency Committee has doubled its seismic preparedness efforts. Yet funding is scarce and there should be more real support from local government agencies. Political decision-makers and all sectors of society recognize that earthquakes are a potential problem. However, there is an incomplete understanding of the real implications of future earthquakes, especially because the vulnerability of the city has increased in the years since the 1976 earthquake hit. In a way, there is a sense of overconfidence or else a sense that other events are still far in the future.

Setting up earthquake scenarios could be a powerful means to increase the awareness of political decision-makers and other sectors that could influence the formal implementation of disaster preparedness.

References

CEPREDENAC, 1989, "Informe Final. Encuentro regional sobre el proceso de reducción de riesgo de desastres naturales", Guatemala, May 8 to 12.

CIFA / DIGI, 1993, "Análisis de Vulnerabilidad Física para la mitigación de Desastres en el Area Metropolitana de Guatemala", Programa de Estrategias de Mitigación de Desastres, Univ.S.Carlos.

USGS, 1976, "The Guatemalan Earthquake of February 4, 1976", U.S.Geological Survey, Professional Paper 1002, U.S.Printing Office.

Selected Bibliography

Gándara G.,J.L. and H.Marroquín (1982). "La vivienda popular en Guatemala antes y después del terremoto de 1976", tomo II. OEA, CRN, Univ.S.Carlos de Guatemala, Editorial Universitaria.

Gándara G.,José Luis (1990), "Desastres Naturales y Zonas de Riesgo en Guatemala. Condiciones y Opciones de Prevención y Mitigación", Vol I y IV. Univ.S.Carlos de Guatemala, CSUCA, International Development Research Centre.

Gándara G.,José Luis (1993), "Desastres, Planificación y Desarrollo. Manejo de Amenazas Naturales para Reducir los Daños en Guatemala", Proyecto DRDE-1024/92, Organizac.de los Estados Americanos.

Instituto Nacional de Estadística (1981). "IX Censo de Población y VI Censo de Habitación"

Monzón D.,H.(1984), "Programa de Cooperación Técnica en Ingeniería de Terremotos - Fase II", GTZ (Germany) and INDE (Guatemala)

SIGLO VEINTIUNO (1993, ago 31), "El Area Metropolitana de Guatemala" in Suplemento "Nuestra Capital", Guatemala.

EARTHQUAKE HAZARD IN EL SALVADOR

ABIGAÍL CASTRO DE PÉREZ
MARÍA LUISA BENÍTEZ
JOSÉ LUIS RIVERA
Ministerio de Educacion
Nueva San Salvador,
El Salvador, C. 4.

1. Demographics

El Salvador covers 21,040 square kilometers, all of which is formed by basins, a situation that has great physical and ecological consequences on the disasters that we try to present here. Most soils found in the Salvadorean territory have their origins in volcanic or fluvial activity.

In the last decade, the urbanization process in the metropolitan areas of El Salvador has accelerated, as a result of rural migration. In some cases, the migration was voluntary, but in others it was caused by the armed conflict. The demand for urban land to satisfy the basic necessities of housing has created a greater chaos than the one before the conflict. This growth has not been planned; consequently, the human settlements do not have basic services such as running water, electricity or drainage. Currently, the housing demand has resulted in construction on any type of soil, even the sides of a volcano and the banks of rivers, and on the same soils that were affected by the last earthquake (October 10, 1986). While the houses consist of only one floor due to the people's culture as well as the seismic activity, hotel, industrial and commercial constructions usually involve multi-story construction.

After the catastrophe of the 1986 earthquake, the government of El Salvador, working in conjunction with groups of builders, decided to create a Culture of Anti-seismic Construction. Unfortunately, this has been executed without rigorous enforcement, due to the lack of systematic supervision of construction in the country. Therefore, there is no guarantee that the approved designs and materials will be used.

2. Seismicity

Currently, El Salvador has inadequate and obsolete telemetric equipment that does not allow the country's seismic activity to be efficiently monitored. Because of this, a project has been initiated to strengthen and diversify the seismological networks of Central America at a projected cost (between 1990-1993) of US$762,800.

As a point of reference for the seismic activity, we can say that Central America has experienced many destructive earthquakes, some of them in the last twenty years. Among these were the earthquakes of Managua—1972, 20,000 dead, 400,000 homeless and total estimated damage of $1B; Guatemala—1976, 26,000 dead, 1.2 million injured, 1.5 million homeless and total damage of $5B; and El Salvador—1986, 1500 dead, 10,000 injured, 200,000 homeless, 40,000 houses destroyed and total lost of $1B.

Even though the elements of plate tectonics are understood, El Salvador does not have detailed knowledge of its seismicity. There are not clearly-defined epicenters, only zones of seismic activity, which can be divided into regional, local, and volcanic. The source of oceanic seismic activity becomes the main seismic zone of El Salvador, which is located between 30-55 km off the Salvadorean coast. This zone has 50-60 km deep foci.

With the available information we can describe the existence of 33 seismic zones within the regional seismicity. Local seismic activity is very common in almost all the territory. The available information indicates the existence of at least 69 seismic zones within the local seismicity.

Various sources have noted 51 earthquakes in the country between 1524 and 1990, and according to the political-administrative division of the country, data shows that 11 of the 14 *departamentos* (or districts) of the country have had earthquake damage.

It is important to mention that the last earthquake, in October 1986, showed the real housing conditions for the people. Prior to the catastrophe, the housing deficit was estimated at 587,000 units, of which 30% was urban and 70% rural. From the total housing units existing after the earthquake, 40% were considered in bad shape, 17% inhabitable. Thereby, the evaluation of the housing damage due to the disaster of 1986 reflected a dramatic reality: a high housing deficit and the poor condition of existing housing units.

The same earthquake caused extensive damage to educational buildings. There were 150 public and private schools that were damaged by the earthquake, affecting around 30,000 students. The communities of these schools were affected for a long time until the structures were replaced and/or rebuilt. Finally, in 1993 the government of El Salvador through the Ministry of Education and international agencies finished the rebuilding of schools damaged by the 1986 earthquake.

The estimated total damage due to seismic activity in El Salvador in the last 25 years is US$9B. However, a closer study would doubtless provide a better understanding of the tragic effects of earthquakes on El Salvador.

3. Emergency Response

El Salvador has a general organization called the Committee of National Emergency (COEN), responsible for response to earthquakes or any other disaster. After the many natural disasters experienced by El Salvador, and their future inevitability, different strategies have been developed to organize emergency response, firstly to reduce the vulnerability, and secondly, to efficiently help those affected by a disaster.

COEN is an organization of the central government, formed by the Ministers of National Defense, Health, Public Works, Agriculture, and the Interior. As a practical response strategy, COEN broadened its services to other national sectors and drove the integration of the Technical Multisectorial Committee for Disasters (COTIDE).

COTIDE is formed by government and non-governmental organizations. In addition to the government agencies mentioned above, there is the Ministry of Education, which is responsible for raising public awareness. Among the non-governmental agencies are the fire department, Salvadorean Red Cross, Salvadorean Green Cross, and some international agencies such as Copperazione Italiana, PRODERE, OPS/OMS, World Vision, and CEPRODE.

INSTITUTIONS	DEPARTMENT	RESPONSIBILITIES
Ministry of the Interior	COEN	National Coordination of all institutions
Ministry of Education	School Program of Emergencies	Runs the School Emergency Program through preparation, prevention and response activities
Ministry of Public Health	Technical disaster preparedness unit	Plans and coordinates emergency preparedness health activities
Salvadorean Red Cross	Volunteer National Direction	Supports disaster victim. Provides technical support for execution of emergency plans.
National Fire Department		Provides technical support
PRODERE	PNUD/ACNUR/OIT/OMS	Provides technical and financial support. Runs Local Emergency Committee
CEPRODE		Investigates municipal school emergency plans
World Vision		Provides technical support for various programs
PanAmerican Health Organization (PAHO)	OPS/OMS	Provides technical assistance for program development
Copperazione Italiana	Health Program	Provides technical and financial assistance to local emergency plans

On a national level, COTIDE does not have an integrated social agent, such as the private sector of the country, although the response from this sector is quickly evident after an earthquake.

Earthquake safety is now being developed, more than ever before, to unify the country at the moment when an earthquake occurs. Primary emphasis is placed on coordinating efforts made by institutions on the same level, so that these institutions do not make (or duplicate) the same technical and financial efforts. The work focuses on *prevention* of disasters and awareness-raising.

Through COEN, some institutions have trained personnel to distribute supplies after a disaster. This has been developed with the cooperation of OPS/OMS.

On a national level, an actualization of the Civil Defense laws was presented before the National Assembly. This calls for the development of smaller geographic units such as the Local Committees for Emergency that function on the community level. Disaster prevention has become a fundamental part of the Basic Education Programs. These programs are supported by bibliographic material that promotes prevention (i.e., Handbook for School Emergency Plans) and practical instruction for teachers, students, and the general community.

On a scientific level, the first studies of risk to metropolitan San Salvador have been made, so that the institutions have enough technical information to prepare the Salvadorean public. Along this line, El Salvador has become a regional participant in the National Committee of the International Decade for Natural Disaster Reduction 1990-2000.

4. Construction

As mentioned previously, the rules of design construction in the metropolitan area were reviewed after the 1986 earthquake and revised to account for a building's vulnerability to seismic activity as well as the quality of its materials. Further, the Builders Association has seriously considered building vulnerability in its anti-seismic construction plan. Universities have also included subjects related to structural engineering and vulnerability in their curricula of Engineering and Architecture, related to anti-seismic construction.

5. Public Awareness about Earthquakes

Raising public awareness requires the active participation of the mass media, so that the scientific knowledge of the committees and universities will reach the population in a suitable form. At the same time, the Associations of Engineers and Architecture must be involved to promote seismically-safe construction and to localize vulnerable construction sites and so better regulate housing. Multilateral meetings at the highest levels should result in regional policy aimed at prevention, preparation and mitigation of natural disasters. The multilateral meetings should also provide an effective procedural base to coordinate among national and international humanitarian organizations.

The Ministry of Education is conscious of the permanent earthquake hazard and learned much from the destructive 1986 event, in which the structures of 11 schools and more than 400 school rooms were destroyed (at a cost of US$265 million), some lives were lost, and more than a thousand students were injured. In preparation for a future disaster, the Ministry is developing programs to prevent the rise of social and environmental vulnerability. This program should contain prevention and preparations for emergency situations that have an impact on the school structures and the installation of emergency equipment (i.e., first aid kits and fire extinguishers).

6. Research

COTIDE is the institution devoted to disaster prevention and therefore is coordinating actions to be taken in the academic fields and urban planning. While the academic field is developed, this is not the case with urban planning, even though the government of El Salvador is making great efforts to improve this situation. Some regulations exist which attempt to define a real urban plan, but an "anti-seismic culture" must be created with the public so that the people who live in El Salvador take an interest in knowing if the construction and locations of their homes are safe.

Currently, the government of El Salvador is working to develop this awareness. International organizations have been responsible for conducting the majority of scientific investigations that have been completed toward this goal. The Centro de Protección para Desastres, a Salvadorean nonprofit institution, has been designated as being responsible for national research.

SEISMIC SAFETY OF THE LIMA METROPOLITAN AREA

JORGE E. ALVA HURTADO
Professor of Civil Engineering
Director of CISMID
National University of Engineering
Lima, Peru

1. Demographics

The city of Lima is located on the western coast of South America, on a desert strip between the Pacific Ocean and the Andean Mountains. The city has the following geographical coordinates: 12°02' south latitude and 77°1' west longitude. The total area of metropolitan Lima is about 500 km². Lima was founded by the Spaniards in 1535, being the site of the viceroyalty of Peru with a port (Callao) near the city. Downtown Lima (Lima-Cercado) is the oldest area, and from that area the city developed and expanded.

The present population of Lima is 6.4 millon (1990), whereas the total population of Peru is 22.3 million. The population of Lima thus represents 28.7% of the national total. The average population density per square kilometer in Lima is 2,281 persons/km². However, twenty of the fifty districts in Lima are over 10,000 persons/km², and three are above 30,000 persons/km². The population growth for the country is 2.5% per year, whereas in Lima that value is higher, reaching 3.5% per year. Table 1 presents information on population growth in Lima and Peru.

Table 1: Population of Peru and Metropolitan Lima

Year	National population in Peru (millions)	Metropolitan Lima population (millions)	Percentage of the Country	Density (hab/km²)
1940	6.207	0.645	10.4	229.4
1961	9.906	1.846	18.6	665.3
1972	13.538	3.302	24.3	1,743.3
1981	17.031	4.600	27.1	1,635.9
1990	22.332	6.414	28.7	2,281.1

Metropolitan Lima is the most important city in Peru. Almost 29% of the population of the country lives in Lima, and 70% of the economic and industrial activities are located there because the city presents advantages for export and industrial activities. In spite of the city's high level of vulnerability to destructive earthquakes, very little is being done to protect it from seismic hazards.

2. Seismicity

The region is a segment of the Circum-Pacific Belt, which is one of the most seismically-active regions in the world. The seismic activity in this region is mainly caused by the subduction of the Nazca Plate beneath the South American Plate. Silgado (1978) has compiled historical information about the most important seismic events that occurred in Peru from the 16th century to the present time. Based on this study and on its reinterpretation, Alva-Hurtado et al.. (1984) have proposed a map of maximum seismic intensity distribution of Peru, which is shown in Figure 1. From this map it can be seen that a Modified Mercalli Intensity of X has been assigned to Lima. The seismic activity in Peru is higher along the coast, although a zone of superficial earthquakes is present in the high jungle, or sub-andean, region.

The most important earthquakes that affected Lima are here briefly described:

• July 9, 1586. The city of Lima suffered an earthquake that caused 22 deaths. Tsunami in the port of Callao. Intensity in Lima of IX MMI.

• November 13, 1655. Strong ground motion that destroyed houses and buildings in Lima. One person killed. Intensities in Callao of IX and in Lima of VIII MMI.

• June 17, 1678. A strong earthquake occurred in Lima. Damages in Callao and Lima. Nine persons died. Maximum intensities of VIII MMI in Lima and Callao.

• October 20, 1687. Two earthquakes struck the area south of Lima. More than 100 persons died. Tsunami in Callao. Ground cracking in Cañete. Intensities of X MMI in Cañete, VIII MMI in Ica and VII MMI in Lima and Callao.

• October 28, 1746. Earthquake in Lima and tsunami in Callao. Maximum intensity of X MMI. Almost total destruction of buildings in Lima and Callao. From a population of 60,000 inhabitants, 1,100 died because of the earthquake in Lima. In Callao from a population of 4,000 only 200 survived.

• March 30, 1828. Earthquake in Lima that caused 30 deaths. Intensity VII MMI in Lima and VI MMI in Callao and Chorrillos.

• May 24, 1940. Earthquake in Lima. Intensity of VIII MMI. The quake caused 179 deaths and 3,500 persons were injured. Magnitude M_s = 8.2. A tsunami was generated.

• October 17, 1966. Earthquake in the northern part of Lima. One hundred deaths. Maximum intensity of VIII MMI. Maximum ground acceleration of 0.4g in Lima. Magnitude m_b = 6.3.

• October 3, 1974. Earthquake south of Lima. 78 persons died and 2,550 were injured. Maximum intensity of VIII MMI.

The existence of zones more susceptible to seismic effects due to local subsoil conditions has been made evident by recent earthquakes that occurred in Lima. Valencia (1940) reported an intensity of VII MMI in Callao, La Molina, Barranco, and Chorrillos districts during the 1940, May 24 earthquake. Silgado (1978) reported severe damages to buildings in Callao and La Molina districts during the 1966, October 17 earthquake. Damage distribution surveyed in Lima after the 1974, October 3 earthquake is the most complete information about local soil effects in

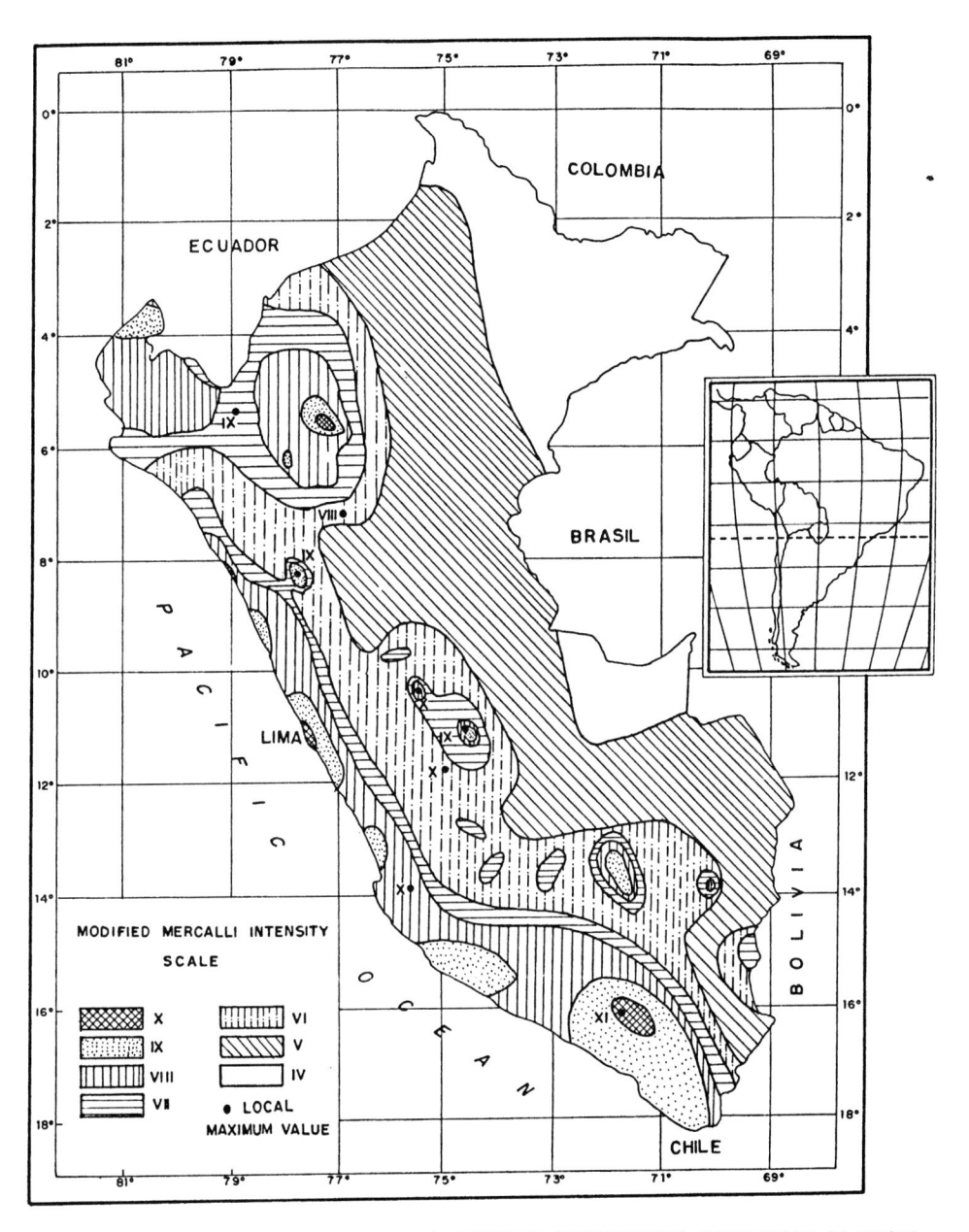

FIGURE 1. DISTRIBUTION OF MAXIMA SEISMIC INTENSITIES OBSERVED IN PERU
(After Alva et al 1984)

the city (Espinosa et al., 1977; Giesecke et al., 1980; Repetto et al., 1980).

Espinosa et al. (1977) assigned a MMI seismic intensity of IX to La Molina district, higher than any other site in Lima, probably due to a focusing effect that might have caused an energy concentration and an interference of seismic waves because of the geometric shape of the valley. Spectacular damages to the Agrarian National University, where reinforced concrete buildings completely collapsed, were reported in this district during the 1974 earthquake. Severe damages due to differential settlement resulting from compaction of loose sands were reported in "Reina de los Angeles" school. (Moran et al., 1975; Sarrazin et al., 1976).

Maximum seismic intensities of IX and VIII MMI were reported in La Punta and Callao, respectively (Espinosa et al., 1977). An important failure on a silo lift tower in the harbor, as well as failures on many buildings were reported in these two districts. Figure 2 shows the official map of intensity distribution for Lima after the 1974, October 3 earthquake, proposed by the Geophysical Institute of Peru (Giesecke et al., 1980).

3. Geological and Geotechnical Characteristics

Lima is located mainly on a fluvio-alluvial deposit of variable characteristics, corresponding to the debris cones of the Rimac and Chillon rivers from Quaternary age. The hills surrounding Lima are mainly intrusive rocks and materials from the Cretaceous period. Most of the city is located on a flat surface. Martinez Vargas (1986) has described the geology of Lima. Figure 3 shows a simplified map of the geological conditions of the city.

The predominant material in the subsoil of downtown Lima is a conglomerate (mixture of boulders, gravel and sand) under loose-to-compact state and interlayered by fine-to-medium sand, silts and clays. This conglomerate is found mixed with colluvial, alluvial and eolian deposits, resulting in erratic soils in the contact zones and surroundings at the north, south and east of the city. Fine soils are found in La Punta-Callao, La Molina, Chorrillos and Barranco districts. Eolian and marine sands appear in the coast to the north and south of Lima. Fills have been deposited on the upper part of the cliffs. Figure 4 shows the soil mechanics map for Lima elaborated by Martinez Vargas (1986).

La Molina district is located to the east of Lima, lying on an alluvial valley of semi-elliptical shape, surrounded by hills and connected to the Rimac valley. It appears that the zone was a lake in the geological past, and was formed because of the obstruction of the outflow. The soil profile is composed by clays, silts and loose sands from the surface up to 8 to 15 m deep, and gravel and dense sands below. The groundwater table is found below 10 m in depth.

The soil profile of Callao is composed by surface fills covering a layer of fine material which consists of sands, silts and clays, and sometimes peat. Below this fine material is found the Lima conglomerate, reaching variable depths, and some very deep, hard marine clay layers in specific areas. In La Punta district, the ground surface consists of a sequence of poor-graded gravel with silts and sands intermixed non-uniformly, with thicknesses ranging from 7 to 15 m. Below this layer are found fine sands with lenses of silts and clays. The conglomerate of Lima is presently at 30 m deep. The depth of the groundwater table varies from 1.5 to 3.5 m in La Punta, whereas in Callao it varies from 2 to 8 m.

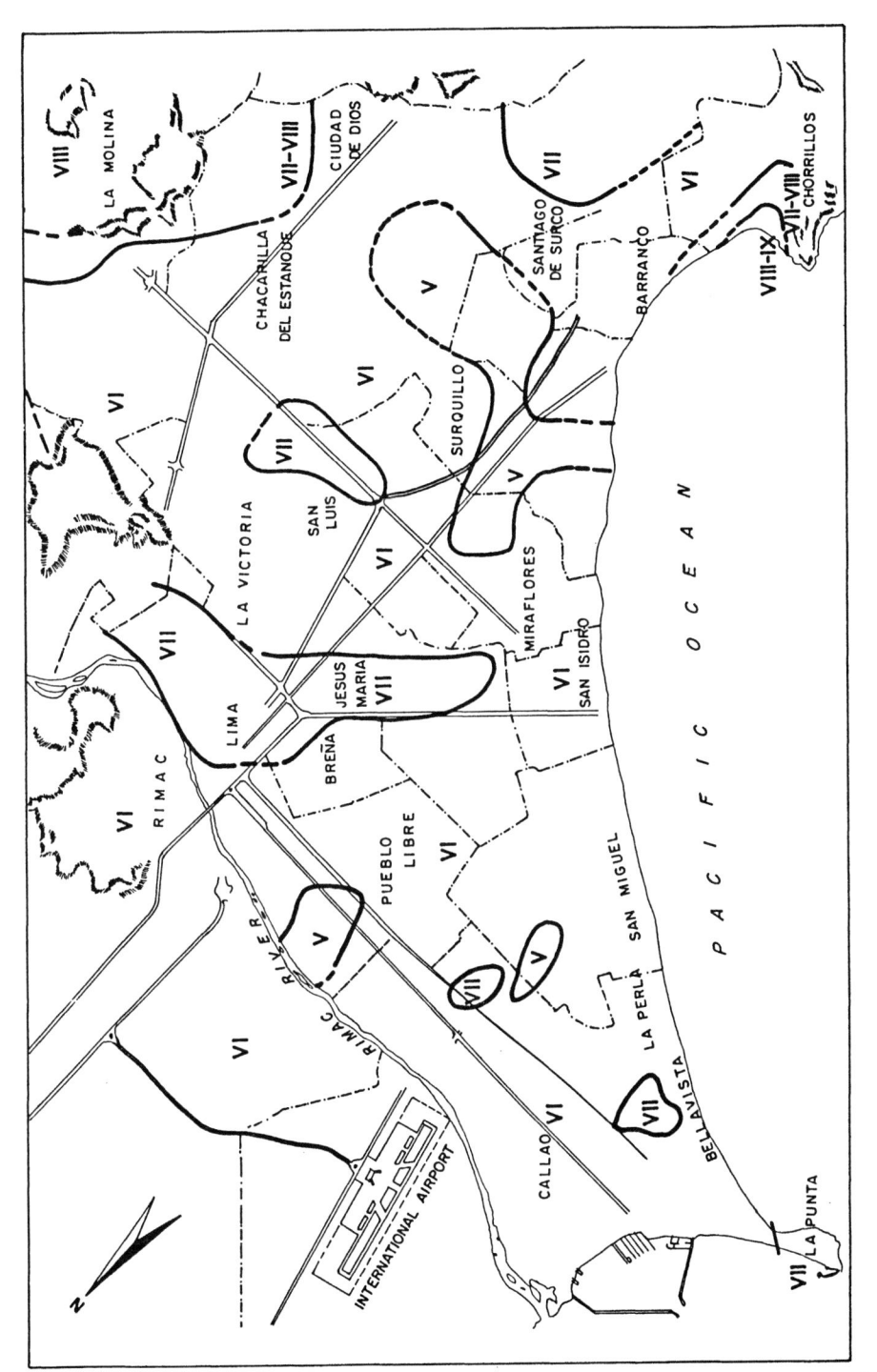

FIGURE 2. MM INTENSITY DISTRIBUTION OF 1974, OCTOBER 3 EARTHQUAKE FOR LIMA CITY
(Giesecke et al, 1980)

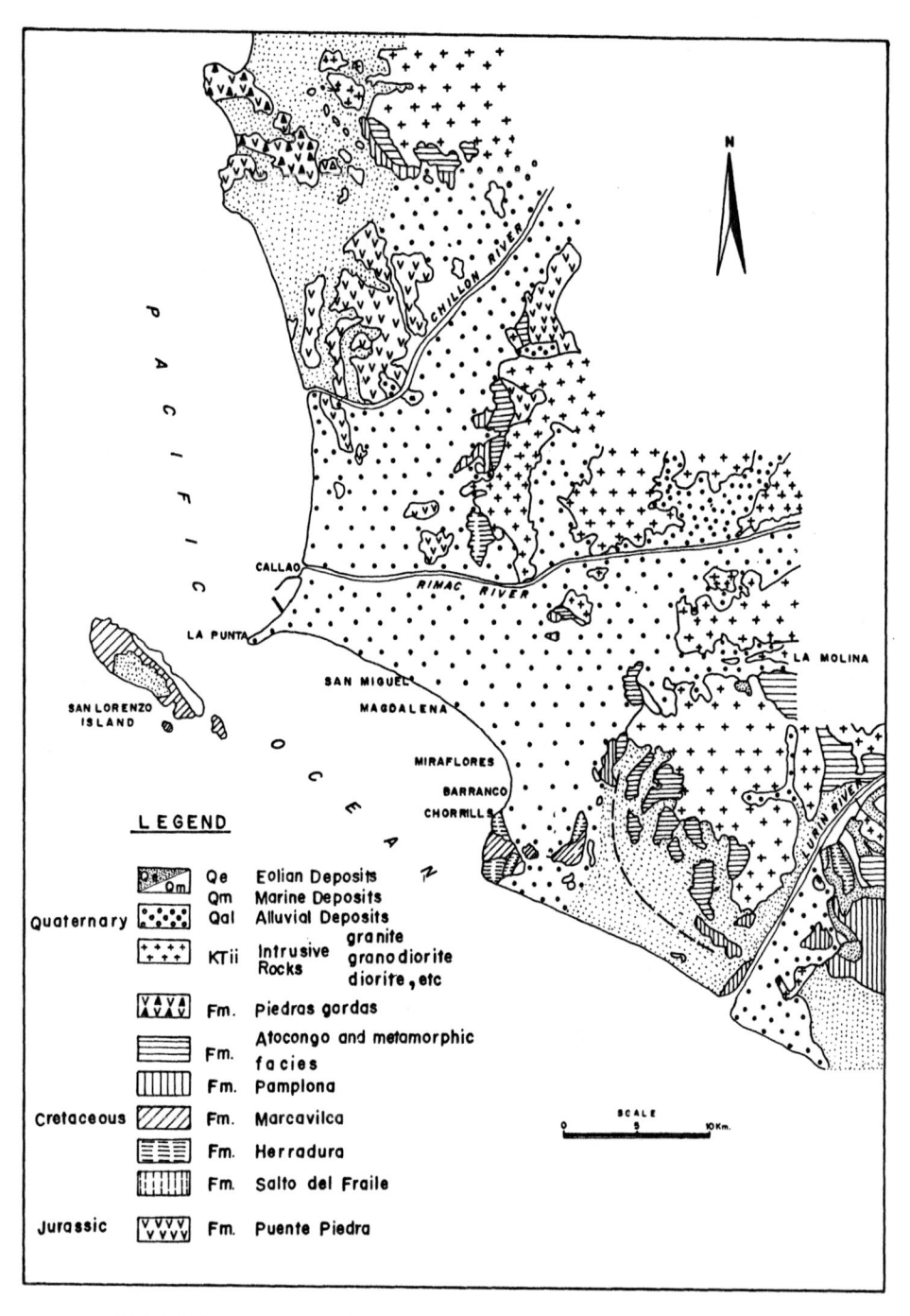

FIGURE 3. MAP OF GEOLOGICAL CONDITIONS FOR LIMA CITY

(Martinez Vargas A. 1986)

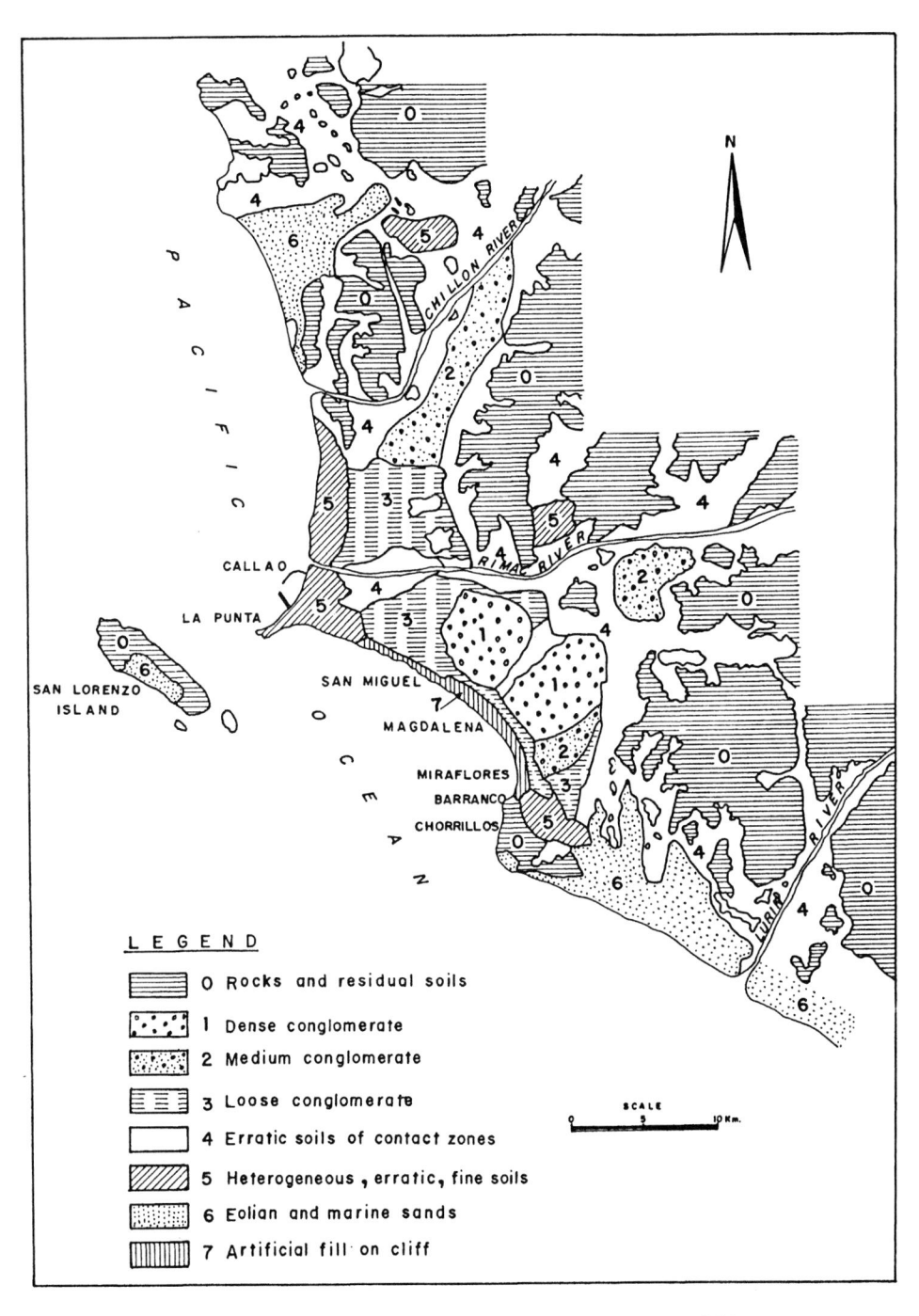

FIGURE 4. MAP OF SOIL CONDITIONS FOR LIMA CITY
(Martinez Vargas A. 1986)

4. Vulnerability

Kuroiwa (1977) from the National University of Engineering initiated the studies on vulnerability of Lima Metropolitan Area and proposed a Seismic Protection Plan for the city. In that study a "most probable earthquake" was selected. The proposed intensity distribution of that earthquake was slightly stronger than that of 1940: IX MMI in La Molina, Chorrillos and La Punta-Callao; VIII MMI in Barranco, La Victoria, Lima Cercado, Rimac; and VII in the rest of the city.

To estimate the damage, different types of buildings were considered: adobe, brick and reinforced concrete, and also subdivisions of these types according to their strength, age, preservation, defects, etc. The relationship between intensity and damage was prepared on the basis of experience in past earthquakes in Peru. The buildings were divided into houses, schools, hospitals and industrial factories. Because of the fact that in Lima Metropolitan Area several thousands of dwellings exist, a statistical survey was undertaken. The city was divided into sections with similar types of buildings and soil conditions. From each section, 5% of the blocks were selected and 10% of the houses of those blocks were evaluated. Similar sampling methods were employed for schools, hospitals and industrial buildings. Studies were also undertaken on lifeline networks and secondary effects, such as tsunamis and fires.

The results were presented in four categories of damage, and for each type the percentages of buildings existing in a section were plotted. The most critical areas were located in the old section of the city, where a significant percentage of adobe construction exists. Most of these constructions are overcrowded and have only a narrow escape. Some old adobe buildings are used for schools. Many brick and reinforced concrete buildings are not earthquake-resistant because the seismic code was officially adopted in 1970. Many of these buildings have the same defects that caused failures in the past (short columns, large eccentricities, resistance in only one direction, etc.). The industrial buildings have costly equipment in inappropriate and weak structures. Little attention was paid to potential earthquakes and fires in those buildings. Some of the hospitals were old and earthquake-damage prone. The study concluded that the results should be considered as preliminary because of the complexity and extent of the problem, and more detailed studies were recommended. To minimize earthquake losses, efforts should be made to improve existing buildings, demolish those in precarious conditions and control the excessive growth of Lima.

The National Institute for Urban Development, INADUR (1983), undertook studies to prepare a Seismic Protection Plan for Lima Metropolitan Area. Figure 5 presents the critical zones for Lima proposed in that study. A large proportion of the housing within the critical zones is highly vulnerable, not only because of the poor resistance of the original structures and materials, but also because of the intensive wear and tear to which they have been submitted by overcrowding during recent decades.

The oldest houses (50 years or more) are large, with heavy roofs and adobe or *quincha* structures with wooden joists. In the past, these houses had relatively high seismic resistance, which explains why they did not collapse during previous earthquakes. However, with time these buildings became vulnerable because of their intensive over-use. In the critical zones there is more recent housing (30 or 40 years old) built of adobe or of provisional material, of one story and laid out in groups of independent rooms (*corralones* or *callejones*). These houses are also vulnerable, but because of their small size and lightweight roof material have lower risk and rapid recuperation.

The change in use of large, old houses to rented tenements for low-income groups is the starting point of the physical deterioration. This change in use is explained by the accelerated demand for

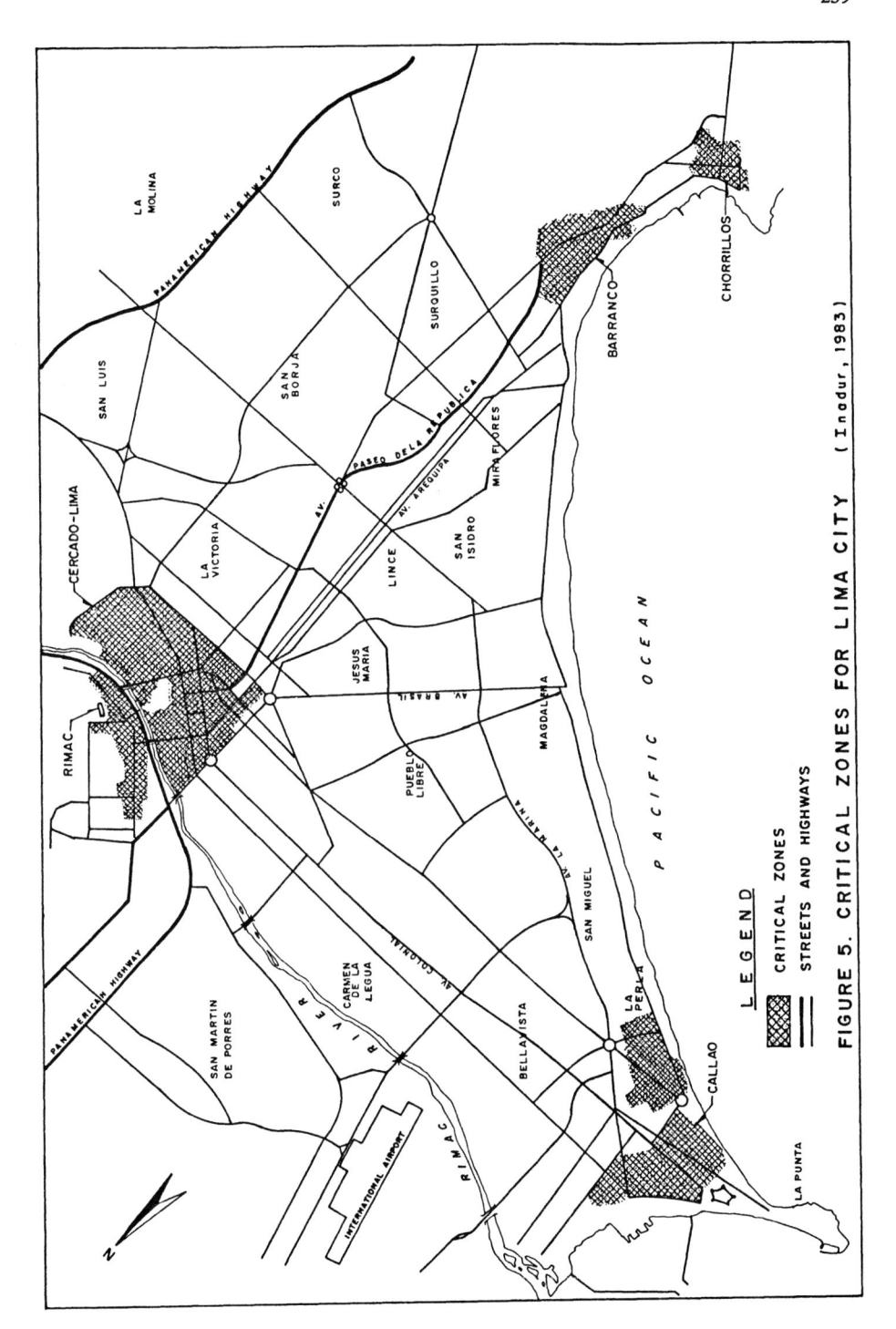

FIGURE 5. CRITICAL ZONES FOR LIMA CITY (Inadur, 1983)

cheap, rented housing in Lima-Cercado by urban migrants and because the state and private sector are not constructing houses for rental. The process of deterioration of the housing in the critical zones produces a highly vulnerable situation in the event of an earthquake. The study conducted by INADUR focused on the critical zones in Callao, Rimac, Barranco, Chorrillos and Lima-Cercado. Figures 6 and 7 present as an example the net density distribution of population and the level of material risk for Lima-Cercado. Other examples were also presented for other critical zones in Lima.

Recently, the National Institute of Civil Defense (INDECI) (1992) finished a Project on Identification, Localization and Qualification of Deteriorated Housing with Collapse Potential in Lima-Cercado. The area under study consists of 273 blocks with a total area of 21.9 km² and a housing area of 11.4 km². The population of the Lima-Cercado is 508,782 with a density of 232 persons/ha and a net density of 444 persons/ha. There are 73,091 houses in the area. In Lima-Cercado there are 18,087 houses under risk of collapse. The population involved is 101,689 persons. In addition to the housing problem there are other problems in Lima-Cercado, such as 41,500 informal merchants selling on the streets and 5,000 public transportation vehicles creating traffic problems.

A more specific study on the vulnerability of buildings in Lima-Cercado was done by Rios (1991) from the Japan-Peru Center for Earthquake Engineering Research and Disaster Mitigation, CISMID, of the National University of Engineering in Lima. The area under study was Barrios Altos within Lima-Cercado. In the research zone, 50% of the area was occupied by housing and the rest by business and industry. From the total housing buildings, 60% corresponds to *corralones* and *callejones*, 27% to old adobe houses, and 13% to modern building apartments. From the total housing buildings, 89% represents adobe and *quincha* materials, 9.6% brick, and 1.4% provisional construction materials. From the 1964 houses surveyed, 58% (1,137) are considered to have a high potential for collapse, 13% (258) can have severe damage, and 29% (571) moderate to light damage under earthquakes. The total population in the area under study was 9,814 persons, with a net density of 832 persons/ha. A moderate earthquake in the area under study can produce 100 deaths, 1,000 injured and 7,000 persons displaced.

5. Conclusions

1) Lima has been exposed to destructive earthquakes with relatively short intervals. In the present century, three earthquakes had intensities of VIII MMI. The last one occurred twenty years ago.

2) The population in Lima has considerably increased in recent years. In 1993 the population of Metropolitan Lima was above 7 million. Almost 30% of the total population of the country is concentrated in the capital.

3) Recent past earthquakes have caused heavier damages in some parts of the city, such as the districts of La Molina, Barranco, Chorrillos and La Punta-Callao, which have different soils than the gravel alluvial deposit of downtown Lima.

4) Based on vulnerability studies in Lima Metropolitan area, the critical zones for damage from an expected earthquake are: Lima-Cercado, Rimac, Callao, Barranco and

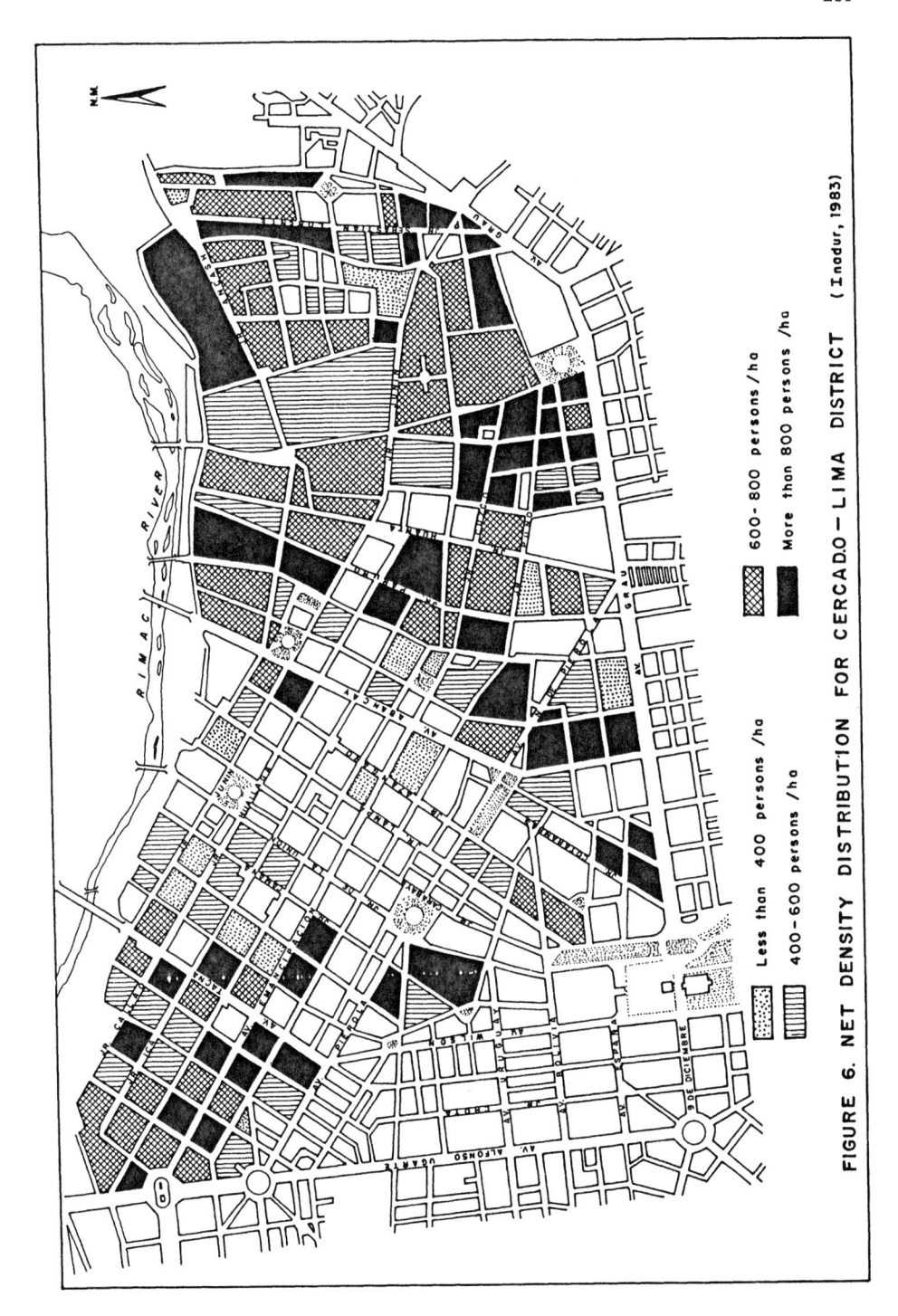

FIGURE 6. NET DENSITY DISTRIBUTION FOR CERCADO-LIMA DISTRICT (Inadur, 1983)

Less than 400 persons/ha
400-600 persons/ha
600-800 persons/ha
More than 800 persons/ha

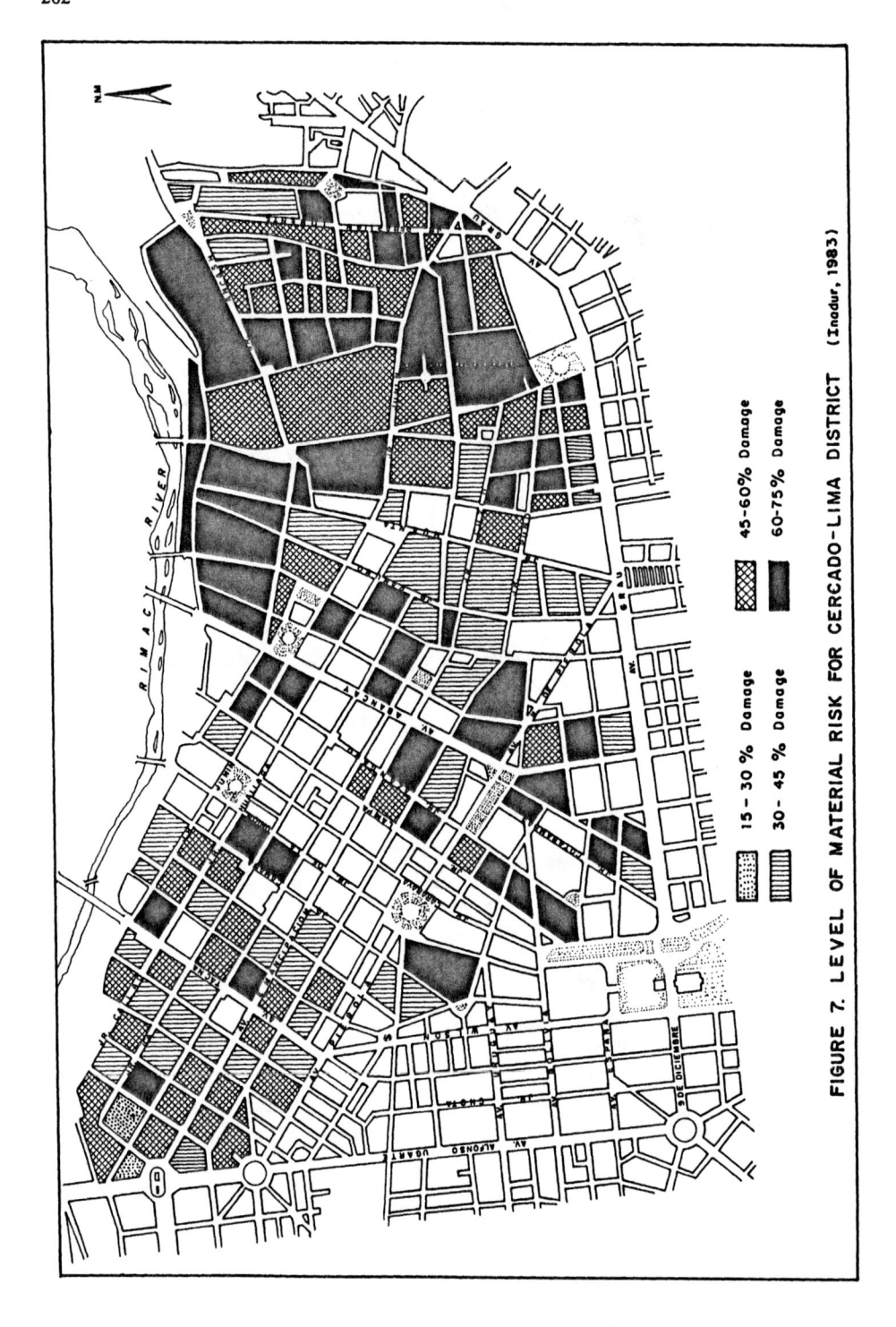

FIGURE 7. LEVEL OF MATERIAL RISK FOR CERCADO-LIMA DISTRICT (Inadur, 1983)

Chorrillos. Those zones correspond to high-density population areas with highly vulnerable buildings.

5) The vulnerability of Lima to destructive earthquakes is very high and has increased dramatically in recent years. The population has also considerably increased, the buildings have deteriorated, and new buildings have occupied lands marginal from the viewpoint of seismic strength.

6) Urgent measures must be prepared to improve the seismic safety of Lima Metropolitan Area.

References

Alva-Hurtado J.E., Meneses-Loja J.F. and Guzman V. (1984), "Distribución de Máximas Intensidades Sísmicas Observadas en el Perú". Proceedings of the V National Conference on Civil Engineering, Tacna, Perú.

Espinosa A.F., Husid R., Algermissen S.T. and De Las Casas J. (1977), "The Lima Earthquake of October 3, 1974: Intensity Distribution", Bulletin of the Seismological Society of America, Vol. 67, No. 5, pp. 149-1439, october.

Giesecke A., Ocola L., Silgado E., Herrera J. and Giuliani H. (1980), "El Terremoto de Lima del 3 de Octubre de 1974", CERESIS/UNESCO.

INADUR (1983), "Seismic Protection of Metropolitan Lima. A SuMMIary of a Study of Vulnerability, Risk and Recuperation Potential in Lima's Critical Areas", Instituto Nacional de Desarrollo Urbano, Ministry of Housing, Lima, Peru.

INDECI (1992), " Identificación, Localización y Calificación de Viviendas Tugurizadas con Riesgo de Colapso en Lima Cercado", Instituto Nacional de Defensa Civil, Ministry of Interior, Lima, Peru.

Kuroiwa J. (1977), "Protección de Lima Metropolitana ante Sismos Destructivos", Faculty of Civil Engineering, National University of Engineering, Lima, Peru.

Martinez Vargas A. (1986), "Características del Subsuelo en Lima Metropolitana", Proceedings of a Seminar on Foundation Design and Construction, Peruvian CoMMIittee on Soil Mechanics, Foundations and Rock Mechanics, Lima, Peru.

Moran D., Ferver G., Thiel C., Stratta J., Valera J. and Wyllie L. (1975), "Engineering Aspects of the Lima, Peru Earthquake of October 3, 1974", Earthquake Engineering Research Institute, CA.

Repetto P., Arango I. and Seed H.B. (1980), "Influence of Site Characteristics on Building

Damage During the October 3, 1974 Lima Earthquake", Report No. UCB/EERC-80/41, Earthquake Engineering Research Center, University of California, Berkeley.

Ríos J.F. (1991), "Estudio de la Vulnerabilidad y Medidas de Prevención Sísmica en el Cercado de Lima", Thesis submitted to the Faculty of Civil Engineering, National University of Engineering, Lima, Peru.

Sarrazin M., Saragoni R. and Monge J. (1976), "El Terremoto del Perú del 3 de Octubre de 1974". Proceedings of Second Chilean Meetings of Seismology and Earthquake Engineering, Santiago, Chile.

Silgado E. (1978), "Historia de los Sismos más Notables Ocurridos en el Perú (1513-1974)". Institute of Geology and Mining, Bulletin No. 3, Serie C, Geodynamics and Geological Engineering, Lima, Perú.

Valencia R. (1940), "El Terremoto del 24 de Mayo de 1940, Sus efectos y Enseñanzas". Journal of the Catholic University, Vol. VIII, pp. 6-7, Lima, Perú.

EARTHQUAKE DAMAGE SCENARIOS IN LISBON FOR DISASTER PREPAREDNESS

L. MENDES-VICTOR[1]
C. S. OLIVEIRA[2]
I. PAIS[3]
P. TEVES-COSTA[1]
[1]*Faculty of Sciences of the Lisbon University*
[2]*Instituto Superior Técnico, Lisbon*
[3]*Municipal Department of Civil Protection, Lisbon City Council*

1. Introduction

Research on the earthquake history of the town of Lisbon has been improved in order to assure the scientific background to launch an effective earthquake disaster preparedness program. The characterization of the seismology of the area as well as its geology were fundamental in preparing a microzonation of the seismic risk of the building stock. Furthermore, to evaluate the seismic risk, the dynamics of the population, the property value and the vulnerability of the building typology were considered and integrated in a seismic impact model. This knowledge has been used to prepare the first version of the seismic risk emergency plan for the town of Lisbon, and many efforts are carried out to improve it in order to assure the adequate disaster preparedness.

Due to its particular situation, Lisbon has been affected by several strong earthquakes in the past; since the 12th century, at least nine earthquakes caused severe damage, and four of them had a Modified Mercalli Intensity greater than or equal to VIII. The last big one struck November 1, 1755 and caused the destruction of a great part of the town (Pereira de Sousa, 1911-1932). In this century, although several quakes were felt—the 28 February, 1969 event reportedly produced a maximum intensity of VII in the Lisbon area— none of them caused severe damage. Strong shaking could be produced, however, at any time by a high magnitude earthquake at a medium epicentral distance, or a moderate magnitude earthquake at a shorter epicentral distance (due to the tectonic activity of some faults close to the town, as occurred in 1531).

For these reasons, it is of primary interest to try and prevent the damage that could occur in the future. The evaluation of the impact of potential earthquakes in the Lisbon area aims at (1) the determination of the areas of higher risks, and (2) the quantification of losses in terms of deaths and structural damage. This evaluation requires the development of different tasks, the most important ones dealing with hazard, microzonation, building vulnerability, and space-time dynamics of the population throughout the day.

As the capital of Portugal, Lisbon has a floating population during the day, with about 1,200,000 people during work hours and about 660,000 residents within the county limits. The building typologies vary greatly (five categories were identified from the total of approximately 60,000 buildings) with an historical center, some old quarters and new zones. However, the coexistence in some areas of different typologies increases the difficulty in predicting building behaviour. If a damaging earthquake occurs, the social and economic impact would be significant.

In 1984, a methodology to predict earthquake losses on a consistent basis was developed, taking into account all the knowledge in the fields mentioned above (Oliveira and Mendes-Victor, 1984). For the town of Lisbon two seismic scenarios corresponding to the two main sources were considered: a strong far earthquake (produced in the Gorringe Bank area), and a moderate near earthquake (produced in the Lower Tagus Valley Fault). A global loss function was defined, taking into account the individual losses of each category of buildings which is a function of their vulnerability, the cost of the construction, the area of the buildings, and the number of buildings of that class. The total damage cost, the total affected population, the density of losses, and the density of affected population were also estimated.

Since its first version, this model has been improved. An example of application to a pilot area will be presented. At the present time, work is being undertaken in order to extend the methodology to the whole town of Lisbon.

The results provided by this model are being used by the Municipal Department of Civil Protection of Lisbon City Council to elaborate the seismic risk emergency plan for Lisbon.

2. Historical Seismicity

Among the earthquakes which caused intensities greater than or equal to VIII in Lisbon, the 1531 and the 1755 were the most recent.

The 1531 earthquake had an estimated magnitude of 7.1, and its source was located at 10-20 km north of Lisbon. Intensities of VIII and IX (MMI) were reported, and the most significant damage was observed in the "downtown" alluvium valley and hillsides. Liquefaction phenomena, landslides, and an anomalous rise in the river level were identified. A sulfur scent was set free. About 25% of the houses were damaged, and 10% suffered total collapse; 2% of the population was killed (Henriques et al., 1988).

On November 1, 1755, Lisbon suffered a large earthquake whose effects were felt throughout Europe (for instance the water in Loch Lamond, Scotland vibrated for one-and-a-half hours with an amplitude of 60 cm). Several authors consider this the biggest earthquake that ever occurred and estimate its magnitude at 8.5 to 9. The focus was located around 200 km southwest of Lisbon. This earthquake caused severe damage: intensities of IX and X were felt in the downtown and central hills, and of VIII in the remainder areas of the town [Figure 1]. The earthquake produced a tsunami which increased the damage on the rivershore zones. Triggered by the earthquake, a big fire lasting five to six days increased the damages in the central part of the town. In all, 32 churches, 31 monasteries, 15 convents, 53 palaces, and 60 chapels were completely destroyed; among the 20,000 existent houses, only 3,000 were saved. About 10% of the population lost their lives (Pereira de Sousa, 1911-1932).

These two earthquakes correspond to the two main seismogenetic sources affecting the town of Lisbon: the Lower Tagus Valley, a near source believed to be responsible for the 1344, 1512, 1531, and 1909 earthquakes; and the Gorringe Bank, a far source believed to be responsible for the 1356, 1755, 1761, and 1969 earthquakes. Figure 2 shows the location of the main historical earthquakes felt in the mainland territory.

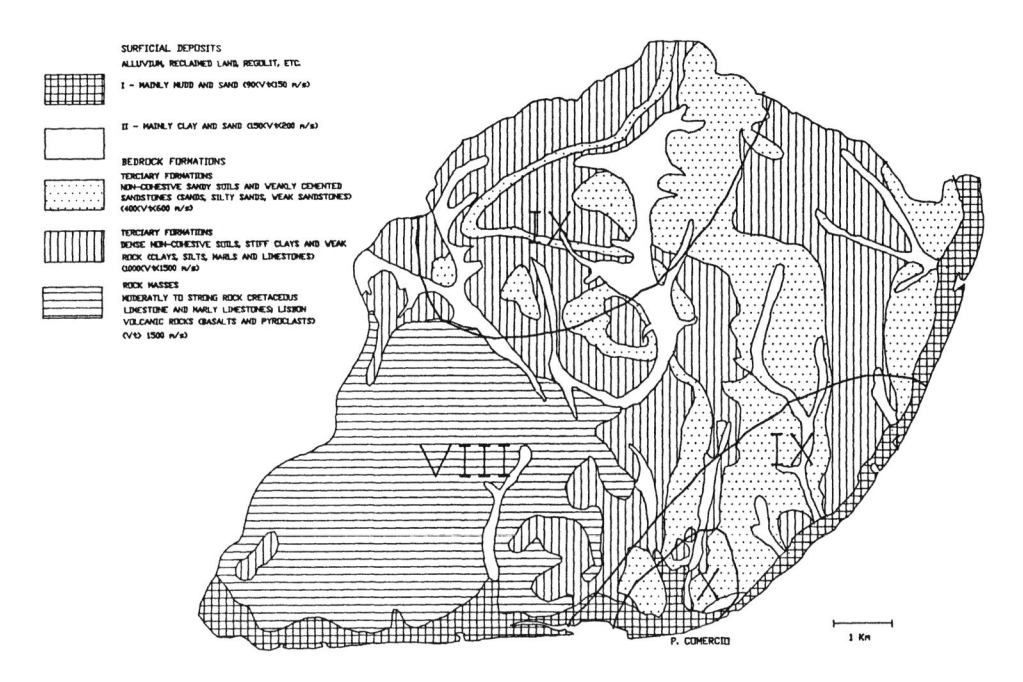

Figure 1. Isoseismals of the 1755 Lisbon earthquake, superimposed on a sketch of the geotechnical map

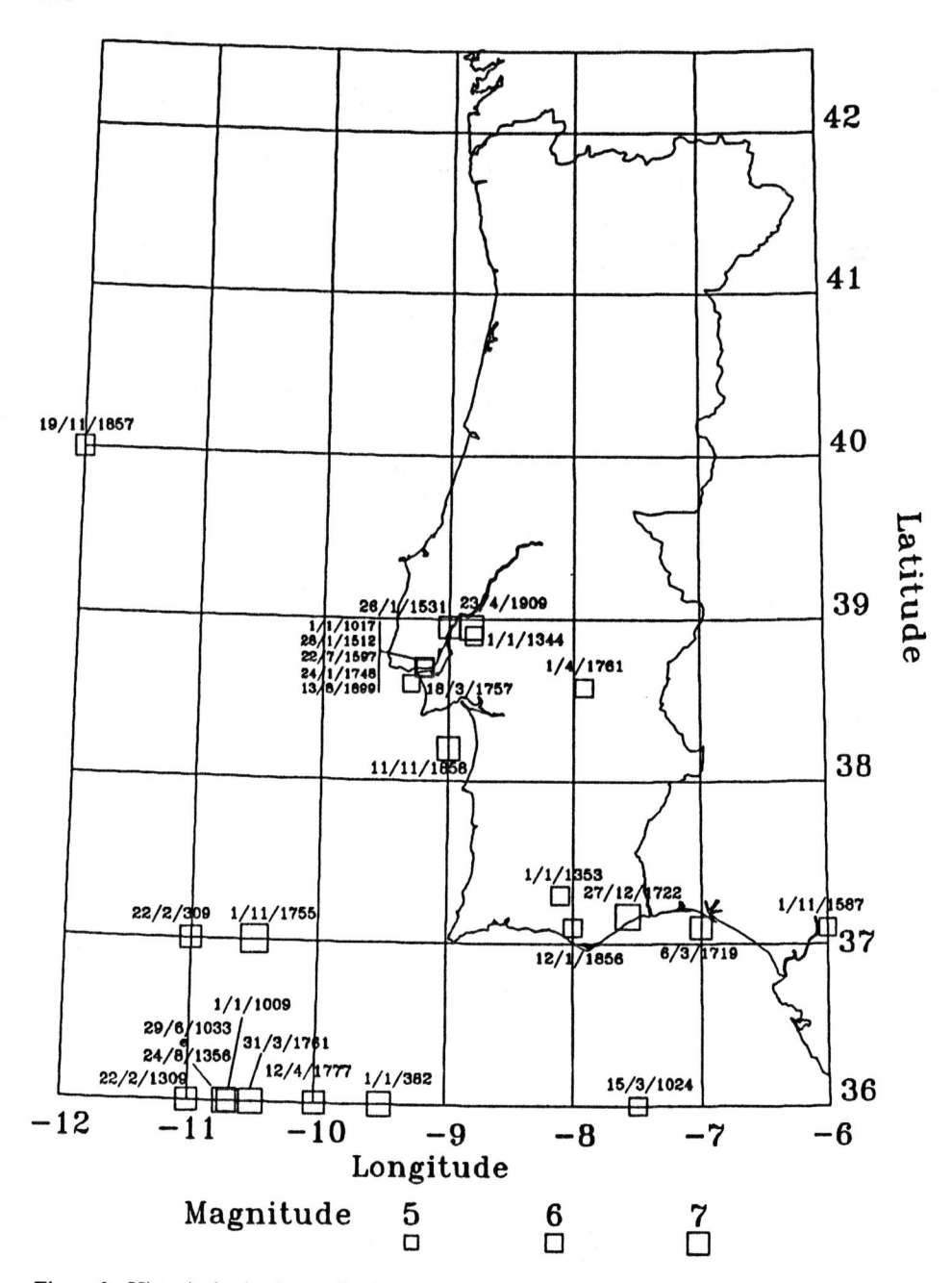

Figure 2. Historical seismicity affecting the mainland territory until 1920 (M5.6) (after Cabral, 1993)

3. Seismogenetic Characterization

The evaluation of the seismic impact on Lisbon of future earthquakes includes the identification and analysis of the activity of seismogenetic sources responsible for the town's historical damaging earthquakes.

The Portuguese territory shows different tectonic features responsible for the observed seismic activity, (Cabral, J., 1993) [Figure 3]. This activity has been studied by several authors (Moreira, 1985; Campos-Costa et al., 1992). Recently, Martins and Mendes-Victor (1993) presented a study of the seismicity of the western part of the Iberian Peninsula, between longitudes of 18°W to 2°W and latitudes of 32°N to 43°N.

Four zones were identified, grouping the seismogenetic sources according to the seismic activity and the energy distribution [Figure 4], and the constants **a** and **b** of the Gutenberg-Richter law were estimated for each one of the zones.

The two zones of main importance for the town of Lisbon are zone Ae (where the Gorringe Bank is located - far source) and zone Be (which includes the Lower Tagus Valley Fault - near source). The return periods corresponding to different magnitude levels were evaluated [Table I].

Table I - Return periods for the different seismogenetic zones

Magnitude	Zone A_e	Zone B_e
5.0	3	5
5.5	7	10±1
6.0	15±1	21±2
6.5	30±2	41±6
7.0	63±7	84±13
7.5	135±17	170±33
8.0	288±42	344±80
8.5	614±105	695±188

Return Period (Years)

4. Geology and Topography of the Town: A Short Description

The geology of the Lisbon area is quite complex, presenting many contrasts as the result of a long tectonic history and the opening of the Atlantic Ocean. The main tectonic processes were the subsidence of the Lusitanian basin (since the Mesozoic Era) interrupted by some compressive episodes, the last of which was the Alpine orogeny (until the upper Cretaceous) (Ribeiro et al., 1990; Wilson et al., 1992). Important volcanic eruptions, mainly in the western part of the town (Monsanto anticlinal), occurred at that time in the Lisbon area.

The surface geology of Lisbon is well described in the geological map of the county, published by the Serviços Geológicos de Portugal at the scale of 1:10,000 (Almeida, 1986). Analysis of this map reveals, in the western zone of the town, a predominance of Cretaceous rock masses essentially composed of basalts, limestones and marbles, while towards the east and south, this structure is covered by progressively thicker Miocene deposits dipping 7-10 degrees. These sediments have differentiated lithology—essentially clays, sandstones, silts and limestones—with

270

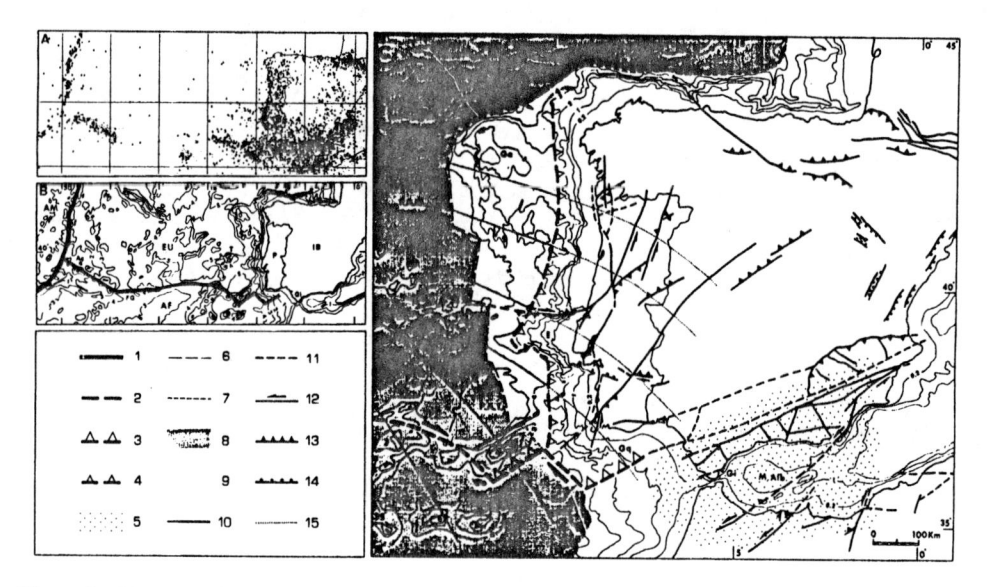

Figure 3. Neotectonic environment of the Portuguese territory according to Cabral's geodynamic model (1993). (A) Seismicity of the Azores-Gibraltar region, 1900-1991. (B) Assumed position of the Azores-Gibraltar boundary. (C) Neotectonic environment of the Portuguese territory showing the main active faults on the Iberian Peninsula and North Africa. Present pattern of the maximum horizontal stress. 1- American (AM), Euroasiatic (EU) and African (AF) plate boundaries in the Atlantic region; 2 - Boundary between the plates (approximated position); 3 - Subduction south of the Gorringe (Go) and Guadalquivir (Gq) submarine banks; 4 - Incipient subduction zone proposed in the continental west Iberia margin; 5 - Diffuse limit of the plates (continental collision); 6 - Approximated location of the boundary between the continental and the oceanic crusts; 7 - item, estimated; 8 - Atlantic Ocean crust; 9 - thinning continental crust; 10 - Active fault; 11 - item, probably trace fault; 12 - Strike-slip fault; 13 - Inverse active fault; 14 - Normal active fault; 15 - Estimated pattern of the present maximum horizontal stress (after Cabral, 1993).

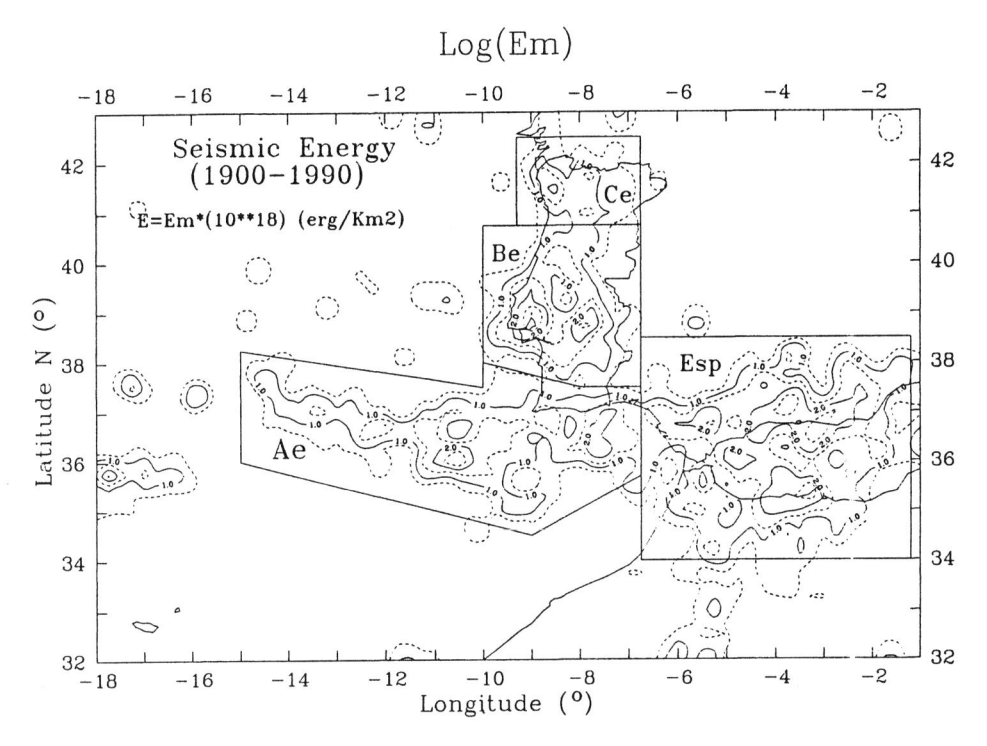

Figure 4. Distribution of the seismic energy and identification of the studied seismogenetic zones.

different geotechnical properties. The valleys in the lower parts of the town are filled with recent, thin alluvial deposits which have a low seismic impedance and may have important consequences for the seismic response.

A sketch of the geotechnical map of the town is displayed in Figure 1 and is based on the surface geology and the geotechnical properties of the different formations (SNPC, 1983).

The topography of Lisbon consists of seven low hills separated by long, narrow valleys, which are the remnants of old streams. Figure 5 displays the topography of the town, showing its great variety.

5. Microzonation—Additional Geotechnical Studies

The scattered destruction reported in historical documents has been considered as a form of clarifying the meaning of seismic risk in the dense urban areas. In order to complement the information provided by the geological map and the geotechnical data, several studies and experiments were undertaken since 1980.

The first experiment, carried out in 1980-81, consisted of the recording, processing and interpretation of the seismic signals produced by chemical explosions in the Tagus riverbed. Several papers already published (SNPC, 1983; Mendes-Victor, 1987; Teves-Costa, 1989) present a detailed description of the experiment. One result of this study was a set of microzonation maps for the town of Lisbon, based on the geotechnical properties of the soils and on the characteristics of wave propagation, evaluated for different seismic scenarios (Oliveira and Mendes-Victor, 1984). Figure 6 displays the microzonation map for the upper bound of the Gorringe scenario.

In 1988-89, several theoretical studies were developed, using the Aki-Larner technique (Teves-Costa, 1989; Bard and Teves-Costa, 1989; Teves-Costa and Bard, 1990; Teves-Costa and Oliveira, 1991). These studies suggested the existence of site effects in the alluvial valleys and in some hills.

In a second experiment, carried out in 1991, using the above-mentioned technique, seismic records were obtained at selected sites, according to the spatial distribution of damage reported in the historical documents and the previous theoretical studies. Published papers emphasized the amplification of the seismic wave energy for certain frequencies atop Castle Hill and in some alluvial valleys (Teves-Costa and Mendes-Victor, 1992). The theoretical results obtained by different techniques (Mota, 1992; Teves-Costa, 1993), showed a quite reasonable agreement with the experimental values.

6. Building Stock Evaluation

The behaviour of buildings during earthquakes is very difficult to predict and depends upon a great number of parameters. In the town of Lisbon, there are buildings of different types, ages, number of stories, and material properties, usually supported laterally by other buildings and showing discontinuities in height and plan. The buildings may be located on a flat or steep area and may be in good structural condition or bad, due to lack of repair.

Lisbon's building stock is composed mainly of modern, middle-size buildings for housing, larger office buildings and a considerable number of old buildings used both for housing and services in

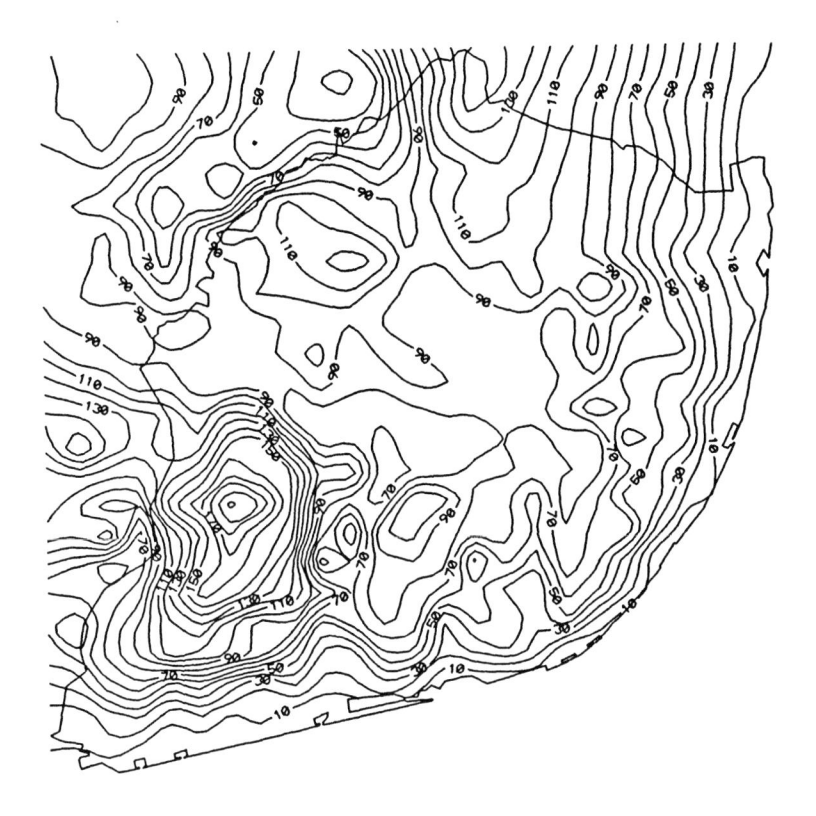

Figure 5. Topography of Lisbon

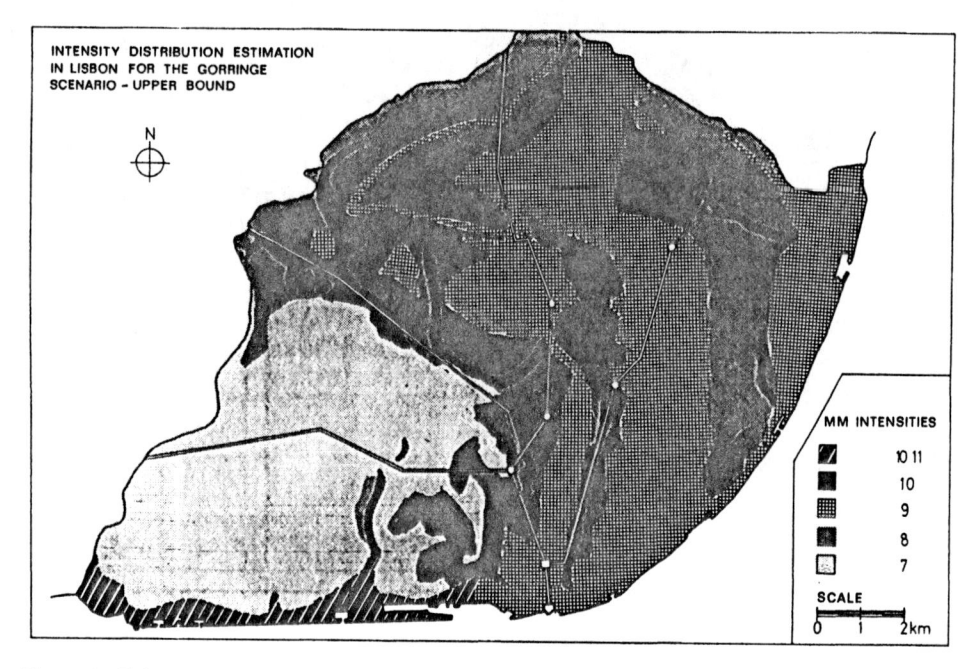

Figure 6. Seismic intensity distribution in Lisbon (MMI). Gorringe scenario - upper bound (after Oliveira et al., 1984)

the city's historical center.

In order to allow for a more precise understanding of their seismic behaviour, according to their main structural properties, Lisbon's buildings have been classified into five different categories (Oliveira and Cabrita, 1985a):

A - Masonry stone buildings prior to 1755, low-rise, most in bad shape (Freq. > 3Hz);

B - Masonry stone buildings constructed during the period 1755-1880 with horizontal ties and in good shape (Freq. > 2.5 Hz);

C - Brick masonry, tall buildings constructed during the period 1880-1940. Wooden floors. (Freq. > 2 Hz);

D - Dual structures with masonry-resistant walls and reinforced concrete (RC) slabs or RC moment resistant frames heavily infilled with brick walls constructed during the period of 1940-1960 (Freq. > 2 Hz);

E - Modern RC buildings constructed after 1960, designed according to modern, lateral load requirements (Freq. < 2.5 Hz).

Figure 7 presents a schematic spatial distribution of the above-mentioned building categories according to expert opinion (Oliveira et al., 1985b).

The structural characterization of Lisbon's building stock was made using several types of questionnaire surveys covering different areas and is presented in Figure 8.

The survey covering the areas of Alameda, Anjos-Pena and the western rivershore area was made under LNEC's (National Laboratory of Civil Engineering) supervision and was specially designed for future seismic impact studies. The survey sheets had 30 different parameters to determine, and trained teams of geographers, architects and engineers were formed to ensure as much as possible the homogeneity of data collection. An example of the output results of this survey is given in Table II.

The survey covering the areas of Lisbon's historical center and other ancient quarters, now under rehabilitation, was made by Lisbon's City Council Municipal Department of Urban Rehabilitation. Most of the parameters contained in this survey are similar to the ones used in LNEC's survey.

In the areas not yet directly studied by the two previous surveys, the Population and Housing Censuses of 1981 and 1991 were used (INE, 1991). The data contained in the Censuses was completed through expeditious field work. This work, covering representative modern urban areas, used sampling techniques to determine the number of stories and the area of land covered.

The vulnerability curves were assigned to each category based on the type of construction, natural period of vibration and on statistics obtained from thirteen worldwide earthquakes [Table III](Oliveira, 1988).

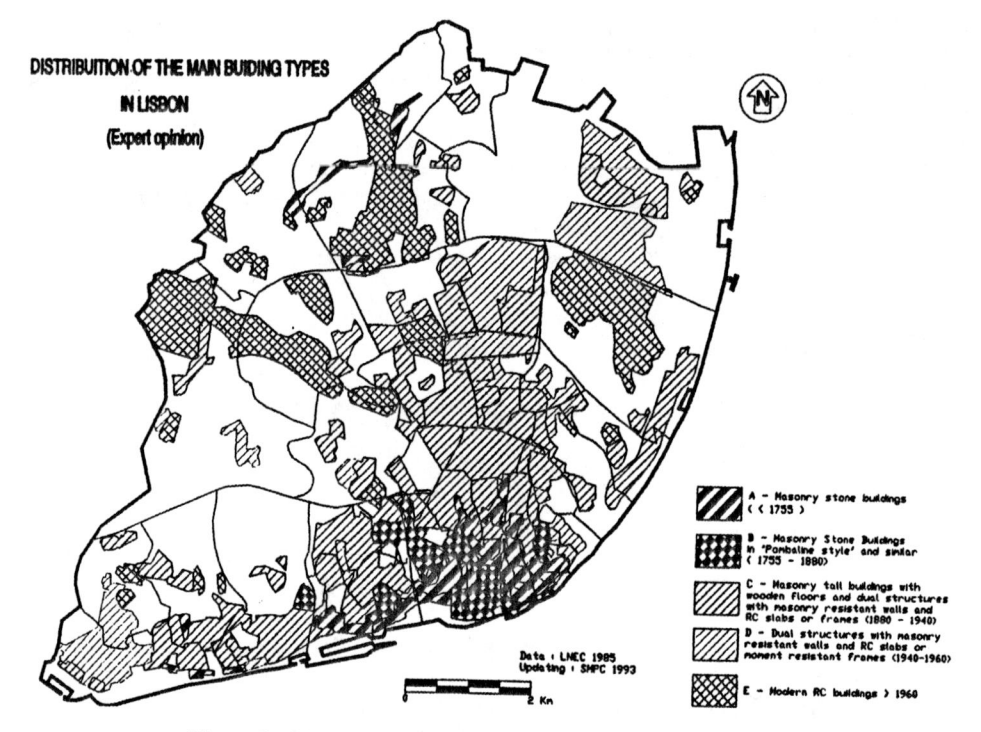

Figure 7. Distribution of the main building types in Lisbon

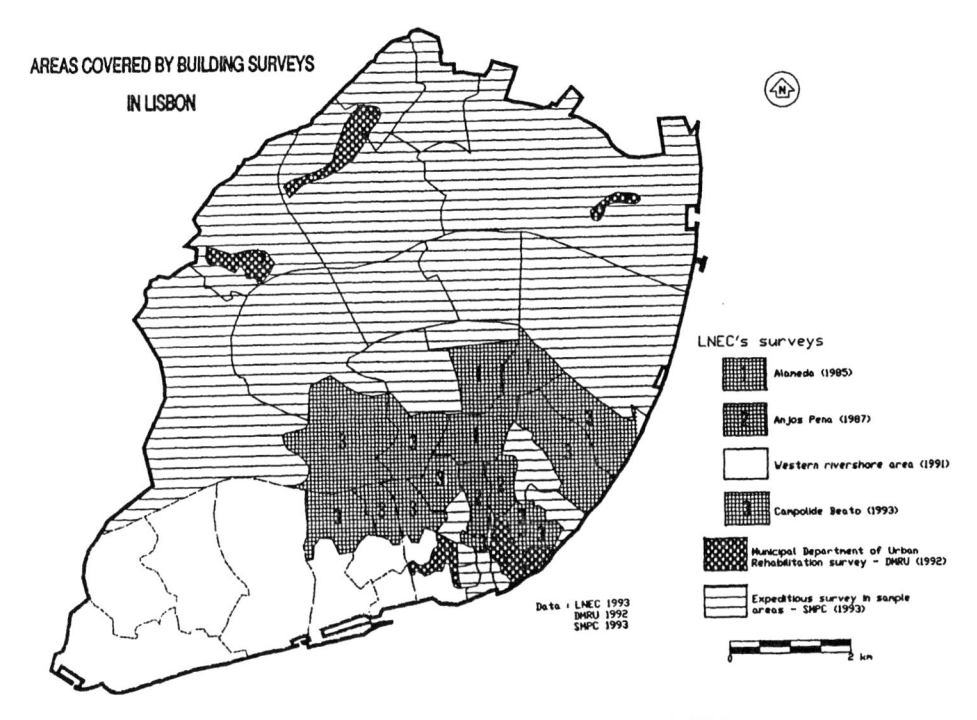

Figure 8. Areas covered by building surveys in Lisbon.

Table II- Example of an output of the LNEC's survey

	Alameda	Anjos	Western Rivershore	Total
Geographic area	286 ha	105 ha	1644 ha	2035 ha
No. of blocks	185	46	729	960
No. of buildings	3273	1877	10355	15505
Total no. of stories	15692	4710	31577	51979
Total no. of dwellings	22264	6411	44445	73120
Average no. of stories	4.32	3.52	3.07	3.49
Predominant epochs	1870-1940: 55%	1870-1940: 47.5%	1755-1889: 37.7%	-
Average area per building	280 m^2	202 m^2	521 m^2	431 m^2
Average area per dwellings	112 m^2	99 m^2	138 m^2	127 m^2
Predominant uses: housing	42.0%	47.2%	71.9%	53.7%
mixed	46.0%	35.2%	20.0%	33.7%
State conservation: average	55.0%	54.5%	43.0%	50.8%
bad	20.0%	28.0%	18.0%	22.0%
Population	64520*	30243*	96848**	-
Employees in the area	47912*	18636*	69812**	-
Occupation: Land covered/ geographic area	0.32	0.36	0.08	0.13
Use: Construction area/ geographic area	1.56	1.26	0.32	0.54

* - Data from Marin et. al. (1983); ** - Data from S.M.P.C. - C.M.L. (1993a)

Table III - Vulnerability variables

| M.M. Intensity | BUILDING TYPES ||||||||||
	A		B		C		D		E	
VII	0.2	1.0	0.05	0.15	0.5	1.0	0.05	0.08	0.1	0.1
VIII	1.0	5.0	0.1	0.3	1.0	2.0	0.1	0.15	0.5	0.5
IX	2.0	10.0	1.0	3.0	5.0	10.0	1.0	1.5	0.8	0.8
X	10.0	50.0	10.0	30.0	10.0	20.0	2.0	3.0	1.0	1.0
Source	s1	s2	s1	s2	s1	s2	s1	s2	s1	s2

Source: 1 - Gorringe; 2 - Lower Tagus Valley

7. Space-Time Dynamics of the Population

The population of the town of Lisbon and its metropolitan area (composed of 18 counties), which covers no more than 3% of Portugal's total area, has been estimated around 2,700,000. Nevertheless, only 660,000 of these are inhabitants of the town of Lisbon, which has lost 18% of its resident population between the 1980's and the 1990's (INE, 1991). The remaining population is confined to municipalities that have become mainly residential areas.

However, the active population in the town of Lisbon represents more than 54% of the total employment in the whole metropolitan area. At present, 520,000 people are employed in the town. About 80% work in the services and in commercial activities (Silva et al., 1993).

Figure 9 presents the spatial distribution of the employment in Lisbon. Although nowadays there is a tendency to disperse this distribution, there still can be seen a concentration of employment in the central area and along the main urban structural axis (north axis and western rivershore axis) that links the center to the peripheral parishes. These peripheral parishes are more predominantly residential.

In terms of job availability, the relation of jobs/inhabitants has increased from 56/100 in the 1980s to 79/100 in the 1990s, and this has brought serious urban and social consequences, such as a decrease in the number of residences, as they are progressively occupied by commercial activities and services; the abandonment of the central area due to population flow towards the periphery; and a strong increase in the number of daily commuters. A special study was developed (Marin et al., 1983) to estimate the population flow throughout the day in each of Lisbon's 23 unit areas. The division into areas was made according to socioeconomic and building characteristics in order to obtain homogeneous areas.

The dynamics of the population throughout the day is shown in Figure 10 (Pais, 1992), taking into account the five periods considered for this study: 0:00-7:30; 7:30-9:30; 9:30-18:00; 18:00-20:00; 20:00-24:00. An updated version of this study for 1993 is now being prepared by the Municipal Department of Civil Protection of Lisbon's City Council (SMPC-CML, 1993a).

As presented in Figure 10, the central areas of the town show a considerable increase in population during the day, especially from 9:30 to 18:00. In fact there are about 250,000 vehicles entering and leaving the town daily (besides the 400,000 registered in the city itself - DPPE-CML, 1993), mainly during the rush hours in the morning (7:30 - 9:30) and in the afternoon (18:00 - 20:00). If we add the railway and fluvial traffic to all this, which also transport a great number of people from the periphery to the city, it can be said that there are about 600,000 people entering the city daily. The current total population in Lisbon on weekdays, from 9:30 to 18:00 is about 1,200,000 people.

Casualties and injuries cannot easily be correlated with building damage and earthquake intensity. The number of casualties will depend in part on the occupancy conditions prevailing at the time of the event. It is sometimes considered that the San Fernando data can provide a reference estimation [Table IV]. These estimates are meant to be median. It seems reasonable to consider the two-thirds probability that actual casualties will occur between 0.5 and 1.5 times the figures in Table IV (Oliveira, 1988).

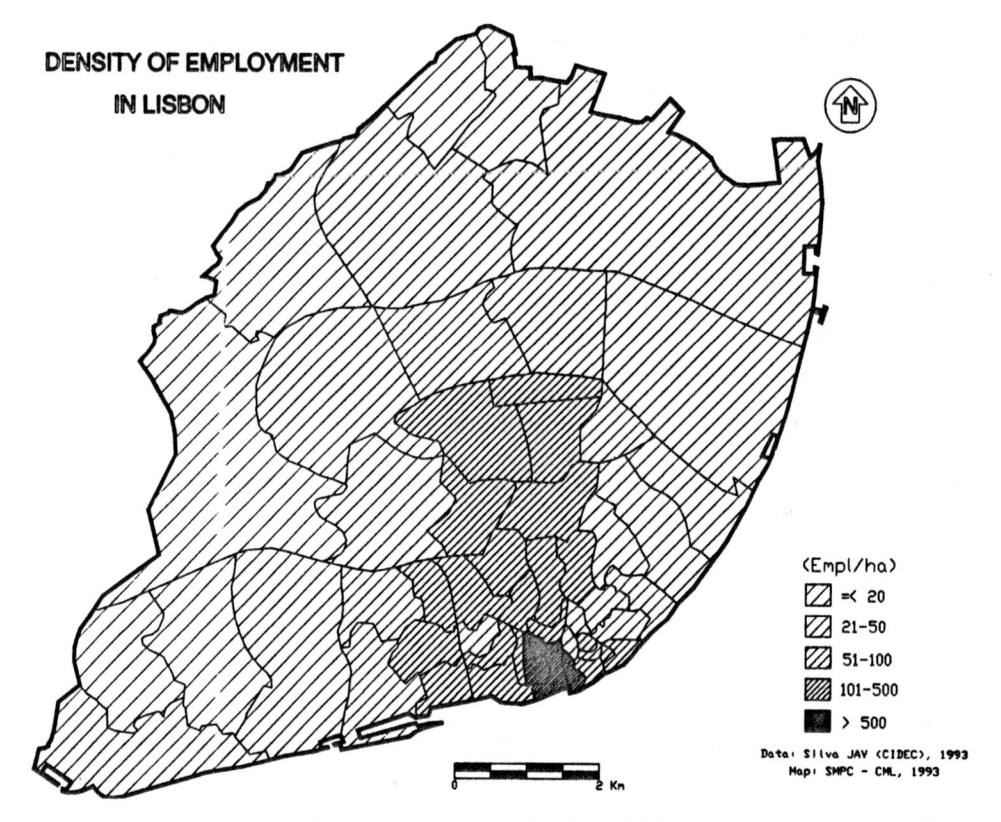

Figure 9. Employment density in Lisbon.

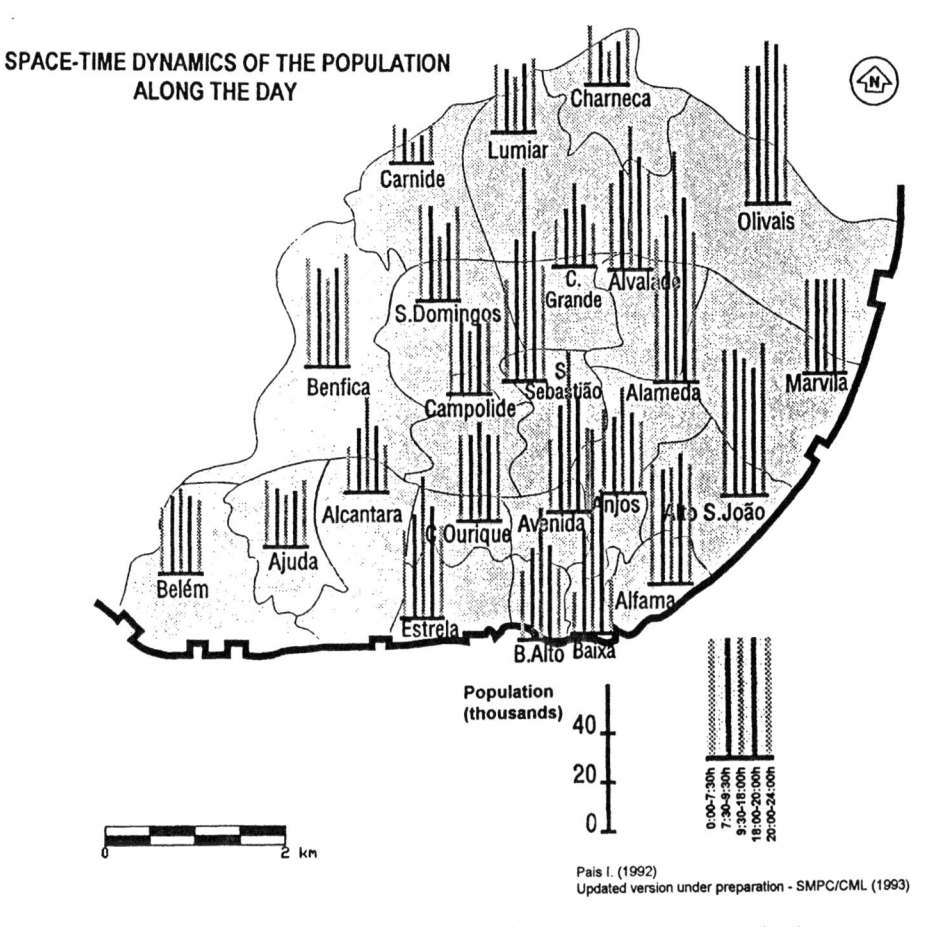

Figure 10. Space-time dynamics of Lisbon's population throughout the day.

Table IV - Population Vulnerability

Intensity	Estimated Human Casualties per 10^6 Inhabitants
VI	None
VII	10
VIII	150
>VIII	500

8. Seismic Impact Model

The seismic impact model takes into account the building stock and the public urban area. Monuments, schools and important, dangerous, or particular buildings are not considered in this model.

The mathematical model to analyze the seismic impact model considers the following:
1. Parameters :
    (a) The city of Lisbon was divided into 53 unit areas (the 53 Lisbon parishes), j (j = 1,53);
    (b) The existing buildings were grouped into 5 categories, i (i = 1,5), each one of them integrating the mean number of stories $n_{i,j}$ and area $a_{i,j}$;
    (c) The seismic scenarios of disaster (source), k (k = 1,2), with associated probability of occurrence $F_k (.)$;
    (d) The local topography and ground behaviour are included into 4 classes of intensity l (l = 1,4), corresponding to the MMI VII to X;
    (e) The time of occurrence of the earthquake is considered, m (m = 1,5) : in Lisbon, 5 periods have been studied;
    (f) The public urban area, defined as the public area not occupied by buildings, is classified in 3 types, r (r = 1,3): urban open spaces (gardens and parks, agricultural and non-cultivated areas); urban areas occupied by road networks and current infrastructures; and particularly sensitive urban areas, such as important lifelines (main roads, tunnels, bridges and viaducts, underground and railway networks, airport and harbour facilities).
2. Variables :
    $V_{i,j,k,l}$ - mean vulnerability for spectrum $S_{j,k,l}(\varpi)$, (in unit j, for source k and intensity l), in buildings of class i, with area $a_{i,j}$ and $n_{i,j}$ stories;
    $VP_{i,j,k,l}$ - mean value of affected population by the building vulnerability $V_{i,j,k,l}$;
    A_j - area of the unit j;
    $N_{i,j}$ - total amount of buildings of class i in unit j;
    $al_{l,j}$ - area of intensity l in the total area of unit j;
    $P_{j,m}$ - total amount of population in unit j during the period m;
    $C_{i,j}$ - mean household value per square meter in buildings of class i in unit j;
    $aep_{r,l,j}$ - area of public urban area of type r in class of intensity l in unit j;
    $Vep_{r,l}$ - mean vulnerability of public urban area of type r in class of intensity l;
    Cep_r - mean public urban value for type r.

3. Functions :
   Individual Loss Function (ILF) :

   $$ILF_{i,j,k} = C_{i,j}\, a_{i,j}\, n_{i,j}\, \Sigma_l\, V_{i,j,k,l}\, (al_{l,j}/A_j)\, N_{i,j}$$

   Global Loss Function (GLF):

   $$GLF_{j,k} = \Sigma_i\, ILF_{i,j,k} + \Sigma_r\, Cep_r\, aep_{r,l,j}\, Vep_{r,l}$$

   Affected Population (AP) :

   $$AP_{j,k,m} = \Sigma_i\, \Sigma_l\, VP_{i,j,k,l}\, (al_{l,j}/A_j)\, N_{i,j}\, P_{j,m}$$

   Density of Losses Index (DL) :

   $$DL_{j,k} = GLF_{j,k} / (\Sigma_i\, C_{i,j}\, a_{i,j}\, n_{i,j}\, N_{i,j} + \Sigma_r\, Cep_r\, \Sigma_l\, aep_{r,l,j})$$

   Density of Affected Population (DAP) :

   $$DAP_{j,k,m} = AP_{j,k,m} / P_{m,j}$$

   The influence of all seismic sources is obtained by the convolution :

   $$RGLF_j = \int GLF_{j,k}\, dF_k$$

This model is similar to the one presented by Oliveira et al.(1988) but made to include the public urban area parameter. At the moment different necessary tasks (building characterization, space-time dynamics, building costs and public urban value) are being developed in order to apply the model to the entire town of Lisbon. The results of this exercise will be used as the main input to the elaboration of the seismic risk emergency plan for Lisbon county (first version). The need to have this first version ("Plan Zero") comes from the urgency to provide Lisbon, as quick as possible, with an adequate instrument to reduce the seismic risk through an effective coordination of rescue operations, and correct planning in order to mitigate vulnerability. Therefore, the above-mentioned tasks are being performed using expeditious field work and simplified estimations.

An example of application was produced for Lisbon's western rivershore area, which is already covered by the LNEC's building survey. This example refers to a scenario similar to the one presented in Figure 6, for the period 9:00-18:00 h. Using the vulnerability functions in Tables III and IV and current real estate prices (Fortuna, 1993), material losses and population casualties were estimated and are presented in Figures 11 and 12. The geographic units used for the calculations were the parishes, and the plot variables were divided into four categories. These results correspond to a preliminary simulation to test the model and analyze the differences between the parishes under study. At the present time, the model is being tested for the entire city using different vulnerability functions and seismic scenarios.

These types of estimations and mapping are very important in the elaboration of emergency plans. But it is also essential to study the seismic behavior of important, dangerous, and particular buildings, such as hospitals, schools, shopping centers, and other highly-populated buildings, as well as lifelines. In the short-term, the seismic vulnerability of such buildings will have to be evaluated, in order to estimate their behavior under different disaster scenarios.

DAMAGE ESTIMATION USING CURRENT REAL ESTATE PRICES IN LISBON WESTERN RIVERSHORE AREA

Damage (thousand escudos)
- < 400000
- 400000 - 600000
- 600000 - 800000
- > 800000

SMPC - CML, 1993

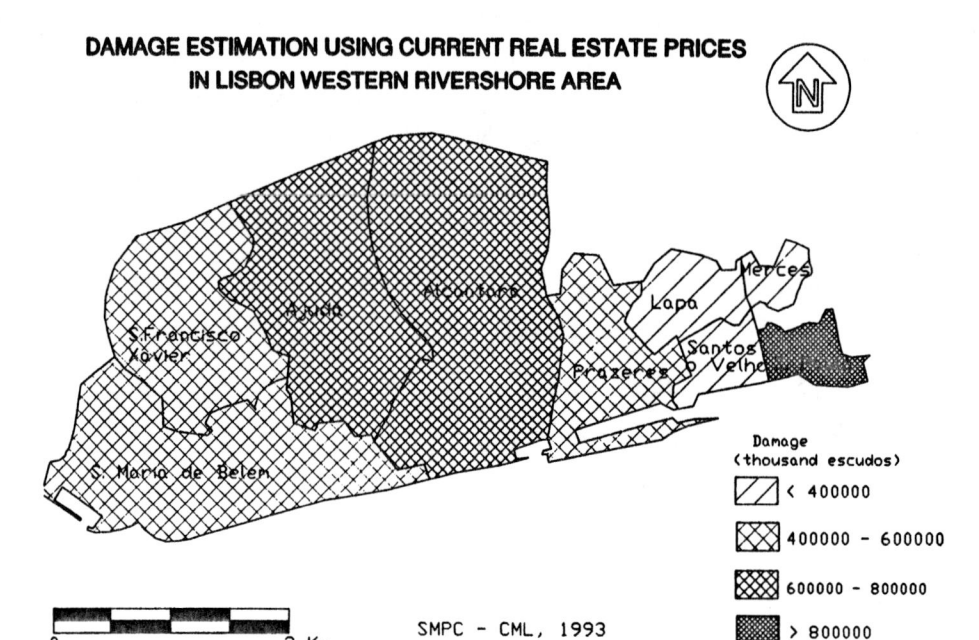

NUMBER OF CASUALTIES DURING THE PERIOD 9:00 - 18:00 H. IN LISBON WESTERN RIVERSHORE AREA

Nr. Casualties
- < 50
- 50 - 150
- 150 - 300
- > 300

SMPC - CML, 1993

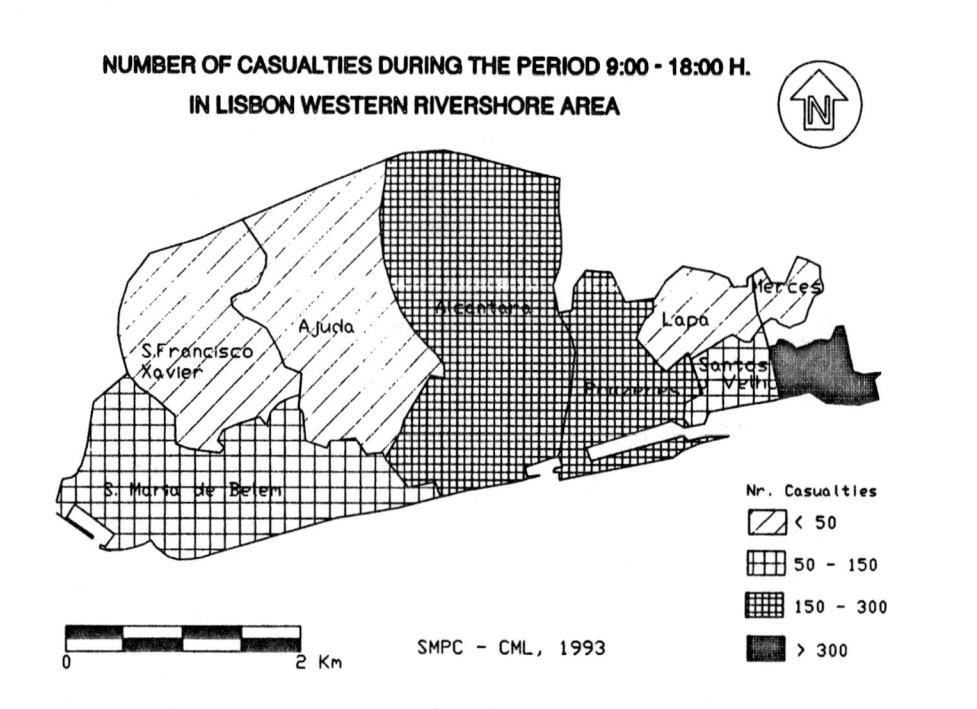

9. Earthquake Damage Scenarios and Emergency Response

The definition of earthquake damage scenarios must provide, as far as possible, a realistic image of the anticipated disaster. Efficient, coordinated emergency intervention and an adequate management of rescue resources depend heavily on a correct estimation of human and material losses and their spatial distribution, as well as of the socioeconomic impact and costs. This estimation also allows for a better oriented prevention policy both at operational and financial levels.

The definition of earthquake damage scenarios in the town of Lisbon is an essential step in the elaboration of the seismic risk emergency plan, which is being prepared by the Municipal Department of Civil Protection (SMPC) of the Lisbon City Council (CML). This department is in charge of the implementation of the civil protection local policy as far as human and material safety are concerned.

The CML is aware of the fact that safety has become a fundamental component of urban development and of the population's quality of living (SMPC-CML,1993c). To achieve these objectives, CML relies on the contribution of all entities and departments involved in the process of production and management of urban space, particularly those belonging to the technical and scientific communities and, of course, those working in the areas of urban safety and security. The civil protection policy of CML is guided by the principle that the safety and well-being of the population must be founded on an attitude of preventive planning, as well as on adequate education and citizen participation. That is why CML gave high priority to the elaboration of a seismic risk emergency plan, in order to provide the city with an efficient response, namely a framework for rescue operations and the reduction of vulnerability, while effectively coordinating resources.

A considerable number of studies have been developed in order to define earthquake damage scenarios in Lisbon. This work was done on the basis of a close and permanent coordination among scientists and technicians, working in the areas of seismology, earthquake engineering, demography, computing, statistics, and others.

The SMPC is in charge of coordinating and operating 24 hours a day the necessary means of rescue in case an accident occurs. If such an accident takes place, every mechanism is operated and the situation is site-managed. Then problems are surveyed, the accident impacts are minimized, and all necessary resources are operated and managed until a normal situation is recovered.

The SMPC also coordinates emergency operations of assistance to the affected population, including post-event field operations and psycho-social support.

The Municipal Commission of Civil Protection, a consultative body of the SMPC, plays a very important role is this process. It is composed of representatives from all departments involved in the security of the city: fire department, security forces, health services, army, transports and communication department, social service (in charge of the supply of shelter, food and warm clothes), as well as the departments responsible for lifelines and logistics.

An efficient communication network connecting the SMPC and the other involved departments is extremely important in an emergency situation. Apart from regular telephone lines, the SMPC also uses a special telephone network and a radio network, both directly connected to those departments.

There are often accidents of small and medium size in Lisbon, mostly of urban nature, such as fires and building collapses. When such situations happen, the SMPC carries out so-called "daily emergency intervention," which not only tackles the actual problem but also serves as an "experimental lab" to test and improve action techniques. They have also proved to be very useful in creating a better and closer relationship among all people working in the process of rescue management, which can be very important and helpful if a major accident (e.g. an earthquake) happens.

In case of a major disaster with great socioeconomic impact, coordination with the Civil Protection National and District Authorities takes place.

10. Earthquake Awareness Programs

Public information and awareness programs are extremely important in an efficient system of civil protection, not only in what concerns decision-makers but also at the population level in general. These programs engender and spread new attitudes and behaviors among people, while adequately reacting to different hazardous situations. They become (individually and in groups) both receivers of and actors in security and civil protection .

The CML has been increasingly investing in this area. About 30% of the budget annually allocated to the SMPC by the City Council is devoted to public awareness and training programs.

Specific awareness programs are developed for different target population groups and for all possible kinds of risk, but special attention is given to the seismic risk.

For instance, a considerable number of programs are being carried out in schools. Schools are considered a priority target group, because young people are exceptionally receptive to the programs and would spread new attitudes (SMPC-CML, 1993b); others programs are being developed in companies, public departments and local authorities.

The insurance sector is investing considerably in prevention and education programs. Companies often sign partnership protocols with the City Council and they are now financing partially or totally a number of projects which are being carried out. This interaction between local authorities and private companies has contributed to a significant improvement in the number and quality of those programs.

11. Building Legislation

In regard to legislation, the most recent national code for structural safety of buildings and bridges dates from 1983 (Decree-law 235/83 - 31st May). Chapter seven of that decree-law refers to earthquake action.

According to that code, Portugal was divided into four zones (A to D) of descending seismicity. The rules defined for zone A apply to Lisbon because it is located in that zone.

The 1993 Municipal Urban Plan, made under the Planning Department of CML supervision, also defines regulations both for building construction and for the location of economical and social activities in risk areas exposed to natural or technological hazards. As for seismic risk, the Municipal Urban Plan defines complementary legislation applying to the most sensitive areas of the city.

Modern seismic codes were first developed in 1958. Since then, approximately 23% of the present building stock was subjected to lateral resisting loads.

CML is responsible for the enforcement of the regulations.

12. Final Considerations

The studies reported in this paper need further development in most of the tasks described. They are already in progress, with the goal of reducing the many uncertainties emerging from the present model. A more detailed study will require larger-scale analysis.

Acknowledgements

The authors would like to thank Dr. Maria Figueirinhas and Dr. Manuel João Ribeiro from S.M.P.C., and Eng. João Caldeira Cabral from I.S.D., for their valuable contributions.

This work was partially supported by JNICT's project nr. STRDA/C/CEN/428/92.

References and Bibliography

Almeida, F. Moitinho, 1986. Geological Map of the Lisbon County. Serv. Geol. Portugal, Lisbon.

Bard, P.Y. and P. Teves-Costa, 1989. Prédictions Numériques et Spectres Reglementaires Adaptés au Site - Un Exemple de Desaccord Typique. Proc. 2ème Coll. Nat. de l'AFPS "Génie - Parasismique et Aspects Vibratoires dans le Génie Civil", St. Rémy-les-Chevreuse, Avril, I:S2-46 - S2-57.

Cabral, J., 1993. *Neotectónica de Portugal Continental.* PhD Thesis, Lisbon University.

Campos-Costa, A., C.S. Oliveira and M.L. Sousa, 1992. Seismic Hazard-Consistent Studies for Portugal. Proc. 10th World Conf. Earthq. Engin., Madrid 19-24 July, p. 477-482, A.A. Balkema.

D.P.P.E. - C.M.L., 1993. *Plano Director Municipal de Lisboa.* Câm. Munic. de Lisboa, Lisbon.

Fortuna, 1993. Caderno Conjuntura (Imobiliário). Magazine Fortuna, Electroliber edition, September 1993, p. VII, Lisbon.

Henriques, M.C.J., M.T. Mouzinho and M.F.F. Natividade, 1988. *O Sismo de 26 de Janeiro de 1531.* Comissão para o Catálogo Sísmico Nacional, Lisbon.

I.N.E., 1991. *Censos 91* (Population and Housing Census 1991). Inst. Nac. Estatística, Lisbon.

Marin A., F. Correia and J. Gaspar, 1983. *Lisboa : Estimativa da População Presente por Zonas e Intervalos de Tempo.* SNPC, Program for Minimizing Seismic Hazard, Lisbon.

Martins, I. and L.A. Mendes-Victor, 1993. A Actividade Sísmica na Região Oeste da Península Ibérica. Energética e Períodos de Retorno. Publ. 20, IGIDL, Lisbon University.

Mendes-Victor, L.A., 1987. Estimation of the Seismic Impact in a Metropolitan Area Based on Hazard Analysis and Microzonation - An Example. The town of Lisbon. Pact, 18: 183-213, Council of Europe, Belgium.

Moreira, V.S., 1985. Seismotectonics of Portugal and its Adjacent Area in the Atlantic. *Tectonophysics*, 117: 85-96

Mota, R., 1993. *Modelação Unidimensional não Linear - Aplicação a Bacias Aluvionares de Lisboa*. Master Thesis, Lisbon University.

Oliveira, C.S., 1986. *A Sismicidade Histórica e a Revisão do Catálogo Sísmico*. Report LNEC, Lisbon.

Oliveira, C.S., 1988. Seismic Risk Analysis and Civil Protection. Proc. Int. Workshop on Natural Disasters in European-Mediterranean Countries, Perugia.

Oliveira, C.S. and L.A. Mendes-Victor, 1984. Prediction of Seismic Impact in the Town of Lisbon based on Hazard Analysis and Microzonation. Proc. 8th World Conf. Earthq. Engin., San Francisco, VIII: 639-647, Prentice-Hall.

Oliveira, C.S. and A.M.R. Cabrita, 1985a. *Tipificação do Parque Habitacional*. Report LNEC, Lisbon.

Oliveira, C.S., J. Gaspar and F. Correia, 1985b. *Levantamento do Parque Habitacional - Vol I - Zona da Alameda*. Report LNEC, Lisbon.

Oliveira, C.S., L. A. Mendes-Victor, J. Gaspar and G. Silveira, 1988. Seismic Impact of Future Earthquakes in the Town of Lisbon : An Example of Application. Proc. 9th World Conf. Earthq. Engin., Tokyo.

Pais, I., 1992. Risco Sísmico : Prevenção Versus Vulnerabilidade Geológica. Proc. 1st Meeting of Lisbon City Council Technicians, Lisbon.

Pereira de Sousa, F.L., 1919-1932. *O Terramoto do 1º de Novembro de 1755 em Portugal e um Estudo Demográfico*. Serviços Geológicos, 4 vols, Lisbon.

Ribeiro, A., M.C. Kullberg, J.C. Kullberg., G. Manuppella and S. Phipps, 1990. A Review of Alpine Tectonics in Portugal: Foreland Detachment in Basement and Cover Rocks. *Tectonophysics*, 184: 357-366.

Silva, J.A.V., E.R. Gonçalves and M.J. Quedas, 1993 - *Estudo de Caracterização e Distribuição Espacial do Emprego na Cidade de Lisboa* - C.I.D.E.C., Lisbon.

S.M.P.C. - C.M.L., 1993a. *Actualização a 1993 das Estimativas da População Presente na Cidade de Lisboa por Zonas e Períodos do Dia*. Internal Report, Lisbon.

S.M.P.C. - C.M.L., 1993b. *Crescer na Segurança*. Internal Report, Lisbon.

S.M.P.C. - C.M.L., 1993c. Documento Contributivo para o Plano Director Municipal. Internal Report, Dep. Planeam. Estratégico, Lisbon.

SNPC, 1983. *Program for Minimizing Seismic Hazard - Final Report* (in portuguese). Serv. Nac. de Protecção Civil, Lisbon.

Teves-Costa, P., 1989. *Radiação Elástica de uma Fonte Sísmica em Meio Estratificado - Aplicação àMicrozonagem de Lisboa*. PhD Thesis, Lisbon University.

Teves-Costa, P., 1993. Geophysical Aspects of the Lisbon Town - Site Effects and the 1755 Earthquake. Presented on the seminar "Les Systèmes Nationaux Face aux Séimes Majeurs - Réponse des Autorités et Vulnérabilité du Bâti", Lisbon 26-28 Nov 1992, CUEBC Ravello, Special Edition (in press).

Teves-Costa, P. and P.Y. Bard, 1990. Site Effects in the City of Lisbon: Response Spectra for Some Alluvial Sites. Proc. of the ECE/UN Seminar on "Prediction of Earthquakes - Occurrence and Ground Motion", Lisbon 14-18 Nov 1988, LNEC, I: 499-515.

Teves-Costa, P. and C.S. Oliveira, 1991. A Study on the Microzonation of the Town of Lisbon: Improvement of Previous Results. Proc. 4th Int. Conf. on Seismic Zonation, Stanford 25-29 Aug, III: 657-664.

Teves-Costa, P. and L.A. Mendes-Victor, 1992. Site Effects Modelling Experiment. Proc. 10th World Conf. Earthq. Engin., Madrid 19-24 July, p. 1081-1084, A.A. Balkema.

Wilson, R.C.L., R.N. Hiscott, M.G. Willis and F.M. Gradstein, 1992. The Lusitanian Basin of West-Central Portugal: Mesozoic and Tertiary Tectonic, Stratigraphic and Subsidence History. In: *Extensional Tectonics and Stratigraphy of the North Atlantic Margins*, p.341-361. Am. Assoc. Pet. Geol. Mem., 46.

GRANADA FACING AN EARTHQUAKE

F. VIDAL[1]
G. DEL CASTILLO[2]
[1]*Instituto Andaluz de Geofísica y Prevención de Desastres Sísmicos.*
P.B. 2145. 18080 Granada. Spain.
[2]*Servicio de Protección Civil. Consejería de Gobernación. Junta de Andalucía.*
C/ Harinas, 9. 41071 Sevilla. Spain.

1. Introduction

In many places of the world it is likely that a strong earthquake will occur, and recently many countries have experienced such earthquakes and the consequent destruction and deaths. It is therefore largely accepted that programs for earthquake hazard reduction be made an urgent priority for the highly hazardous regions; however, this frequently makes us forget that in regions of moderate seismic activity, such as Granada's basin, the application of these programs is also of great need and could achieve spectacular results, particularly in minimizing the number of injured and killed.

The reduction of damage due to earthquakes is a difficult but attainable goal, and so the decade of the 1990s has been named the United Nations' "International Decade for Natural Disaster Reduction." In this spirit we celebrate the present Workshop, which will be an effective way to analyze and present some guidelines useful for the mitigation of seismic disasters and applicable on a worldwide scale to towns with a high or moderate seismic hazard.

Another efficient way to reduce consequences of earthquakes, in addition to urban planning controls and earthquake-resistant construction, is the development of studies of earthquake damage scenarios which evaluate the incidence of an earthquake in an urban area, bearing in mind the demographic, political and social aspects in relation to the effects produced by a great shock.

2. Granada: History of Construction and Demography

Granada, a small town in the south of Spain with a population of about 280,000, is located in the eastern part of Andalusia on the Depression of Granada, a neogene basin of the Betic Region. The town lies on Quaternary and Plio-quaternary sedimentary deposits. Geologically two types of structures are present. One is characterized by compact conglomerates (Alhambra Formation) of Pliocene age; this zone includes the hills whereon lies the old part of the town. The second geological type, where lies the remaining part of the town and area of modern development, is covered by recent alluvial sediments (clay, water saturated sands, etc.) of Quaternary age, a product of the Genil River deposits.

2.1 HISTORY OF CONSTRUCTION

Granada was the capital of the last Islamic kingdom on the Iberian Peninsula, a kingdom which left a deep architectonic and urban imprint on the town, with historical quarters like the Albaycin or important monuments like La Alhambra and El Generalife. In Roman and Arab times, the town was situated on the hills where the Albaycin and the Alhambra currently are, growing steadily from the 11th to the 15th centuries. After the conquest in 1492, Granada was a rich and flourishing town where the construction of a great number of public and private buildings took place, constituting a very important historical and cultural heritage, like the Cathedral, the Hospital Real, churches, and palaces of the nobility. From the 16th to the 18th centuries, the town spread towards the alluvial deposit plain of the rivers Darro and Genil, and then remained roughly the same size from the 19th century to 1940 [Figure 1]. After 1960, however, there was a great increase in construction work, and the area of the town's development expanded as much as 300% [Figure 1].

Structures built before the 1960s typically had fewer than four stories, and the construction was of brick masonry, sometimes using a mixture of brick masonry reinforced with concrete frames, but without specific anti-seismic design. The vulnerability of these buildings is of class B on the EMS scale. (The EMS scale consists of six classes of decreasing vulnerability, A-F.) The floor space of the dwellings was generally larger than those of the present day.

After 1960, reinforced concrete was used, and the buildings usually had between three and nine stories. They were designed by architects and the taller ones by structural engineers, though until 1974 without any seismic consideration. Since 1980, however, the seismic codes have been taken into account, with an occasional exception.

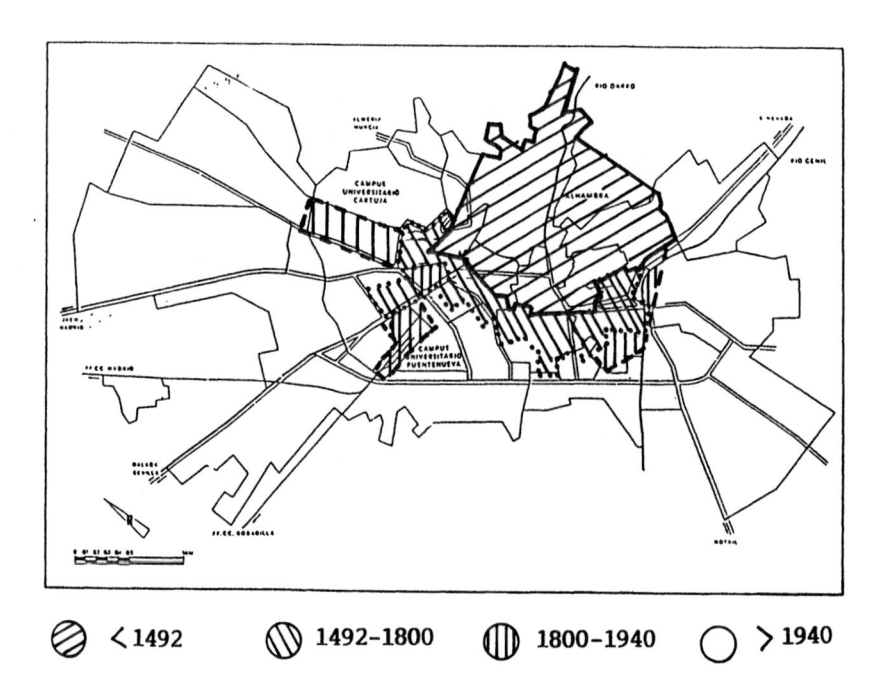

Figure 1. Evolution of Granada's Development

Between 1960 and 1980, the structures were very heavy, with great cross-sections in the beams and columns and with deficient steel reinforcement and occasionally an inadequate base (isolated shallow bolsters); the buildings were very regular and symmetrical in their dimensions. Their vulnerability is generally of class C and sometimes of class B. After 1980, the structures were lighter and more elastic, with more steel reinforcement. The buildings were also more irregular in plan and elevation, with inadequate building separation. Their vulnerability is of class C or D (EMS scale) [see Figure1].

In 1956 urban planning did not take seismic phenomenona into consideration. In 1973 a general urban plan was created, and in 1974, the seismic code PDS-1 went into effect. The new urban plan created in 1985 followed the PDS-1 code, and this seismic building code still stands today. The maximum intensity expected by the seismic code PDS-1 for Granada is IX (MSK scale) and the maximum horizontal acceleration 0.3g. Both the plans of 1973 and 1985 laid down the rules for the future construction of blocks of flats.

In 1983 a new Seismic Building Code was started, of which the last version appeared in 1992, but because it still has not been legally approved it is not valid. This code requires a horizontal base acceleration for building designs of 0.24 g and estimates an intensity IX for the return period of 500 years.

Because of Granada's long, multi-layered architectural history, there are also many differences in the construction of roads. In the Albaycin, the streets are very winding and narrow, with a width between 1.5 m and 5.0 m, although there are some narrower than 1.5 m and a few wider than 5.0 m. Moreover, in many cases they are very steep and they even have parts with steps, which makes vehicle access impossible. In the rest of the town's older section, built before this century, the streets usually range between 6 and 10 m in width, with some of them are slightly wider; normally they are straight and level. Until 1960 the urban structure was similar, though with a more regular geometry, but with only two main streets. In the section built after 1960, the streets are wider, straighter and more level, and there are many main streets with a width of between 20 and 40 m. Also there is a circular road around Granada with many entrances to and exits from the town.

2.2 DEMOGRAPHY

According to the census of 1986, Granada's population numbers about 280,000 within its urban nucleus. But there are 28 villages within a radius of 10 km, forming a metropolitan area with approximately 130,000 inhabitants; these people have a close relationship with the urban nucleus in things like work, studies and services.

From 1960 to 1986, the growth rate of the population was on the order of 78.5% in the town of Granada and 70.0% in the villages of the metropolitan area. This rhythm of growth has been maintained since 1986 in the town and has been slightly higher in the surrounding villages. In the last five years, the population has increased 6% due to births and 8.7% due to immigration; 5% of this immigration is from the Andalusian region, generally from rural zones.

The districts with the highest density of population are El Zaidín (655-1271 inh/km^2), Los Vergeles, La Redonda and Fuentenueva (454-655 inh/km^2) [Figure 2]; half of Granada's population lives in these quarters, where there are more than 100 or 150 dwellings per hectare, depending on the quarter.

⊜ 450-1300   ⦀ 380-450   ⊘ 140-380   ○ 40-140

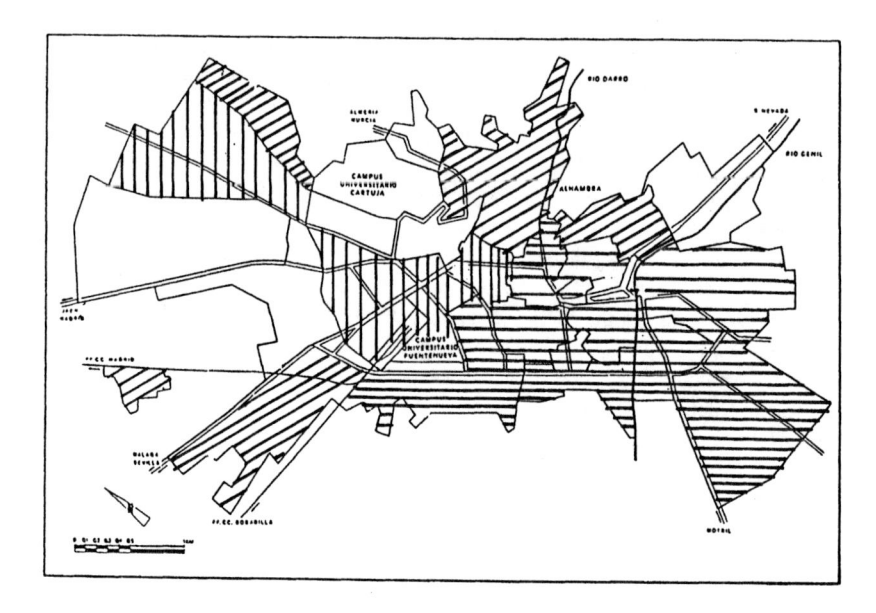

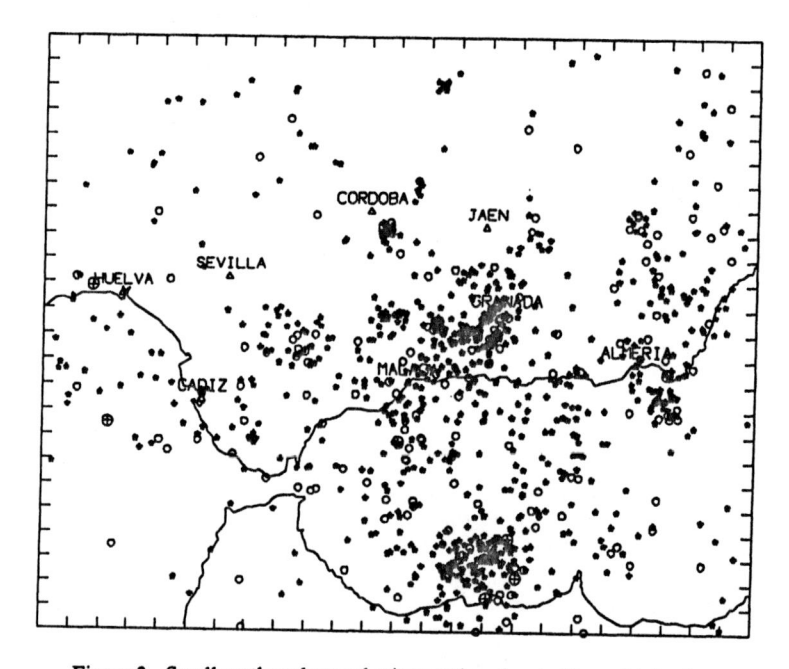

Figure 2. Population Density (inhabitants/km2) in different districts

Figure 3. Small earthquakes and microearthquakes in Granada's basin.

3. Seismicity and Seismic Hazard

The Mediterranean region is seismically active. Its eastern part, the Aegean and Balkan zones, has the highest seismic hazard. In the western part, the seismic activity is concentrated north of Africa and on the southern Iberian Peninsula, and then it spreads toward the west where it meets the Mid-Atlantic Ridge. Between 10° W and 20° W, large earthquakes occur (for example, the 1755 event, which had magnitude greater than 8.0).

Granada's basin, in the Beticas-Alboran region, is the area of the greatest seismic activity in Spain. The neogene basin is formed by fracture systems with directions N10-30E, N40-60W and N90-100E (Vidal et al., 1984), with four sedimentary and detritic depocenters (Cubillas, Chimeneas, Genil and Granada) which have a depth of between 2600 m and 2800 m (Morales et al., 1990).

In the past, the eastern part of Andalusia, and consequently the province of Granada, was affected by the occurrence of destructive earthquakes. Since the 15th century, the more important earthquakes with intensities equal to or greater than IX (MSK scale) were those of 1431, 1504, 1518, 1522, 1531, 1680, 1804, 1806, and 1884; the epicenters of the earthquakes of 1431, 1806 and 1884 were within Granada's basin. Moreover, a series of damaging earthquakes occurred in this century in 1910 and 1956 near Granada, though with relatively small radii of damage.

Granada's basin experiences numerous small earthquakes and microearthquakes that show a strong tendency for clustering [Figure 3]; the activity occurs mainly in seismic series and swarms (Peña et al., 1993). The activity is mainly superficial with hypocenters shallower than 20 km; however, there are some as deep as 80 km, and a few very deep ones (630-650 km), such as the earthquakes of 1954, 1973 and 1990. Towards the south of Granada's province there is seismic activity with hypocenters of up to 120 km in depth (Vidal, 1986). The earthquakes or seismic series which occur inside the basin have return periods of 10-20 years for magnitudes of up to 5.5, and of 200 years for magnitudes between 6.0 and 7.0.

3.1 SEISMIC NETWORKS

Because of the high seismic activity, the Observatorio de Cartuja (Granada), consisting only of one seismic station since 1903, started a project in 1979 for building a seismic network: the Red Sísmica de Andalucía (RSA) (Andalusian Seismic Network). The Observatory designed and built the seismic instruments which have since been used by the Observatory and other institutions. The RSA records in analog form, but it also has a seismic detection system which stores in digital form and evaluates a series of parameters of the events automatically.

The first belt of stations around Granada's basin began operating in 1983, and currently the Observatory is installing a second belt; there is also an array in Almeria province which belongs to the RSA. So a network of 21 stations over central and eastern Andalusia has been established, with the recording of the two belts centralized in the Observatory (Alguacil et al., 1990). Furthermore, there exists another microearthquake network run by the Real Instituto y Observatorio de la Armada (Royal Institute and Observatory of the Navy), in the west of Andalusia, using the same instrumentation and detection-analysis programs as the Observatory of Granada. Finally, the National Seismic Network, a division of the IGN (Instituto Geográfico Nacional), has a series of seismological stations in Andalusia with their recordings centralized in Madrid and an automatic location system.

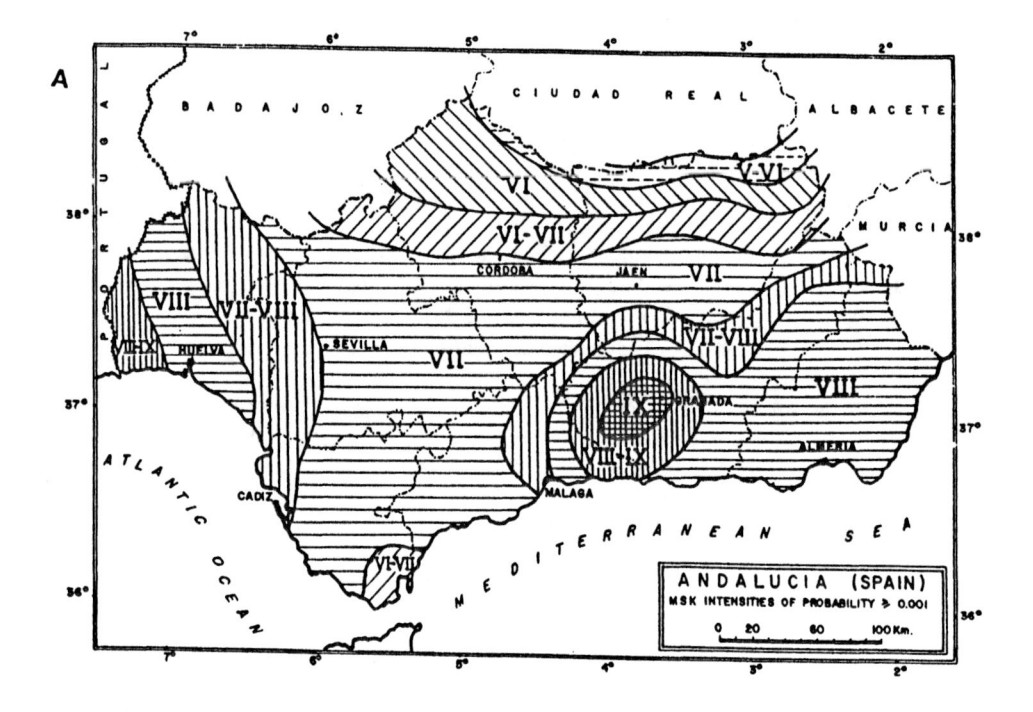

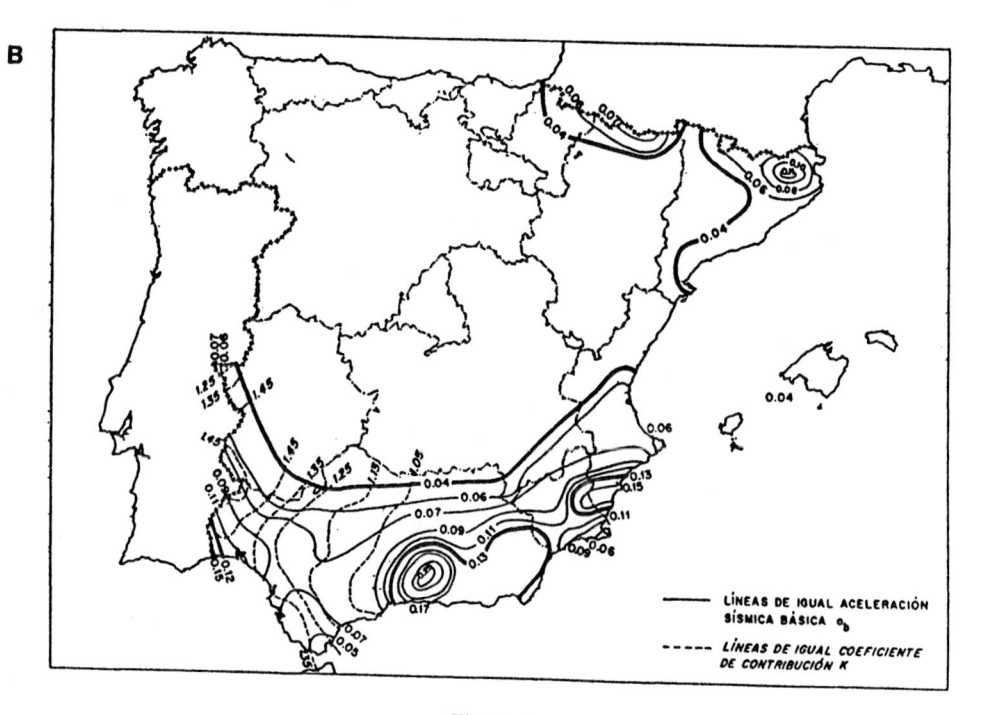

Figure 4.

3.2 GRANADA'S SEISMIC HAZARD

The work done by different authors on seismic hazard (Muñoz, 1984; Martín, 1983), using probabilistic methods, shows that an intensity IX has, for the town of Granada, a return period of about 500 years [Figure 4]; this intensity has been taken into account in the new Spanish Seismic Code of 1992.

The active seismological fractures situated near Granada are potential sources of such intensities in the town; these fractures delimit the more depressed parts of the basin. Among them, the most important affecting Granada are those at the edges of Sierra Elvira, the Pinos Puente-Santafé, and Albolote-Atarfe seismic areas. Another important source is the area of Alhama de Granada, at about 40 km from Granada, where the last large earthquake of the region ocurred in 1884. With less intensity we have the remaining active zones of the basin, like those of Padul-Dúrcal, Loja, and Chimeneas.

The effects of the ground shaking in Granada's basin extend over a smaller area than in the rest of the Andalusian region, due to the existence of a high seismic attenuation with values of Q smaller than 100 (Ibañez et al., 1990; De Miguel et al., 1992). The influence of the geological and local soil conditions on the intensity of the ground shaking and consequently on the associated damage to buildings, has been observed in the historical earthquakes of 1806, 1884, 1911, 1956, and 1979. Liquefaction phenomena have taken place in the first three events (Vidal, 1986).

Moreover, through studies of the seismic noise and spectral analysis, important phenomena of amplification have been characterized for frequencies ranging from 1.5 to 3.0 Hz (Morales et al., 1991, 1993).

In the town of Granada, certain areas are more prone to earthquake damage due to their characteristics: firstly the geology and topography of the hillsides of the Albaycin and the Alhambra; secondly the alluvial deposits in the plain; and, more importantly, in the south and southeast of the town, where there are areas with a high water content.

4. Emergency Response

In 1989-1990 the Civil Defense Office lay down a national plan for seismic emergency, specifying it for different provinces of Spain. A main Operative System coordinates the different departments and emergency response to be achieved by those responsible at local, provincial, regional, and state levels. The particular earthquake emergency plan for Granada has a structure of Managing and Executive Offices. The Managing Office consists of a director and council; a supporting department, including a Center for Earthquake Information; a Surveillance Department; and a Department for Coordination with the Armed Forces. The Executive Office consists of Departments of Work Coordination, a Health Group, Social Action, Technical Assistance, and Rescue (Castillo, 1993).

Presently the state and regional administration are publishing basic guidelines for special plans concerning earthquake risk, using the basic code of Civil Defense. These plans must include: 1) characterization of the earthquake risk prior to the earthquake occurrence, including seismic hazard, vulnerability, seismic risk and an estimation of material damage and number of victims, and also information about earthquakes once they occur, including a rapid evaluations of their effects; 2) intervention measures in case of a seismic catastrophe, including rescue of victims and actions related to building security, lifelines, and transport; and 3) establishment of the various stages and situations, the initial step being to control and handle the emergency. There exists a seismic emergency plan of the city drawn up in 1987 and revised in 1990 which is coordinated by the civil governor.

Granada has three groups of Civil Defense, each dependent on a different administration (local, regional and state), with a staff of fifteen people. The city also has two police forces, a local one (with 380) and a state (with a force of 722); two fire brigades with 136 firemen, one located in the north and the other in the south of the city; five public hospitals with 2,340 beds and two private ones with 80 beds. Apart from the telephone there are other means of communication used frequently by the Civil Defense, police, hospitals, etc.

5. Earthquake Awareness

In Andalusia there is great awareness of the more frequent hazards, such as fires, inundations and industrial accidents. La Junta de Andalucía, the Government of the Andalusian region, is highly concerned about the above-mentioned catastrophes and annually updates all its emergency planning with human and economic resources. Being conscious of the social and economic responsibility of earthquake preparedness, La Junta de Andalucía and the University of Granada created the "Instituto Andaluz de Geofísica y Prevención de Desastres Sísmicos" (IAGPDS) (Andalusian Institute of Geophysics and Prevention of Seismic Disasters) in 1990, which they support economically. The center works in research and prevention. Since 1984, the Junta de Andalucía has partially financed some basic seismic studies and the implementation of the seismic Andalusian network (RSA).

Only the Andalusian population living in the basin of Granada was aware of the earthquake hazard, as they had frequently felt the effects of earthquakes. Due to the continual demand for information about hazard, prevention and self-protection from an earthquake, the IAGPDS is constantly preparing different programs on seismic knowledge at the university level (seminars and courses) as well as training of professionals (engineers, architects, geologists, geophysicists, members of the Civil Defense, administrative staff) and the general public through conferences and other widespread activities (pamphlets, radio, television, newspapers, expositions, technical informs). All this has changed the builders' attitudes and ensured that the organizations concerned with urban planning and construction supervision pay greater attention to the fulfillment of the seismic code.

6. Final Comments and Conclusions

In the overall local preparation for an earthquake damage scenario, an input earthquake should be considered, and the vulnerability of the buildings and systems estimated, as well as the characteristics of the population and the expected damage to structures, lifelines and lives.

For the seismic input in a specific place, it is necessary to obtain the design earthquake in the form of accelerograms from seismic records (mainly from strong motion or broadband sets) and the design accelerogram at the ground surface from geotechnical or instrumental data. Another way is by calculating, from the small earthquakes recorded at the site, a synthetic accelerogram at the bedrock and from this the shaking characteristics at different places.

There are two ways to find the structural response and, consequently, the damages of the buildings: one by means of instrumentation or by structural analysis; and the other by a qualitative method through experiences of past earthquakes at the site or in other places with similar architectonic and urban characteristics. These two ways could be applied for the damage quantification. The instrumental analysis is doubtless the more realistic one, but it is difficult to apply nowadays to every city with a seismic hazard.

Nevertheless, a qualitative or semi-qualitative estimation of the expected damage is necessary, even though many inaccuracies are involved. The technicians doing the analysis for the estimation should be coordinated by seismologists and by the Civil Defense, so the different phenomena associated with the seismic shaking can be carefully taken into account and the results applied by the policy-makers.

In Andalusia some guidelines have been drawn up so each village can prepare its own emergency plan. In these guidelines, apart from the organizational aspect, special emphasis has been placed on recommending that studies of the population, buildings, local lifelines, structures of high risk, economic and sociocultural activities take place.

A seismic scenario provides technicians and public authorities with actual, detailed information about the effect of the seismic shaking on the population, buildings, and lifelines; emergency response; and available means and resources for dealing with earthquakes. The study of seismic scenarios leads to the discovery of existing "black points," for example, inhabited buildings with high vulnerability, stretches of roads susceptible to collapse, insufficiency of main roads and streets needed for rescue action, hospitals without easy access or emergency generating power. The knowledge of these severe deficiencies leads to 1) the "whitening" of the black points (that is, to an increase in the security measurements); and 2) to an urge to incorporate earthquake damage prevention into urban and architectural planning.

The rapid expansion of the town of Granada makes it likely that it will grow towards the Vega, an alluvial zone bordering the Genil River; this expansion will lead to soil amplification and soil failure phenomena. Therefore, the decision-makers must seriously consider the following: 1) innovative and seismically-safe urban planning, including investigation into seismic microzonation to provide consistent earthquake risk management and to regulate the design and construction of buildings, lifeline systems, and the remaining structures; and 2) programs on earthquake preparedness, including information, education and training of the inhabitants in order to avoid or reduce casualties, injuries and disorganization in earthquake response, and also to reduce economic and social consequences.

Finally, the earthquake damage scenario, apart from facilitating improved emergency response, has a positive impact on transportation and urban development. This Workshop should make a recommendation to the authorities of earthquake-threatened countries, regions and important towns to carry out studies of seismic scenarios.

Acknowledgements

The authors are grateful to J. Martin Marfil and his wife Bernadette for their valuable help and useful comments during the preparation of this work.

References

Alguacil, G.; Guirao, J. M.; Gomez, F.; Vidal, F.; Miguel, F. de (1990) - "Red Sísmica de Andalucía (RSA): A Digital PC-Based Seismic Network." *Cahiers du Centre Européen de Géodynamique et de Sismologie, 1, 19-27.*

Canas, J. A.; Miguel, F. de; Vidal, F.; Alguacil, G. (1988) - "Anelastic Rayleigh wave attenuation in the Iberian Peninsula." *Geophys. J. of the R.A.S. Vol 95. 391-396.*

Castillo, G. del (1993)- "Protección Civil: Estructura y objetivos." *Curso sobre Prevención Sísmica". Granada. 14 pp.*

De Miguel, F.; Vidal, F.; Alguacil, G. (1989) - "Spacial and energetic trends of the microearthquakes activity in the Central Betics." *Geodinámica Acta. 3, 87-94.*

De Miguel, F.; Ibañez, J. M.; Alguacil, G.; Canas, J. A.; Vidal, F.; Morales, J.; Peña, J. A.; Posadas, A. M.; Ibañez, J. M.; Guzman, A.; Guirao, J. M. (1992). - "1-18 Hz Lg attenuation in Granada Basin (Southern Spain)." *Geophys. Journ. Int. 111, 270-280.*

Morales, J.; Vidal, F.; Miguel, F. de; Alguacil, G.; Posadas, A. M.; Ibañez, J. M.; Guzman, A.; Guirao, J. M. (1990) - "Basement structure of the Granada Basin. Betic Cordillera, Southern Spain." *Tectonophysics. Vol 177, 337-348.*

Morales, J.; Seo, K.; Samano, T.; Peña, J. A.; Ibañez, J. M.; Vidal, F. (1991). "Site effect on seismic motion in Granada Basin, Southern Spain, based on microtremors mesurement and site-dependence Q_c value." *Earthquake Seismology - Earthquake Engineering. 38, 15-44.*

Peña, J. A.; Vidal, F.; Posadas, A. M.; Morales, J.; Alguacil, G.; Miguel, F. de; Ibañez, J. M.; Romacho, M. D.; Lopez-Linares, A. (1992).- "Hypocentral clustering properties of Betic Earthquakes." *Tectonophysics, 221.*

Vidal, F.; Miguel, F. de; Sanz de Galdeano, C. (1984) - "Neotectónica y sismicidad de la Depresión de Granada." *Energía Nuclear. Núm. 149-150. Madrid. 267-275.*

Vidal, F.; Alguacil, G.; Miguel, F. de; Castillo, G. del; Medrano, V. (1986) - "Medidas de protección frente a terremotos." *Imprenta Ave María. Granada. Granada. 14 pp.*

Vidal, F. (1987) - "Sismotectónica de las Béticas y Mar de Alborán." *Tesis Doctoral.* Universidad de Granada. Granada. 457 pp.

SEISMIC EXPOSURE AND MITIGATION POLICY IN NICE, FRANCE

PIERRE-YVES BARD[1]
FRANÇOIS FEUILLADE[2]
[1]*LGIT, Grenoble Observatory*
BP 53X, 38041 Grenoble Cedex, France
LCPC, 58 Bd Lefebvre
75732 Paris Cedex 15, France
[2]*Ville de Nice, Service Voirie*
11 rue St-François-de-Paule
0600 Nice, France

1. Introduction

The Nice Côte-d'Azur area is a well-known destination for domestic and foreign tourists. The city is located along the Mediterranean Sea, very near to the southern Alps and Italy. Its pleasant geographical and cultural environment, together with its mild climate, have contributed to the important and steady growth of the whole coastal area since the end of the 19th century, growth which is no longer based only on tourism, but also on service industries attracted by the transportation and communication facilities.

It is also known that the area, including the town itself, has been hit in the historical past by several damaging earthquakes, ranking it in the highest seismicity zone of continental France [Figure 1]. Although the actual level of hazard is certainly lower than in many other parts of the world, such as California, Japan or Greece, the existence of somewhat unfavorable geotechnical conditions (the city center is built on young unconsolidated alluviums), combined with the substantial growth of the city in recent decades, has resulted in a significant level of risk, which is probably high enough to justify the establishment of an earthquake damage scenario. Past research supports the idea that Nice's risk is increasing. Some simulation work was done ten years ago that focused on an area located north of Aix-en-Provence (about 200 km west of Nice), which was hit by a magnitude 6 event in 1909. The purpose of the study was to estimate the amount of damage that would be caused in present times (i.e., in 1983) by a repetition of the 1909 event. Although the area under study did not include large cities, this simulation study led to casualty estimates between 400 and 1000—i.e., ten to twenty times larger than the 1909 death toll of 43—as well as to loss estimates of about five billion F (one billion US$). As the same kind of event might be expected for the Nice area, the casualty and loss estimates would certainly be significantly higher.

Given the average low-to-moderate level of seismic hazard in France, and despite the above-mentioned previous study, the French earthquake engineering community has very little experience in such scenario studies and will greatly benefit from the lessons learned by foreign teams on real cases. In order to help in the evaluation of the guidelines proposed in this workshop, the next sections will briefly outline the particular context of the Nice area and will tightly follow the suggestions of the workshop organizers: we will consider Nice's demographic and economic

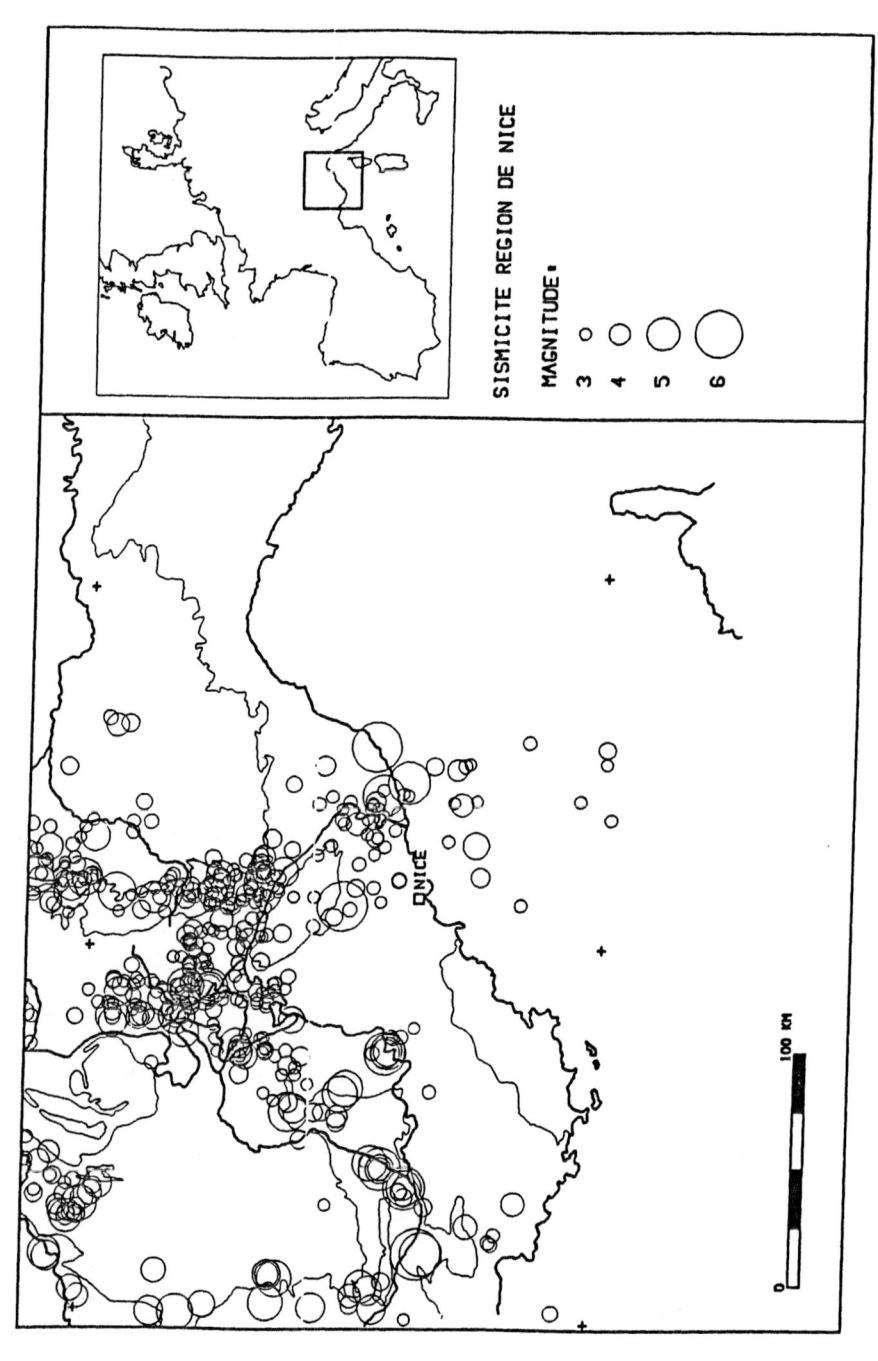

Figure 1: *Seismicity of the larger Nice area for the period 1700 - present. The assumed epicenter of the most destructive event Western Alps, which occurred in 1564 north of Nice, is also indicated (courtesy to Sismalp).*

characteristics, seismological and geotechnical contexts, administrative organization for emergency response, building park and construction code, the information policy and earthquake awareness, and finally the research facilities and programs.

Some details will be added throughout the text to help in understanding the peculiarities of the French administrative system, since the large differences observed in that respect from one country to another should certainly be accounted for in the establishment of broad-use guidelines. Let us simply mention in this introductory part that there exist two intermediate administrative and representation levels between the municipal level and the federal level: Nice is part of the Maritime Alps *Département* (the equivalent of which in the US administrative system is approximately the County), where exist both a board of elected local representatives (*Conseil Général*), and federal services placed under the authority of the *Préfet*. This *Département* is included in a larger regional unit, named "Région Provence - Alpes - Côte d'Azur," with again federal regional services and a regional board of elected representatives (*Conseil Régional*).

2. Demographics

2.1 HISTORICAL DEVELOPMENT

Originally a Greek trading post, and then a Roman one, the town of Nice first developed at the end of the Middle Ages and during the Renaissance. Dependent upon Savoie County, it was initially a settlement around the fortress atop small, rocky "Chateau" hill, just along the sea. The historical center around this hill was first built during the Renaissance.

Up until 1850, the town developed through commercial activities based on its harbour, located just east of Castle Hill. Then, because of the difficulties in competing with Marseille and Genoa, larger cities with larger port facilities, the residential capacities were progressively developed, launching a rapid development of tourist activities by the end of the 19th century.

The design of the urban tissue developed in the first half of the 20th century reflects these residential and touristic functions: the town grew steadily and rapidly on both its flat, alluvial plains and its rocky hills.

Nowadays, in addition to the still-important tourism—including the "business" tourism of international conferences—the growth of the town is also linked with the development of service industries. The center of the town has moved progressively westward, across the last, flat, virgin parts of the Var Valley, which were not previously developed because of flooding hazard. The outskirts of Nice now meet with the suburbs of other growing cities; as a consequence, the whole coast ("Côte d'Azur") now appears as a continuous urban area starting in Cannes and ending along the Italian border in Menton, with its center in Nice. It thus shares some similarities, despite its much smaller size, with the San Francisco Bay area.

2.2 POPULATION

Fifth amongst French cities, the city of Nice *stricto sensu* counts 350,000 inhabitants (compared with 16,000 in the 18th century, and 136,000 in 1911). Due to the tourist activity, the population swells during the summertime, so that Nice is classified as a town exceeding 400,000 inhabitants. The entire Côte d'Azur, from Cannes to Menton, has more than one million inhabitants, with the same large seasonal variations: the population is nearly twice as large in the summer.

Since the total area covered by the city is 72 km², the average density is 4860 hab/km². However, due to the contrasting morphology, with hills covering more than half of the town area, this density varies considerably within the town: the last census in 1990 indicates that it ranges from 200 hab/km² in some parts of the hilly area to 29,800 hab/km² in the densest districts of the flat, alluvial parts. Around 70% of inhabitants live in the alluvial plains that cover 40% of the whole territory, generally in residential buildings having a maximum height of 20 m; the remaining 30% live in the hills, in individual housing or small buildings.

The population is still growing slowly: the difference between the last two censuses of 1982 and 1990 amounts to 11,000 inhabitants. The increase comes mainly from external arrival, since the natural balance between birth and death is still negative (the age pyramid is deformed, containing a higher proportion of older people). The attractive power of the Nice urban area derives not only from its mild climate and location between sea and mountains, but also from its numerous communication and transportation facilities (for instance, the Nice airport ranks second in France for the passenger number), and these factors favor the development and establishment of service industries. Moreover, this population growth is much larger in the smaller, neighbouring villages, where new arrivals generally prefer to settle. Since Nice is the administrative center for the whole Côte-d'Azur urban area, the municipality is forced to maintain and modernize large facilities such as administrations, hospitals, and airports. Simultaneously, the growing population in the hilly parts requires a continuous extension of lifeline facilities (roads, gas, electricity, water supply, etc.), along with the implementation of more stringent zoning and construction regulations, for safety as well as for environmental purposes.

3. Seismicity

3.1 GEOLOGICAL BACKGROUND

The Western Alps area is a young, still-active mountain range resulting from the meeting of the African and Eurasian plates. The exact plate boundary is not clearly identified, and the whole area is characterized by a diffuse seismicity pattern, with a combination of thrust, strike-slip and normal faulting, not yet very well understood from a geodynamical viewpoint. In such a context, the return period for "characteristic" events may largely exceed one thousand years, and classical seismology should certainly be complemented with detailed paleoseismology and neotectonic studies; such studies are just starting now, and are not yet advanced enough to bring significant breakthroughs in the understanding of the local geodynamical pattern. The present section is thus limited to information gained from "classical" seismology.

3.2 HISTORICAL SEISMICITY

There exists a fairly good national seismicity catalogue, which compiles information up to the 19th century. It is, of course, far from being complete over such a long period for moderate events, and it contains a lot of uncertainties, especially as many of the reported earthquakes have apparent epicenters located either to the north of Nice in very sparsely-inhabited mountains, or to the east of Nice, off the Ligurian coast.

However, the information this catalogue provides is valuable and may be summarized as follows [Figure 2 and ref. (1)]:

- For the whole Maritime Alps *Département* in that period, there was one event (1564) with intensity X (MSK scale), four events with intensity IX (1348, 1494, 1644, 1887), three with intensity VIII (1517, 1618, 1854), and three with intensity VII (1612, 1618, 1818). The average return period for really destructive (I > VIII) events within the *Département* is about one century, but is significantly longer in Nice (only two events with local intensity reaching VIII, both in the 19th century).

- The earthquakes responsible for the highest intensities are located to the north-northeast of Nice, but those responsible for highest intensities along the coast (including Nice city) are Ligurian events. For both categories, it is very difficult to obtain magnitude estimates from intensity data, because damage (and intensity) may have been locally largely increased by large landslides, or because the offshore character of the events prevents any estimate of the epicentral intensity. However, the general belief is that, in this area as in all Western Alps, the maximum historical magnitude is about 6-6.5.

- There has not been in the *historical* period any event with a magnitude exceeding 7 within 200 km of Nice. This does not necessarily mean that it never happened, however, nor that it is impossible for the future.

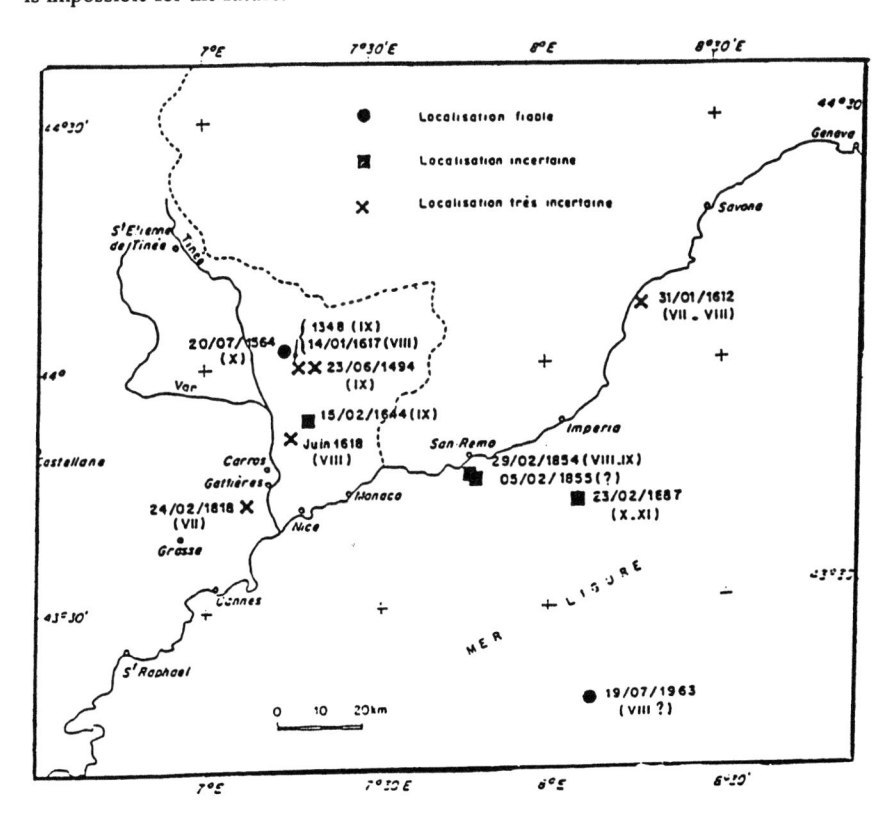

Figure 2: *Historical seismicity of the Nice area. Symbols for epicenters differ according on their reliability: solid circles are reliable, solid squares uncertain, and crosses highly uncertain (from ref. [1]).*

The last damaging event, which occured in 1887, was also the most damaging one in the city's known history. It was located about 70 km east of Nice, 20 km south of Imperia, and caused extensive damage, principally in Italy (640 deaths from Savona to Menton). Magnitude estimates lie between 6 and 6.5+. According to reference (2), the intensity reached VIII in Nice, where damage, however, was limited to a few areas, corresponding to young alluvium. Only 200 houses out of a total of 5484 (for the whole town) needed serious repairs. There were also some indications of liquefaction near the Paillon river, as well as some observations of a small tsunami. Similar, consistent observations outlining the importance of site effects (landslides, liquefaction, and amplification on soft sediments or on elevated topographies) were also made during that earthquake for many cities and villages located along the Mediterranean coast.

It must also be mentioned that, on the basis of this known historical activity, there have been several attempts to obtain probabilistic hazard estimates for the town of Nice [ref. (3)]. Presenting their results would be far beyond the scope of this paper, and it is simply worth noting here that they are significantly different from one another. This was to be expected, since it depends a lot on the delimitation of source zones, which is still a very subjective procedure. The intensity estimates range between 6 and 8 for a return period of 100 years, and between 7.5 and 10 for a return period of 1000 years.

3.3 INSTRUMENTAL SEISMICITY

A local seismological network consisting of seven single-component stations has been in operation since 1977, allowing more than one thousand events to be located, with a maximum magnitude of about 5, and for the identification of several different source zones, located north, northeast, east and southeast of Nice.

As the number of stations was too small to allow precise epicentral locations, two other networks are presently being implemented: the first one, already installed, consists of about forty, single-component stations covering the whole French Alps, while the second, not yet in operation, will consist of ten three-component, high resolution broadband seismometers, and will be coupled with a similar network on the other side of the Italian border. These networks should allow, in the medium term, a more precise assessment of the fault activity. However, the geological context is rather complex and results in a diffuse seismicity; it is not *a priori* obvious that ten or twenty years of precise monitoring will allow identification of all active faults, nor reliable quantitative estimates of their earthquakes' maximum possible magnitudes.

One of the most interesting results of these networks concerns two magnitude 4 events that occured in 1989 and 1990 and were clearly felt in Nice: they could be rather precisely located offshore, only 25 to 30 km from Nice. The question then arises: is the corresponding fault able to produce events with magnitudes equal to, or larger than 6? Definite answers to this question are presently beyond our reach, which underscores the difficulties in choosing the "scenario earthquake."

3.4 GEOTECHNICAL CONDITIONS

Observations performed during recent earthquakes (re)drew attention to the importance of geotechnical conditions.

Downtown Nice is mainly built on two parallel alluvial valleys [Figure 3], corresponding to the present and old beds of the Paillon river. These valleys are filled with young, unconsolidated sediments having a thickness regularly increasing from north to south, where it probably exceeds 50 m along the sea for the central valley. A few standard penetration test (SPT) data are available from a borehole file maintained at the municipality; no other *in situ* measurements (e.g., crosshole, down-hole, and refraction) or lab tests (e.g., resonant column and cyclic triaxial) relevant for seismic analysis have been performed until now. According to some simple preliminary computations [ref. (4)], these sediments should very probably amplify ground motion in a frequency range located between 1 and 4 Hz, and instrumental studies are currently under way to measure this local amplification. These sediments also include, locally, sand lenses that may liquefy for intensity VIII [ref. (4)].

To the very west of Nice, the Var valley is larger and thicker, but its geotechnical characteristics are only very poorly known. The airport is located partly on those deposits, and partly on land gained from the sea with carefully compacted and drained fill. However, the major risk for this airport is probably not related to amplification of ground motion or liquefaction, but rather to its location very near to the continental talus; as it is very steep and covered with very thick, unconsolidated sediments brought by the Var river, large submarine landslides might occur during severe shaking. This could not only seriously affect the stability of runways, but could also induce a very strong and damaging local sea wave along the whole beach (as was already experienced some years ago without any triggering earthquake).

The hills that constitute the rest of the city area are mainly composed of conglomerates, with a variable cementation degree. Although the slopes are in general rather gentle, there are from place to place some cliffs or very steep slopes where landslides or rock falls cannot be ruled out.

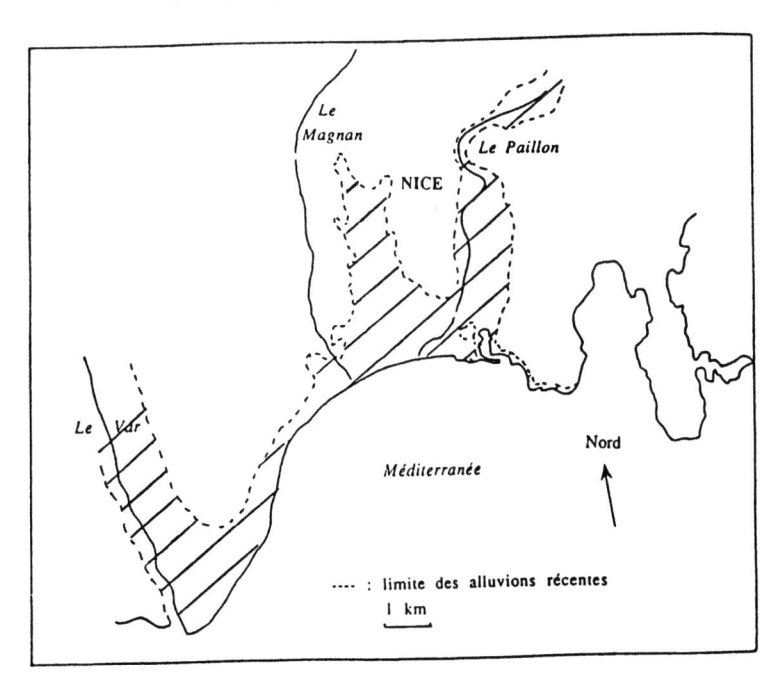

Figure 3: *Simplified geological structure of Nice city: hashed area corresponds to young alluvium, white area to hill zones. Nice dowtown is located in the twin alluvial between Magnan and Paillon rivers.*

4. Emergency response

In the French administrative organization, two persons are responsible for emergency response at the local level: the *Préfet* (representing the federal government at the *Département* level) and the city mayor. Their shared responsibility demands training and coordination of the various relevant services under their specific authority.

There exist various pre-established plans (ORSEC plan, red plan), none of which is specifically dedicated to respond to earthquakes. Nevertheless, the ORSEC plan is designed in principle to allow a rapid response to any kind of accidental situation, such as large landslides, forest fires, explosions, tsunamis, and earthquakes. The choice and activation of a given emergency plan are decided by the *Préfet*, generally in close contact with local authorities, particularly the mayor.

The different services that intervene in these various plans are summarized in the following table. Some of these services, the police for example, depend both on the Mayor and the *Préfet*; others have a single, simple hierarchical level.

Responsible Authority	*Organizations*
Mayor	Municipal Services
	Gendarmerie
	Police
Préfet	Urban Civil Security
	S.D.I.S. (Fire Department)
	D. D. A. F. F. (Agriculture and Forests)
	D. D. E. (Public Works & Housing)
	D. D. A. S. S. (Health and Social Welfare)
	D. R. I. R. E. (Industry and Environment)
	Geophysical Observatories

The S.D.I.S service (fire department) is placed under the authority of the *Préfet*. It consists of 3000 people at the *Département* level, including 600 people alone for the city of Nice. They have an annual budget of 160 million F (around 30 million US$), corresponding to about 300 F ($50) per inhabitant. Particularly relevant for earthquake hazard is the existence within the Municipal Fire Department of one of the most important units of dog handlers in France, specialized in search and rescue after earthquakes. This unit intervened in Mexico City (1985) and in Armenia (1988).

There is a total of 1,200 policemen for the city, most of them under the authority of the *Préfet*, but this total includes a staff of 250 policemen in the Municipal Police placed directly under the authority of the mayor.

There are seven hospitals in the town, with a total staff of 6,000 medical workers, and a total annual budget of two billion F (around 350 million US$).

The coordination between the various "federal" and municipal organizations is ensured through regular meetings: for instance, the police and fire departments meet every month and organize training exercises, an example of which would be to check the common radio systems.

From another point of view, each of these various organizations tries regularly to improve its capacity to respond to emergency situations.

The D.D.A.S.S., for instance, which represents the Ministry of Social Affairs and the Ministry of Public Health at the *Département* level, recently issued an emergency housing plan. Its present capacity allows 270,000 people to be temporarily housed in public buildings and various touristic facilities, and 106,000 people to be fed. The same organization also prepared a plan in close cooperation with both the local representatives of the Ministry of Interior and the municipal services in charge of health, for rapid evaluation of damage to the water supply and sewage systems, in order for them to make the right decisions concerning the possible emergence of contagious diseases.

In another example, the technical services of the town and of the State (D.D.E.) are cooperating to analyse the ground transportation system, in order to decide the retrofitting priorities for first aid. That is, how best to improve access to it in each part of the city, as well as to hospitals.

In summary, although there does not exist a specific earthquake emergency response plan, several procedures have been already implemented in some services, and their coordination is ensured by regular meetings, which should significantly improve the efficiency of the "standard," general-use emergency plan in case of an earthquake.

5. Construction

5.1 CONSTRUCTION CODE

France is too small and too centralized a country to have several competing construction codes. There is therefore only one national code that prescribes the *minimum* design level to adopt, depending on the construction class and the seismic zone. At present, these regulations are applicable only to new buildings, or to buildings for which the use is changed; the actual seismic safety of built areas thus depends both on the age of constructions and on the history of regulation codes.

The first earthquake regulations to be applied in France were promulgated in 1955 after the Orléansville (Algeria) earthquake. They were then improved in 1962 and 1964, and modified in 1969. Since that date, earthquake regulations for new constructions have been gathered in a code entitled "PS69" which is still in application, with some improvements made in 1982 following the 1980 El Asnam (Algeria) event, and in 1985 to account for new seismic zoning. The application of those regulations was, however, limited to two specific classes of new buildings: the high-rise buildings (taller than 28 m) and those which "receive public" (e.g., schools, hospitals, stadiums, museums, and railway stations).

Very recently, in May 1991 and July 1992, two new texts were promulgated. Although they did not change the technical part of the regulations, they considerably extended their application field to all residential buildings, beginning in August 1993, and to individual housing starting in August 1994. Furthermore, completely new technical regulations are now under preparation, based on the recommendations issued by the French Association for Earthquake Engineering, and are sometimes used for particular structures. According to these texts, Nice is located in the highest seismicity zone of continental France, the "moderate seismicity" zone or Zone II. These texts now apply to all new buildings (apart from individual housing units, which will be addressed next year), and to existing buildings in cases of rehabilitation works.

Being aware of the inadequacy of the national *minimum* regulations, the city of Nice has anticipated, whenever possible, the extension of their application field. For example, public buildings such as schools within the city territory have had to conform to the PS69 code since 1972, while at the national level this was not required until 1977. Similarly, in 1979, the city council proposed to extend their application field to *all* buildings (whatever their use), and a special committee was set up in 1980 for that purpose. By December 1980, this led to an official suggestion to the Prime Minister to extend the application field from public buildings to private buildings. The city of Nice thus helped accelerate the application of earthquake regulations to a larger and larger set of buildings.

A detail of the French law must be mentioned: until 1982, no local authorities (including the mayor) were entitled to introduce more severe technical regulations concerning buildings in land-use planning documents than those existing in the national regulations. The decentralization law promulgated in 1982 allowed the mayor to delimit some particular zones with higher hazard level, and to indicate them in land-use planning documents. This is aided through the establishment of Risk Exposure Plans (PER), which correspond to microzoning maps as far as seismic hazard is concerned. However, it was not possible for the mayor to make use of this new, expanded authority until now, given the long time that was needed to put this important law into effect, the lack of a uniformly-accepted methodology for seismic microzoning (especially for deciding the exact limits between the various microzones), and the simultaneous improvements of the national regulations.

5.2 CONSTRUCTION TYPE

The City of Nice comprises about 200,000 housing units (flats or individual homes), including about 12% non-permanent housing (for weekend, summer, or winter holidays). As in most French or Mediterranean towns, the housing park in Nice is very heterogeneous, due to both the history of its economical development, and the heterogeneity of its geological conditions.

The "old" (i.e., pre-1900) town is located in the flat, alluvial area surrounding the Paillon river. It was built with local materials such as stone, wood, sand, and mortar. The use of reinforced concrete started around 1930, and became quasi-systematic after 1945 for public and residential buildings. The mean height of those buildings is about 20 m, which is also the limit fixed in recent land-use planning documents. Most of these structures are built in the flat, alluvial areas. Habitations in the hills, on the other hand, are mainly individual houses made of masonry and/or reinforced concrete.

5.3 CONTROL AND ENFORCEMENT

The design of all 20th century constructions—which represent the vast majority of buildings, given the important growth of the city since 1900—has been done by engineers or architects, with the exception of individual houses in a total inhabited area smaller than 200 m^2, for which it is not legally mandatory.

All new construction projects, as well as modifications to existing buildings, must be registered and cannot start before the delivery of an official administrative approval. These forms are delivered after inspection by two special departments of the City and the State, which comprise a total of about 300 people. 2,100 new housing units are built every year, corresponding to a total floor area of 175,000 m^2.

For constructions that fall in the application field of earthquake regulations (i.e., public buildings before 08/93, and all buildings now), the city requires that the project managers have both the design and the construction of the building checked by a registered control office.

Given the history of earthquake regulations briefly outlined in the previous section and the age and nature of buildings in Nice, it is estimated that almost 95% of the existing buildings have been designed without consideration for earthquake threat. The above-mentioned recent dispositions allow for reasonable optimism if the next earthquake hits far enough in the future for most of the existing buildings to be replaced by new, earthquake-resistant structures. A very valuable output of scenario studies might thus be to help in assessing the actual vulnerability of existing buildings and to assess priorities for retrofitting work.

6. Earthquake Awareness

Local authorities have long been aware of the earthquake threat, as exemplified by the early decisions of the city council concerning the application of building code (see section 5). More recently, in addition to the coordination meetings between city and *Département* services for better implementation of emergency response plans, this awareness has also resulted in the setup of a special committee at the *Département* level which issued a list of first-priority actions, some of which are currently being implemented. The city also became a member of a newly-created association, based in Marseille, and dedicated to risk management in mid-size cities.

The general public has now a rather fuzzy perception of earthquake threat: on the one hand, sensationalist announcements regularly appear in news media, sometimes quoting famous "show-science" personalities; on the other hand, there does not exist any document or pamphlet for dissemination to the general public, describing the present state of knowledge about the seismic hazard, including the uncertainties, together with the mitigation policies presently adopted at the local and state level. The only existing brochure has been issued by the local agency of the state emergency response services, and contains the basic safety instructions as to the way to behave during an earthquake.

To fill this gap, a booklet is now in preparation under the authority of the *Préfet* (*Département* level), whose aim is to provide general information about all natural hazards in the town of Nice. The technical department of the city services has been asked to lead a working group in charge of the publication of the section concerning seismic hazard, and in the longer run, to direct the public awareness. The intentions of this working group are to organize conferences for specifically concerned technical staff, such as health and construction workers, and to give presentations in schools to raise awareness.

The long-term objective, once the information and de-dramatized sensitization of general public will be achieved, is to establish new mitigation procedures, that should then be well understood and accepted by everybody.

To conclude this section, it seems reasonable to say that there exists a fairly good awareness of the earthquake threat among local government officials, and they are willing to improve the mitigation measures. However, to improve these measures requires that the general public be informed of the earthquake threat. To do so, the officials must maintain an uneasy balance between a sensationalist over-dramatization and a complacent minimization, attitudes that may both find some scientific legitimacy, given the present uncertainties as to possible magnitudes and proximity of potential earthquake sources. Further, the officials must also take into account the importance of tourism in the economic life of the city.

7. Research

Before presenting the main research facilities and ongoing programs, some background is necessary. In the academic field too, France is still a highly centralized country; decisions on creating new positions to start or develop a given academic department are officially made in Paris, and do not systematically reflect the wishes of the local authorities or local academic staff. Nevertheless, as regional authorities are responsible for the university buildings and may also help in research funding, there does exist coordination between federal and regional levels.

7.1 STAFF

Broadly speaking, as far as earthquake engineering is concerned, regional research activities are carried out mainly by three institutions: the university, the local technical agency of Ministry of Housing and Public Works (CETE), and the engineering seismology department of BRGM, located in Marseille.

The University of Nice has been only recently involved in activities related with earthquake risk. Despite the existence of a rather strong geophysical department (specializing in marine geophysics and geodynamics), the interest for studies on the Nice area began only in 1990, when a position of Professor of Seismology was created, rapidly followed by two full-time research positions. In addition to the development of a broadband local network (see section 2), this new, young seismological team has also successfully involved their colleagues from neotectonics—who usually worked on very active zones like the Caribbean or the Philippines—in local studies. The team also maintains close contacts with larger seismological teams (Paris, Grenoble, Genoa, Napoli) who are ready to work on the Nice area.

CETE, whose main fields are soils, constructions and road engineering, is presently trying to develop its earthquake engineering activities. A small engineering geology team has been working on microzoning studies for about fifteen years, and they issued the first microzoning study in France, which was mainly a methodological study on the example of Nice. Since their inception, they have gained much experience in instrumental measurements of site effects (and more particularly in urban areas), as well as in liquefaction and slope stability studies. Another new branch of CETE that is slowly developing investigates vulnerability, using both a detailed approach for specific buildings, such as communication centers or hospitals, and a more general, mainly typological approach to estimate the vulnerability of blocks and districts.

The BRGM team comprises more than ten engineers who specialize in earthquake and landslide hazards and geotechnical engineering. In addition to the maintenance of the national seismicity catalogue, they perform numerous microzoning studies as well as seismic analysis of many industrial sites.

BRGM and CETE are used to working together, and CETE and Nice University also have close contacts.

This enumeration shows that, apart from rather good competence in earth and soil sciences, a deficiency exists in the structural earthquake engineering field—the civil engineering department of Nice University is very weak. There are, however, several private consulting companies and control offices that demonstrate competency in that domain. In early 1993, a regional chapter of the French Association of Earthquake Engineering was also established, which should favour contacts between soil and structure people.

7.2 RESEARCH PROGRAMS AND LOCAL STUDIES

Several studies focused on the Nice area have been completed in the last decade by CETE and BRGM: risk exposure plans, including seismic microzoning for about fifteen small cities of the vicinity; a detailed microzoning study of Monaco; investigations on the vulnerability of some mountainous roads north of Nice; and vulnerability estimates for several communication centers.

In addition, new programs are now being implemented with local, national and European funding: field investigations on the neotectonics of Southern Alps; installation of a 3D broadband seismometer local network (see above); starting of a small strong motion network; risk estimates for a restricted part of downtown Nice, involving updating of regional hazard, site effect investigations, and vulnerability estimates for old constructions; and methodological investigations on vulnerability estimates for hospitals.

There also exist some more ambitious proposals, including an eathquake damage scenario study, and the creation near Nice of an earthquake research center grouping the University and CETE teams, but they are still in a discussion phase and not yet funded.

8. Concluding remarks

Although for simplicity's sake this short presentation has focused mainly on the city of Nice, it should be emphasized that a comprehensive earthquake scenario study should certainly consider the whole Côte-d'Azur urban area, since damage within Nice would certainly affect the daily life of people leaving from Cannes to Menton, and, vice versa, damage to vital transportation facilities (motorways, railroads, etc.) even outside Nice would affect the city's economy.

The Nice urban area possesses unique characteristics that distinguish it from most of the other cities considered in this workshop. The seismicity is probably lower—although large events cannot be definitely ruled out—while its economy is mainly based on service industries, tourism, and transportation facilities. Any earthquake damage scenario study should therefore keep in mind these details, which may result in slightly different approaches concerning the management and final outputs of such studies.

References

Bilan, Des connaissances et des actions à mener pour le département des Alpes-Maritimes en matière de séisme et de risque sismique - Plan d'action , Rapport CETE / BRGM pour le Conseil Général des Alpes-Maritimes *(in French)*.

Dadou, C., P. Godefroy and J.-M. Vagneron, 1982. Evaluation probabiliste de l'aléa sismique régional dans le Sud-Est de la France, Document du BRGM n° 59 *(in French)*.

Goula, X., 1981. Analyse du rapport du BRGM concernant le Rique Sismique dans le Sud-Est de la France. Comparaison avec les études réalisées à EDF et au DSN, Note technique CEA/DSN/SAER 81/257 *(in French)*.

Mèneroud, J.-P., and P.-Y. Bard, 1983. Etude méthodologique de microzonage dans les Alpes Maritimes. Compte-rendu de travail, CETE Méditerranée - Laboratoire de Nice, 113 pages *(in French)*.

THE CITY OF PATRAS-W.GREECE: A NATURAL SEISMOLOGICAL LABORATORY TO PERFORM SEISMIC SCENARIO PRACTICES

A. TSELENTIS[1]
A. KARAVOLAS[2]
C. CHRISTOPOULOS[3]
[1]*University of Patras, Seismological Center*
[2]*Mayor, the City of Patras*
[3]*Chamber of Engineers*

1. Introduction

Greece is the most earthquake-prone country in Europe, with about 2% of the whole world's seismic energy release. This takes place within an area accounting for only 0.009% of the world's total area. The cost of repair and reconstruction after the earthquakes in the last decade in Greece is estimated to be around 700 million European Currency Units per year. The indirect cost of earthquakes, due to loss in GNP, social problems, etc., is even higher.

The city of Patras is the capital of Western Greece, the most seismically-active region in the country and Europe (cf. Strong Motion Earthquake Instrument Arrays, Proc. of the International Workshop, Honolulu 1988, W.D. Iwan, Editor, in which Patras was proposed for the deployment of a strong motion instrument array—along with 27 other locations worldwide—as the most suitable location in Europe).

Despite its relatively small size, the city of Patras has been considered as a second international site, after the city of Quito, to perform an earthquake scenario experiment. It is a common engineering practice to employ models in order to test theories or construction practices. Thus, it is much easier to construct a damage scenario of a relatively small city which shares all the interesting factors of an earthquake-prone megacity (variability in building construction, various geological regimes, high seismic risk, etc.), and then, after having tested all the required research tools and developed new methods for seismic hazard mitigation, to try to generalize their application to larger cities. The recent catastrophic earthquake makes Patras a favorable test city to develop and finetune seismic scenario practices.

2. Seismotectonic Regime

The seismicity of the region reflects the complex geodynamics of Greece, which are determined by the convergence of the African and European lithospheric plates [Figure 1]. Furthermore, the Patras seismotectonic region is even more complex since it lies at the junction of two different structural trends within the current neotectonic extensional regime of an approximately N-S direction. The first is the WNW-ESE zone of extension defined by the Gulf of Corinth graben; the second is the NE-SW faulting associated with the Rio graben which has been interpreted as a transfer (transtensional) fault zone linking the Gulf of Patras graben in the southwest with the Corinth-Trichonis zone of extension in the northwest [Figure 1].

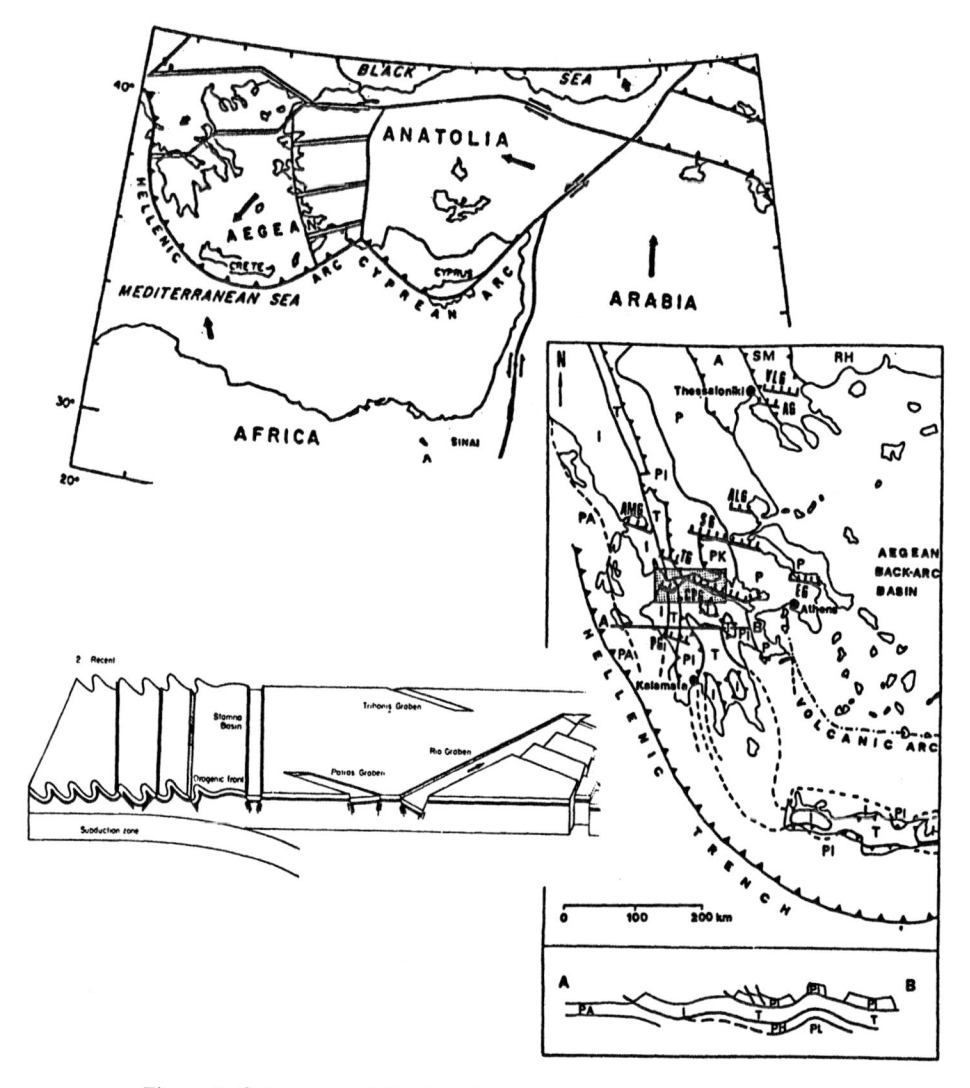

Figure 1. Seismotectonic Regime (modified from Doutsos et al. 1988)

2.1 HISTORICAL SEISMICITY IN THE PATRAS AREA

Macroearthquake catalogues and earthquake studies show that destructive earthquakes have occurred during historic times. The city of Patras has been destroyed many times in the past by earthquakes. Many of these earthquakes also triggered tsunami. The most famous event was that of 373 B.C. (with an estimated magnitude greater than 7.5 on the Richter scale), during which the ancient civilization of Helice totally disappeared. Even today one can observe in a nearby roadcut the face of the fault which was responsible for the aforementioned event. In Table1 we have collected from the existing literature all the available historical data while Figure 2 depicts the events with magnitude greater than 6.

(year	TIME date	hr min)	LAT (°N)	LON (°E)	DEPTH (km)	MAG (Ms)
	BC					
600			38.3	22.6	s	6.8
373	WINTER	NIGHT	38.2	22.2	s	7.0
348			38.4	22.5	s	6.7
279			38.4	22.6	s	6.8
	AD					
23			38.3	22.0	s	6.5
551	JUL 7		38.4	22.4	s	7.2
996			38.3	22.4	s	6.8
1147			38.50	22.50	s	6.25
1402	JUNE		38.1	22.4	s	7.0
1580			38.4	22.3	s	6.7
1660	MAR		38.3	22.5	s	6.4
1714	JUL 27	6 0	38.2	21.7	s	6.6
1742	FEB 22		38.1	22.5	s	6.0
1748	MAY 27	15 0	38.2	22.2	s	6.8
1753	MAR 8		38.1	22.6	s	6.2
1785	JAN 31	4 0	38.2	21.7	s	6.6
1804	JUN 8	3 0	38.2	21.7	s	6.6
1806	JAN 23	.	38.3	21.9	s	6.3
1817	AUG 23	8 0	38.2	22.1	s	6.5
1861	DEC 26	6 30	38.2	22.3	s	6.7
1862	JAN 1		38.25	22.25	s	6.0
1870	AUG 1	-0 41	38.5	22.5	s	6.8
1876	AUG 6	11 0	38.25	21.75	s	4.70
1883	NOV 14	0 45	38.25	21.75	s	4.70
1885	FEB 18	14 30	38.50	21.75	s	5.8
1885	JUL 14	0 4	38.25	21.75	s	4.70
1887	OCT 3	22 53	38.1	22.6	s	6.3
1888	SEP 9	15 15	38.1	22.1	s	6.2
1889	AUG 25	19 13	38.3	20.1	61	7.0

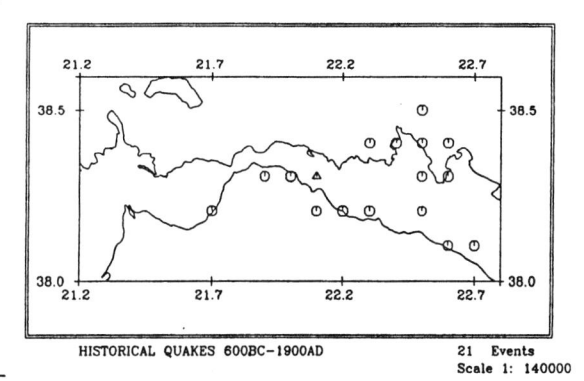

s: shallow earthquakes (depth ≤ 60km)

HISTORICAL QUAKES 600BC-1900AD   21 Events
Scale 1: 140000

2.2 PRESENT SEISMICITY

Figure 3a presents the seismicity of Greece during the last century. The concentration of earthquake epicenters towards the west is characteristic. Even the short-term seismicity of the region depicts a similar pattern. Figure 3b shows the microearthquake activity in the region as it was evaluated by the region's recently-established microearthquake network of the University of Patras. A map showing the average intensities experienced in Western Greece during the present century is presented in Figure 3c.

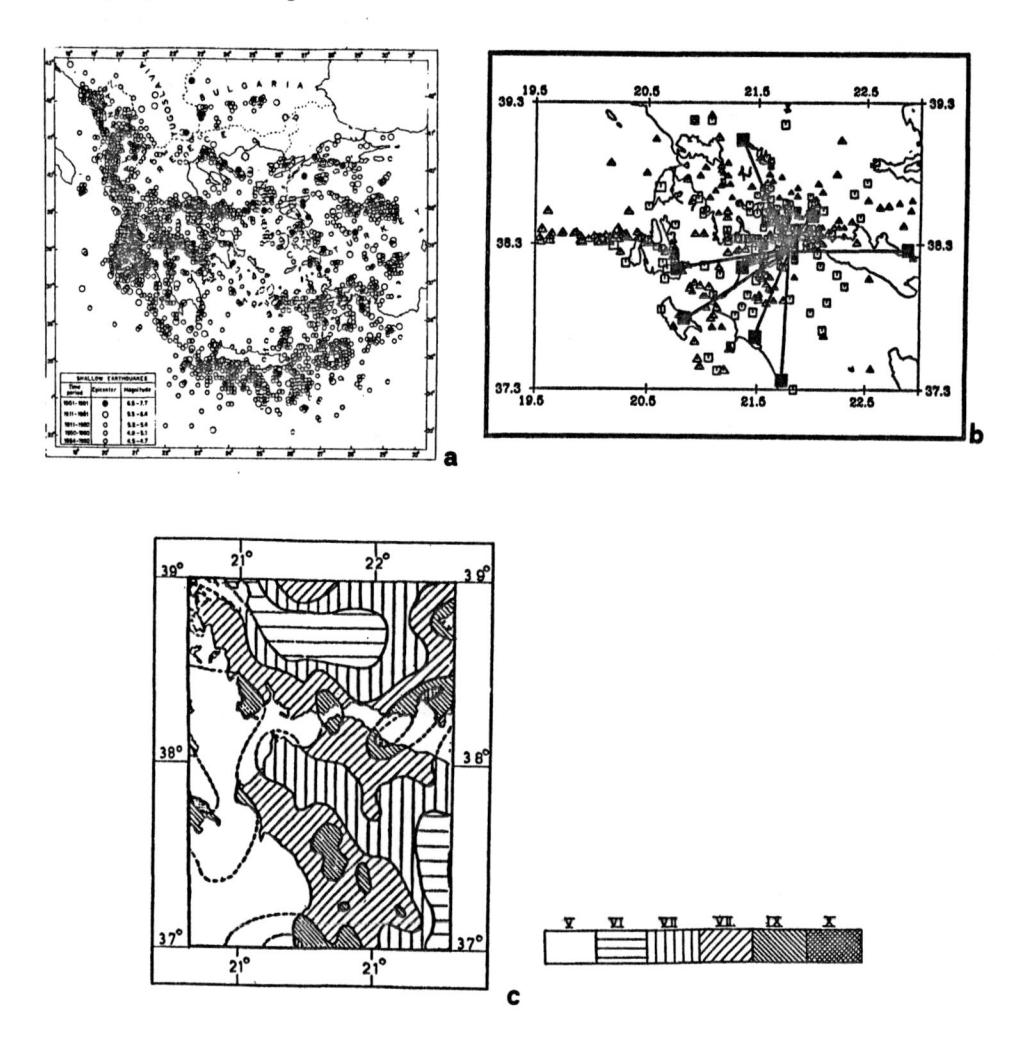

Figure 3:(a)Seismicity of the region during the the 19th century, (b)Short-term (4 months) seismicity, (c)Average intensities.

3. Geological Regime

The broader area of the city of Patras is emplaced in the gulf of Patras basin. The basement rocks, which outcrop in the mountainous area eastward to south-eastward of the city, consist mainly of flysch formations and of thin-bedded limestones and radiolarties. The plio-pleistocene sediments of the basin, exhibiting gentle morphology (elevation up to 200 m) and dipping to the south are mostly covered by old quaternary and recent alluvial deposits. Two horizons can be distinguished in these sediments, namely a lower fine-grained one with a thickness greater than 150 m, and an upper coarse-grained one usually of 50 m thickness.

In the narrow area of the city, which has a medium to low relief, the plio-pleistocene sediments developed consist of dark-grey marls, silty clays and sandy silts in the lower horizons, while in the upper ones alternations of clays, sands and gravels, conglomerates and clayey silts predominate. According to the data of the boreholes carried out, the total thickness of the plio-pleistocene sequence exceeds 300 m and in the coastal zone is covered by younger quaternary deposits.

There are parts of the city built on top of alluvial deposits and sands (region 1 in Figure 4a) with high liquefaction potential. There are also parts of the city built on top of hard rock (region 2 in Figure 4b), or on the top of low thickness sediments. Thus there exists a great variety of geological conditions, even interesting topographic surface and bedrock features.

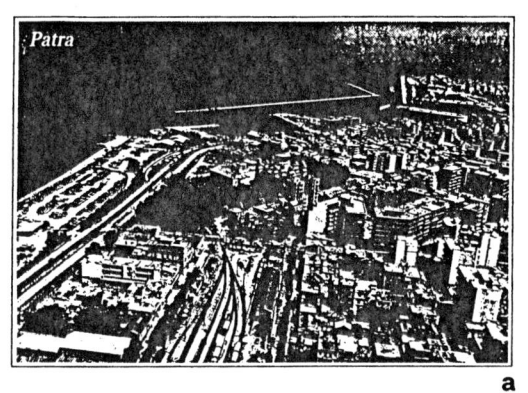

Figure 4:(a)NE view of the city, (b)SW view of the city.

3.1 ACTIVE FAULTS

Recent investigations by the Geology Department of Patras University have revealed the existence of about 500 active faults with length greater than 1.5 km in NW Peloponnesus (Figure 5a). As indicated by slickensides, the kinematic picture of the region is characterized by NNE trending movements along WNW trending faults. The total geological offset across the WNW trending faults near the earth surface was estimated by addition of the maximum thickness of the sedimentary prism, which was deposited in the hanging wall block of the fault, and the high of the fault scarp. Figure 5b gives total offset estimations for 41 major active faults of the region.

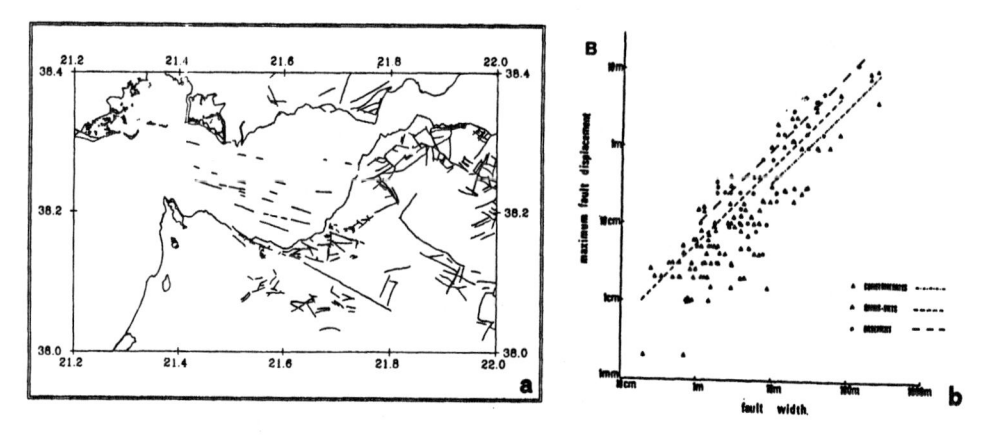

Figure 5:(a)Active faults in the region, (b)Offset estimations (from Doutsos and Poulimenos 1993).

3.2 AN ACTIVE FAULT THROUGH THE CITY

Patras is characterized by a unique geological phenomenon: an active fault runs through its most heavily populated part, resulting in considerable damage to buildings and lifelines.

Patras was struck on August 31, 1989 by an earthquake of small magnitude which caused serious damage to new multi-story and old two-story buildings in a limited area of the city. The main earthquake of magnitude 4.8 with a focal depth of about 3 km, occurred at an epicentral distance of about 5 km from Patras. This earthquake was followed by a series of shocks (Ms=4.5-5.2) in 1990, with epicenters located inside the Gulf of Patras and an epicentral distance from the city of less than 10 km. It is important to mention that the structural damage of buildings was observed in a narrow elongated zone, about 1500 m long and 50 m wide (Figure 6), along a surface rupture related to the reactivation of a normal fault which has been recognized in old airphotos (years 1945 and 1960). This normal fault has an observed length of more than 4 km, one of its edges being at the coast and the other at the mountains. It strikes N70E and is parallel to one of the main groups of faults affecting the plio-pleistocene deposits in the border area of Patras.

The main part of the rupture observed was extended in free ground, causing damage in road pavements and lifeline systems, whereas some part was extended to a heavily-populated area of the city of Patras and caused serious damage to many buildings. This damage was closely related with the surface rupture, whereas the discontinuity of the ground resulted in an unavoidable break in certain parts of the structures along the fault trace. It is obvious that this fault provides a unique opportunity to study seismic scenarios as related to active faults in cities.

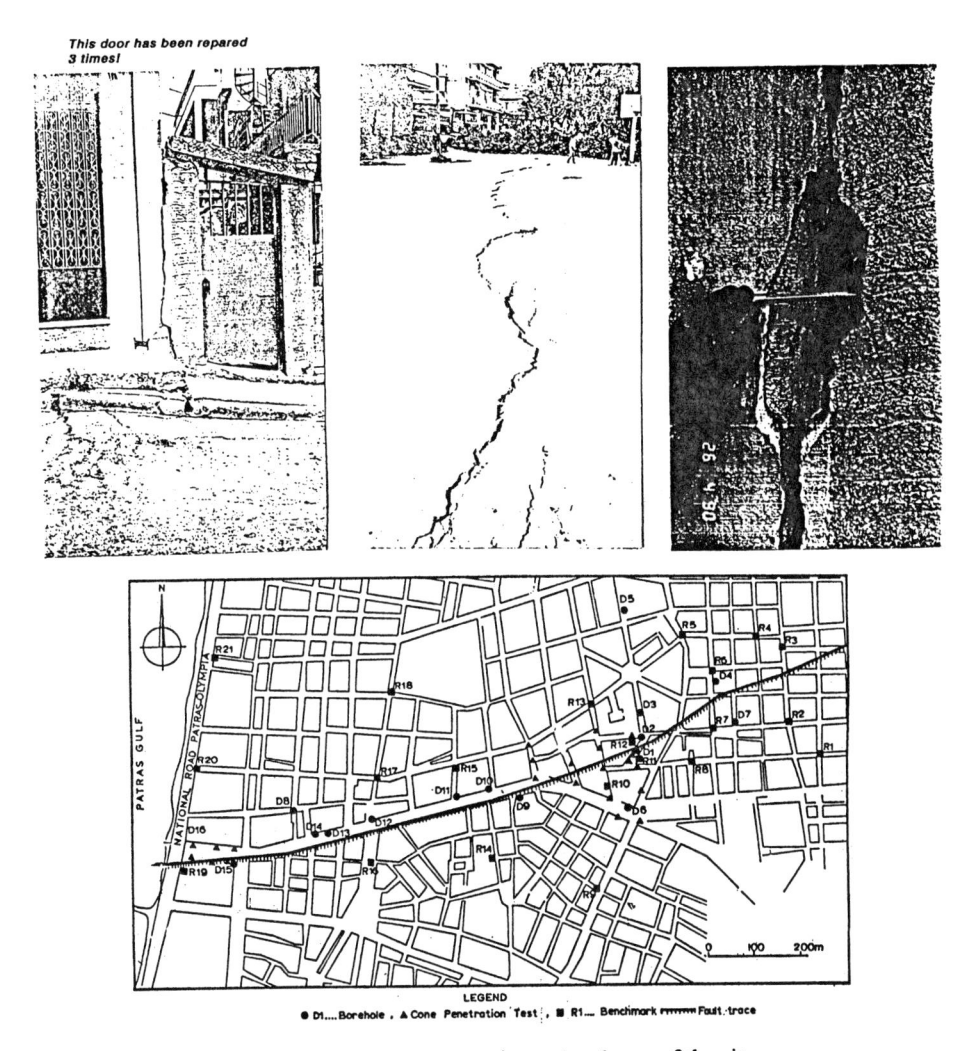

Figure 6: An active fault crosses the most heavily populated part of the city.

4. The July 14, 1993 Earthquake

Just after the decision to develop a seismic scenario in the city, a catastrophic earthquake occurred at the outskirts of the city. Despite its relatively small magnitude (Ms=5.5), it caused considerable damage to the city's buildings, particularly the old ones. Forty-five percent of the city's buildings were damaged, and 12% of them will be demolished. An increase in damages was observed in regions characterized by soft soils (coastal zone, old buried channels). The extremely high degree of damage experienced even by modern buildings in certain parts of the city highlights the importance of site effects.

5. Earthquake Awareness

5.1 SEISMOLOGICAL CENTER

The University of Patras has recently established in the region an eight-station radiolink microearthquake network (Figure 3b, Figure 7). Despite its relatively small time of operation, the analysis of the obtained data has resulted in the delineation of the major earthquake zones and the understanding of the seismotectonic regime.

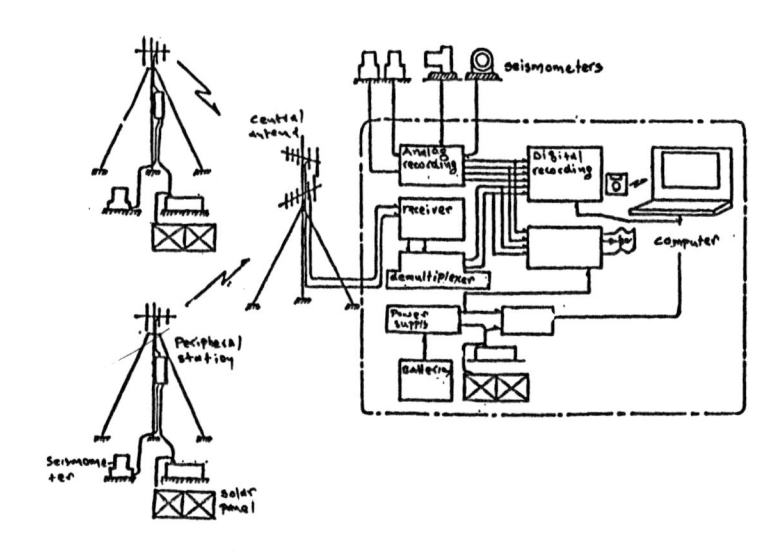

Figure 7:Organization of Patras microearthquake network.

When an earthquake of considerable size occurs, the seismological center informs the local authorities of the spatial and temporal evolution of the aftershock sequence, assessing throughout an algorithm of the probability of occurrence of a greater event or a major aftershock.

In addition to routine seismological monitoring, an earthquake prognostics laboratory has been established for the investigation of earthquake precursors. The prediction strategy is to continuously monitor as many physical parameters as possible (Figure 8), compare the results with the seismic activity, build up a model valid for explaining the phenomena, and provide information on future research activities (Figure 8). In certain cases, where there are serious indications of the existence of precursor signals, the local authorities are informed.

The fact that the Earthquake Prognostics Center operates within a region which is characterized by extremely high seismicity makes it a natural earthquake prediction laboratory, where various research teams from many countries have expressed their interest to perform experiments.

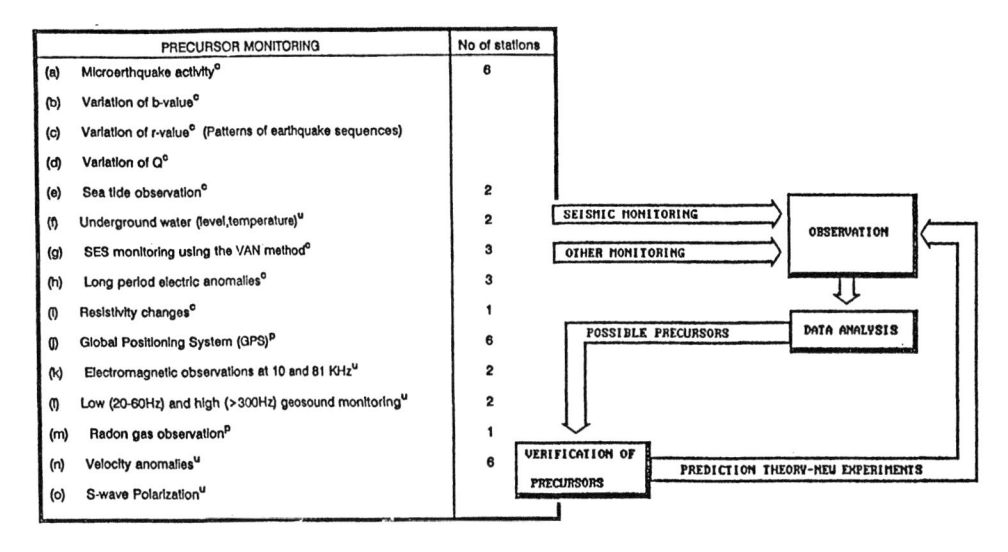

	PRECURSOR MONITORING	No of stations
(a)	Microearthquake activity[c]	6
(b)	Variation of b-value[c]	
(c)	Variation of r-value[c] (Patterns of earthquake sequences)	
(d)	Variation of Q[c]	
(e)	Sea tide observation[c]	2
(f)	Underground water (level,temperature)[u]	2
(g)	SES monitoring using the VAN method[c]	3
(h)	Long period electric anomalies[c]	3
(i)	Resistivity changes[c]	1
(j)	Global Positioning System (GPS)[p]	6
(k)	Electromagnetic observations at 10 and 81 KHz[u]	2
(l)	Low (20-60Hz) and high (>300Hz) geosound monitoring[u]	2
(m)	Radon gas observation[p]	1
(n)	Velocity anomalies[u]	6
(o)	S-wave Polarization[u]	

Fig.8:Earthquake Prognostics Strategy   C=combleted, U=under development P=planed.

5.2 SEISMIC SCENARIO

The Municipality of Patras has realized the high seismic risk of the city and has recently decided to support the development of its seismic scenario.

This will be a pilot project partially financed by the Municipality of Patras, the Chamber of Civil Engineers and GeoHazards International. A later stage of the project, which has been submitted for approval to the European Community, deals with the generalization of the project's results to other Mediterranean cities. Figure 9 depicts in detail the various stages of the work, which is planned to start in early 1994 and is expected to last two years.

The recent catastrophic earthquake of July 14, 1993 which caused serious damage to the city's buildings, provides us with a unique chance to judge the scenario's results and finetune the methodologies developed.

After the completion of the project, the obtained results will be integrated into the city's Geographic Information System and will be disseminated to all interested parties. For example, civil engineers will have the opportunity to link via modem to the main database and obtain information about the response spectra at each construction site, the local authorities will be able to perform earthquake exercises triggering various earthquake scenarios, etc.

CHARACTERIZATION OF SITES

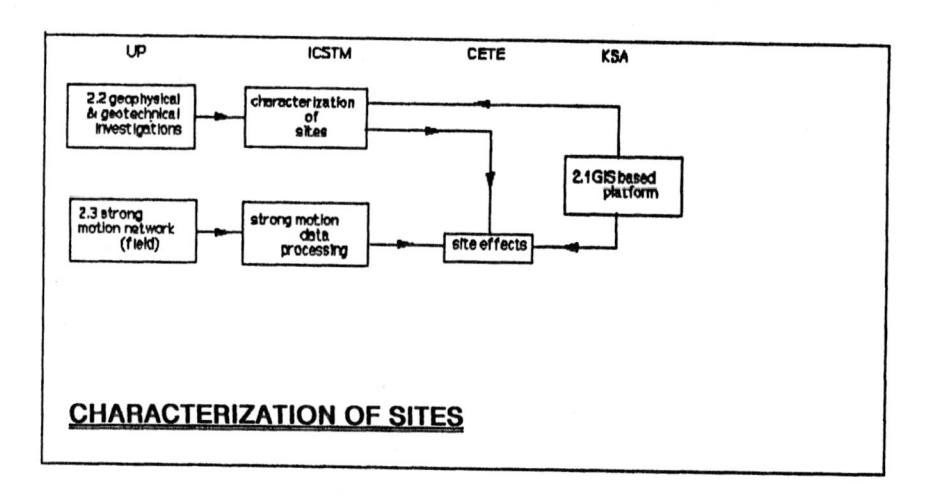

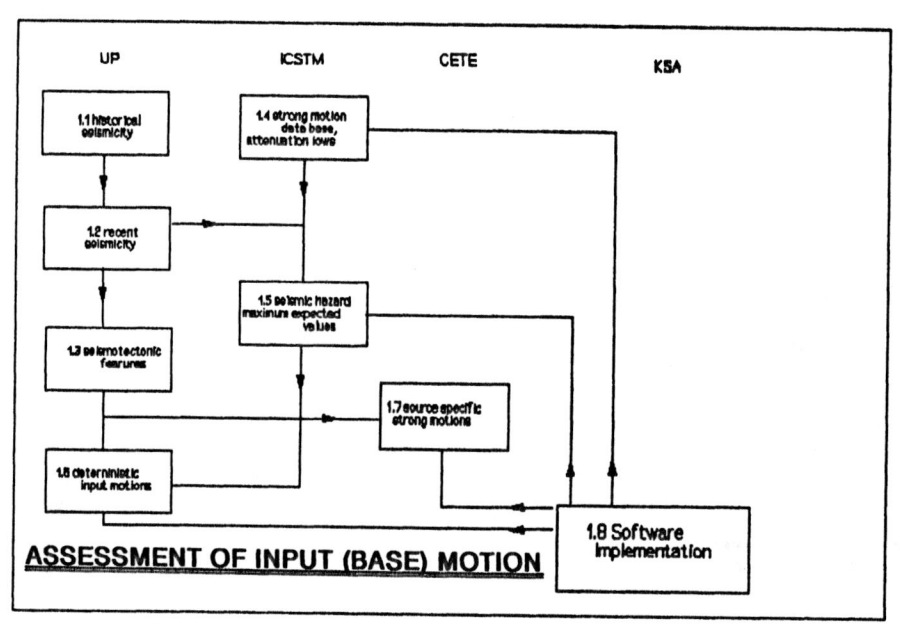

ASSESSMENT OF INPUT (BASE) MOTION

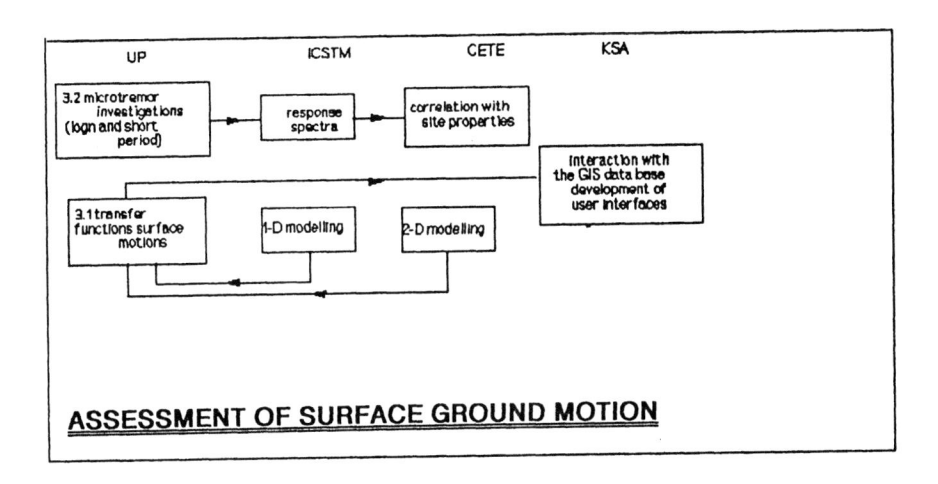

ASSESSMENT OF SURFACE GROUND MOTION

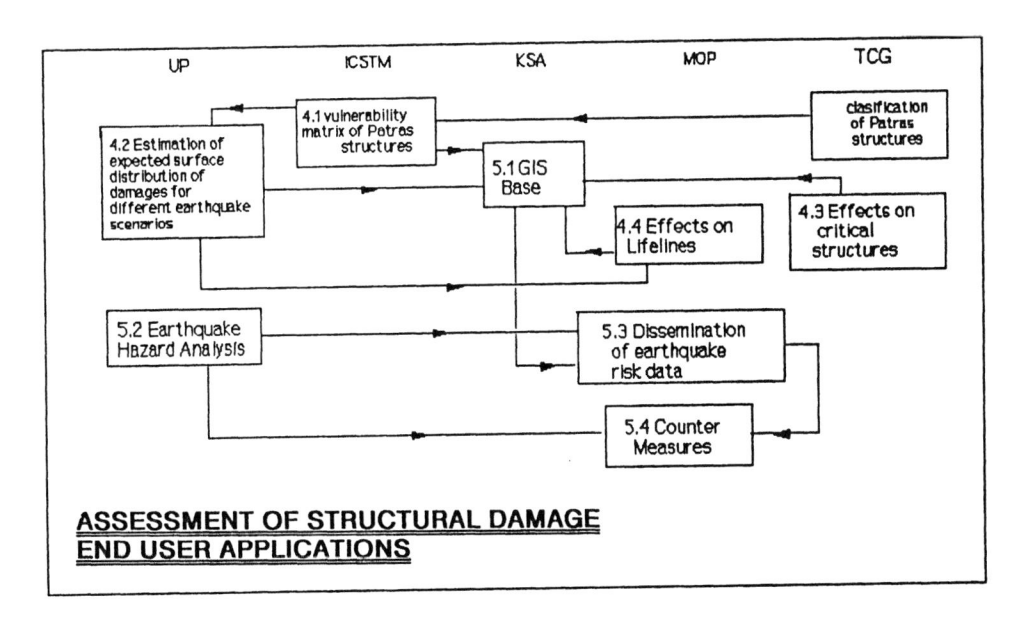

UP | ICSTM | CETE | KSA

- 3.2 microtremor investigations (long and short period)
- response spectra
- correlation with site properties
- Interaction with the GIS data base development of user interfaces
- 3.1 transfer functions surface motions
- 1-D modelling
- 2-D modelling

ASSESSMENT OF STRUCTURAL DAMAGE
END USER APPLICATIONS

UP | ICSTM | KSA | MOP | TCG

- 4.1 vulnerability matrix of Patras structures
- classification of Patras structures
- 4.2 Estimation of expected surface distribution of damages for different earthquake scenarios
- 5.1 GIS Base
- 4.4 Effects on Lifelines
- 4.3 Effects on critical structures
- 5.2 Earthquake Hazard Analysis
- 5.3 Dissemination of earthquake risk data
- 5.4 Counter Measures

References

Doutsos, T., N. Kontopoulos and G. Poulimenos, 1988. "Neotectonic faults in SW Greece," Basin Research, Vol.1, 177-190.

Doutsos, T. and G. Poulimenos, 1993. "Active faults in N. Peloponnese." Jr. Str. Geology, Vol.14, 6, 689-699.

Iwan, W. D., 1988. "Report on the meeting of the IAEE/IASPEI Joint Working Group (Steering Committee) on the Effects of Surface Geology on Seismic Motion." Proceedings, 2nd Workshop, Tokyo.

INDEX

A

Applied Technology Council (ATC), 42, 79, 83, 86
Armenia, see Yerevan
Avalanche, 21

B

Benefit-Cost Ratio (BCR), 28
BICEPP, 8, 19

C

Collateral hazards, 79, 82, 86, 90, 91
Conseil d'Affaires Sociales et Economiques (CASE), 18, 107

D

Damage estimation, 42, 83, 105

E

Earthquake damage scenario, 11, 12, 17, 29, 30, 89, 103, 104, 105, 106, 119, 126, 158
Earthquake hazard, 128
Earthquake hazard reduction, 5, 103, 105
Earthquake risk, 27, 35
Earthquake risk mitigation, 154
Earthquakes, historic
    Alaska, 80
    Armenia, 3, 24, 125, 161
    Coalinga, 25
    Erzincan, 26, 125, 145, 152, 161
    Imperial Valley, 80
    Kwanto, 25, 29
    Loma Prieta, 3, 16, 24, 26, 29, 34, 45, 48, 52, 65, 80, 125, 161
    Managua, 160
    Mexico City, 16, 21, 26, 44, 86, 125, 160
    Niigata, 51
    North Sulawesi, 36
    Northridge, 2, 5
    Philippines, 25
    San Fernando, 25, 80, 160
    San Francisco, 47, 80
    Tangshan, 25

EDS, see Earthquake damage scenario
El Salvador, 247
    Construction, 250
    Demographics, 247
    Emergency response, 248
    Public awareness, 250
    Research, 250
    Seismicity, 247
Emergency response, 103
Escuela Politécnica Nacional, 42

F

Fatalities, 3, 88, 145
Fault rupture, 21, 82, 87, 106, 131
Federal Emergency Management Agency (FEMA), 19
Fire, 82
Fundamental frequency, 81

G

Geofian Scale, 83
Geographic Information System (GIS), 115, 127
GeoHazards International, 11, 16, 18, 34, 41, 106
Gorontalo, 39
Granada, 291
    Demography, 291
    Emergency response, 297
    History of construction, 291
    Public awareness, 298
    Seismicity, 295
Ground displacement, 61
Ground motion, see Ground shaking, 127, 137
Ground shaking, 21, 25, 26, 31
Ground-shaking severity, 79, 80, 90
Guatemala City, 237
    Construction, 243
    Earthquake preparedness, 246
    Emergency response, 245
    Geography, 240
    Geologic setting, 240
    Natural hazards, 237
    Public awareness, 246
    Research, 246
    Seismicity, 240

I

Impedance ratio, 43
Insurance, 29
International Decade of Natural Disaster Reduction (IDNDR), 34
Istanbul, 125
    microzonation, 137

J

Japan, see Saitama prefecture
Japan Meteorological Agency (JMA) Scale, 83

K

Kamchatka, 215
    Emergency response, 217
    Research, 217
    Secondary dangers, 215
    Seismic hazard, 215
Kathmandu Valley, 183
    Construction practice, 192
    Demographics, 186
    Emergency Response, 190
    Public awareness, 193
    Resources, 195
    Seismic hazard, 187

L

Landslide, 64, 68, 82, 87, 106, 131
Lima, 251
    Demographics, 251
    Geology, 254
    Seismicity, 251
    Vulnerability, 258
Liquefaction, 21, 51, 82, 86, 106, 130
Lisbon, 265
    Building legislation, 286
    Building stock, 272
    Earthquake damage scenario, 285
    Emergency response, 285
    Geology, 269
    Historical seismicity, 266
    Microzonation, 272
    Public awareness, 286
    Seismic Impact Model, 282
Los Angeles, 8, 12, 13, 14, 16, 19
Loss estimation, 87, 92

M

Maximum Credible Earthquake, 21
Medvedev-Sponheuer-Karnik (MSK) Intensity, 83, 142
Microzonation, see Zonation
Modified Mercalli Intensity, 42, 43, 83, 140

N

National Earthquake Information Service (NEIS) Database, US, 36
Nepal, see Kathmandu Valley
Nice, 301
    Construction, 309
    Demographics, 303
    Emergency response, 308
    Public awareness, 311
    Research, 312
    Seismicity, 304
North Anatolian Fault, 125

O

ORSTOM, 42, 119
OYO Corporation, 41, 116
OYO Pacific, see GeoHazards International

P

Patras, 315
    Earthquake damage scenario, 323
    Geology, 319
    Public awareness, 322
    Seismicity, 315
Petropavlovsk, see Kamchatka
Probability models, 27
Public policy agenda, 18

Q

Quito, 121
Quito Earthquake Risk Management Project, 9, 11, 12, 31, 41, 47, 106, 115, 121

R

Rossi-Forel Intensity, 83

S

Safety Evaluation Earthquake (SEE), 21, 30
Saitama prefecture, 199
   Construction, 203
   Damage estimation, 206
   Demographics, 199
   Earthquake awareness, 203
   Earthquake damage scenario, 206
   Emergency response, 203
   Seismicity, 199
San Andreas Fault, 11, 12, 27, 30
Seiche, 21, 82
Seismic design codes, 80
Seismic hazard, 35
Seismic Intensity Distribution (SID), 43, 116
Seismic risk evaluation, 36
Seismic Safety Commission, CA, 19, 25, 28, 109
Seismological Society of America, 24
Shear waves, 44, 47, 130
Site effects, see Soil conditions
Slope stability, 65
Soil conditions, 39, 45, 130
Specific Risk, 147
Stanford University, 29
Strong motion recording, 26, 32
Structural characteristics, 79
Structural damage, 92, 106, 117

T

Tsunami, 21, 82

U

U. S. Geological Survey, 84
Uniform Building Code, 28, 45
United Nations Disaster Relief Organization (UNDRO), 7, 35
Urban earthquake risk, 1, 158

V

Vancouver, 221
   Construction practice, 232
   Demographics, 223
   Emergency response, 228
   Ground motion, 228
   Liquefaction, 228
   Public awareness, 228
   Research, 235
   Seismic risk, 224
   Tsunami, 228
Vulnerability, 35, 91
Vulnerability analysis, 140
Vulnerability curves, 43

Y

Yerevan, 167
   Construction practice, 174
   Demography, 168
   Earthquake preparedness, 175
   Emergency response, 176
   Seismic hazard, 169
   Seismic risk, 177
   Seismicity, 169

Z

Zonation, 44, 46